T0146327

SOME WRITERS ON CONCRETE

Published by
Whittles Publishing,
Dunbeath,
Caithness KW6 6EG,
Scotland, UK

www.whittlespublishing.com

ISBN 978-184995-050-3

Printed by Studio RBB, Riga, Latvia.

Some *Writers* on Concrete

THE LITERATURE OF REINFORCED CONCRETE, 1897–1935

Edwin A. R. Trout

Whittles Publishing

Thanks are due to

Hurst Peirce + Malcolm

CONSULTING CIVIL & STRUCTURAL ENGINEERS

for their sponsorship towards the publication of this volume

Contents

The Birth of a Specialist Literature

A dozen years ago [in 1897] reinforced concrete was still in its infancy as a building material and its applications in structures bore large the imprint of an experimental character ... It fell to the share of the promoting enthusiast, who in most cases was also the contactor, to make as brilliant a display of the achievements of reinforced concrete as he possibly could to convince the buyer and his engineer of the wisdom of their investment.

The competition with steel structures had made it however, apparent to the reinforced concrete interests that it was not sufficient to be able to point to a number of successfully built structures which sustained great loads to the convince the engineer ... the builder of reinforced concrete had to explain the behaviour of his structures by assigning to each element its due share in sustaining loads. He had to furnish sufficient data to enable the buy engineer to design and specify his structure with a practical certainty of its safety and fitness to his purposes.

At the same time the engineering profession had come to realize the great possibilities of the new building material, and, surprised by the boldness of some of the structures built with it, began to devote much attention and serious study to the new problem. Numerous articles thus made their appearance in the engineering periodicals...

Gradually the results of carefully prepared and observed tests conducted by experienced experimenters began to appear in the proceedings of the engineering societies and in the technical press, furnishing material for extended discussions and for verifying and correcting the formulas in vogue ... Periodicals specially devoted to reinforced concrete were created, several with good success.

All this ever-increasing mass of information was spread like a floating sea over many journals, transactions, reports and trade publications and the task to condense and filter it down to a book which should form one organic whole, where consecutive thought should separate the wheat from the chaff, required superhuman efforts. Only engineers who attempted to meet this demand even half way, know the amount of labor and time required for it. But the demand for books on reinforced concrete was growing more urgent as the advantages of reinforced concrete were more recognized.

Leon Solomon Moisseiff
The Engineering Digest, 1909

Preface

The inspirations for this study are threefold and a brief exposition of them will go some way to explaining its format. Edwin Sachs – in the introduction to the first issue of his new journal, *Concrete & Constructional Engineering* (*C&CE*) in 1906 – lamented the lack of British books on the subject of reinforced concrete and undertook to harvest the fruits of foreign writing for the benefit of British construction. Having written previously about Sachs and his journal (*Concrete Publications Ltd and its legacy to the construction industry*, *Construction History*, 2005), I wondered if I could quantify the literature, assess whether Sachs was justified in his view of 1906, and measure his success in stimulating home-grown interest over the succeeding years. With this as motivation, I was taken with the example of Spackman's 1926 book *Some writers on cement*, and decided that its blend of bibliography and biography would provide an interesting approach to outlining the emerging literature on concrete. The present title suggests this as a companion volume. Then its scope and definition was influenced by a short passage in Barry Barton's book – *Water towers of Britain* – in which he described the early decades of the 20th Century as a period of discovery and experimentation, in which designers attempted to demonstrate the possibilities of reinforced concrete before practice was standardised, and to a certain extent constrained, in the early 1930s. How did the literature reflect and shape practice at this time?

The literature surveyed here is the specialist literature of concrete, and reinforced concrete in particular. The literature of cement has been outlined admirably by Spackman, and so I have not presumed to duplicate. Neither have I attempted to include general works on roads, bridges, construction and civil or structural engineering – even if they happen to include concrete in their content. The period covered is the early 20th Century – the years in which a distinctive literature emerged. I take 1897 as a starting point – the time that saw construction of the first complete building of reinforced concrete in the UK – and conclude in 1935: the year in which the institutionalisation of concrete practice culminated with the establishment of the Cement & Concrete Association (C&CA).

In limiting the period in such a manner, I follow Spackman's earlier treatment of cement – the story of discovery, patents, proportioning, clinkering, the addition of gypsum and development of testing – through which the British cement industry achieved its modern state: consolidation, in the commercial entity of APCM in 1900 (the Associated Portland Cement Manufacturers, later Blue Circle); technological maturity with the advent of the rotary kiln (from 1898); and the standardisation of output under BS12 (1904). This is not to say cement manufacturing and cement technology was perfect – of course it continued to develop, to mechanise, consolidate, improve quality and strength, incorporate additions – and the literature reflected this. Plenty of authors have written about cement in the 20th Century, but the point is that the technology was mature, recognisably modern: the 'heroic' or 'pioneering' age of discovery was over.

Much the same applies to concrete. The literature mushroomed after 1935, but the basic technology was understood, codified and institutionalised – and these are the themes of this study. By 1935 concrete technology and the concrete industry could be said to have reached maturity. Though pre-stressed concrete, fibre reinforced concrete, self-compacting concrete and countless other technological advances were yet to come, the basic concrete technology was established. With the huge expansion of the precast concrete industry in the 1920s and 1930s, and the introduction of ready-mixed concrete in 1931, the production of concrete was taking its modern form. Practice was reflected in the emerging codes – from the RIBA reports of 1907 and 1911, the London County Council Building Regulations of 1915, and the Concrete Institute's recommendations and codes of 1918 and 1920, to the first national Code of Practice and its accompanying handbook of 1934.

A similar progression can be seen in the emergence of specialist organisations aiming to develop the use of concrete. The Concrete Institute was set up in 1908 and the Concrete Utilities Board (CUB) established in 1916. At the end of the period, the Reinforced Concrete Association was founded in 1933 and the CUB merged with the British Portland Cement Association to form the Cement and Concrete Association in 1935. A parallel pattern unfolded in America, with the establishment of the Portland Cement Association (PCA) and American Concrete Institute (ACI). These, or their successors, continue to prosper, and the collective literature of the period is assembled in the libraries of the institutions, the PCA and The Concrete Society. *Concrete* – the linear descendant of *C&CE* that partly inspired this survey – is still the vehicle for promoting and reviewing the literature, as well as publishing original contributions to the ongoing body of knowledge. And individual authors' influence has survived from this period. With the publication of books by Reynolds and Lea in 1932 and 1935 respectively – editions of which are still in print – the literature also had reached a point that remains relevant today.

My emphasis is on authors – writers – but we must acknowledge that many pioneers of concrete were practitioners and did not write, and that some writers were teachers or journalists rather than technical innovators. Some of the literature is in the form of reports by corporate bodies – professional institutions, trade associations, commercial companies – that often did not credit the writers by name. Though I do not treat them individually, I review the collective output of – significant reports to and recommendations by – the Concrete Institute, Institution of Civil Engineers, Royal

Institute of British Architects, London County Council, Building Research Station, Portland Cement Association, American Concrete Institute, *etc.*

Much of the most advanced work, of course, was published first in journals and proceedings, which again lie beyond the scope of this survey. No attempt has been made to identify the numerous authors of technical articles in the professional or trade press, though key journals will be reviewed separately as corporate sources, and some author entries may comment on periodical output alongside their published books as a reflection of the individual author's interests and professional achievements.

Authors here will be defined as those whose work was issued in book form by a recognised trade publisher – 'publicly' published on the open market rather than circulated as reports, papers or recommendations within the more closely defined membership of a professional or academic institution. These will have had the greatest influence in consolidating and disseminating knowledge, even if not always taking the lead in advancing it.

My interest is principally in the development of a technical literature in the English language, which can be divided into British and American practice, though many titles were published jointly London and New York. In addition, a small number of British Empire works were published in India – notably by the presidency government of Madras and later (in the years around 1930), the Concrete Association of India. British titles are covered comprehensively here while American coverage is a little more selective as reflected by availability in the UK. That said, many American books were reviewed in the British press and the British institutions (of Civil and Structural Engineers, for instance) maintained extensive libraries.

Much of the literature in the early years was dominated by continental writers, as reflected in the relatively high number of German works reviewed in *C&CE* before the outbreak of war in 1914. The names of several American authors too, may indicate post-war migration from Europe. But generally the pioneering work of French, Austrian, German, Dutch and Italian writers appears in this survey only when translated into English. It must be acknowledged that from the point of view of technological development, this survey therefore cannot be regarded as complete, but it does reflect how most engineers in this country will have come to understand the use of concrete as new ideas filtered through American, and then British, writing.

The author entries in the main body of this volume are arranged chronologically by date of first publication, to suggest the unfolding of history and to indicate the gradual accumulation of knowledge available to the readership over time. This does not of course, automatically lead to a smooth biographical progression; different authors will have written at different points in their careers. Some were passing on the fruits of a lifetime's experience, others publicising early research to launch their careers.

Books written from these different motives fulfilled different functions. Monographs often developed new ideas; textbooks summarised existing knowledge for students; others offered guidance – manuals and handbooks for designers and 'the practical man'. Yet others reported the results of testing or inspection programmes, while some promoted commercial concerns or simply the idea of concrete construction.

At the end of the volume an appendix traces the parallel chronological sequences of British and American titles, and another identifies the main publishers and assesses their relative importance. A bibliography sets out fuller details of the books arranged alphabetically by author surname and a reading list points to sources.

David Yeomans – in his study of engineering textbooks and manuals of the interwar period – has demonstrated the value of a bibliographical background to professional development, and so for anyone interested in the progress of technological knowledge in the use of concrete, or the history of the industry it underpinned, it is hoped this present survey will be of interest – that it will help engender an appreciation of how 'concrete' arrived at its present state and provide a memorial to the pioneers of our discipline.

Edwin A. R. Trout
Reading

ACKNOWLEDGEMENTS

No creative enterprise takes place in a vacuum and although responsibility for any errors remains mine alone, I am pleased to acknowledge the help and support extended to me from several quarters:

- The Concrete Society for unfettered access to its collection and copying facilities.
- Colleagues and counterparts in the libraries or publishing operations of various concrete-related organisations for the supply of, and permission to use, portrait photographs as credited individually. They include: Connie Field (PCA), Martha de Jaeger (C&CI), Rob Thomas (IStructE), Steve Adam (*Quarry Management*), Geoffrey Ellis, Jo French (T P Bennett) and Ron Burg (ACI).
- Nick Clarke (BRE) for his comments on the text and assistance in seeking a publisher.
- My wife and daughter for assistance in checking specific data on the Web and my whole family's support during the year or so spent researching and writing.

THE AUTHOR

Edwin Trout is a librarian by profession, and having worked the libraries of the Royal Geographical Society and Institution of Chartered Surveyors, joined the cement concrete sector in 1995.

He is currently the Manager of Information Services at The Concrete Society, having previously fulfilled similar roles for former British Cement Association and Concrete Information Ltd. He has also handled enquiries on behalf of the Concrete Block Association and Cast Stone Association, and between and 2009 supervised the National Concrete Helpline for Concrete Centre.

He also acts as Secretary of the Cement Industry Suppliers' Forum, and in early 2011 was appointed Executive Officer of the Institute of Concrete Technology. Both are now based at The Concrete Society.

With responsibility for the Society's specialist library – an extensive resource founded in 1937 – Edwin has developed an interest in the history of concrete construction and the evolving cement and concrete industries. He has written articles on historical and commercial topics for *Concrete* magazine, *Global Cement*, and *Quarry Management*, and gives talks on the history of the cement industry, cement and concrete transport, and early concrete construction.

This book combines his professional interest in the literature of concrete, with an enthusiasm for the history of his industry. It builds upon his earlier work on publishing history, with several articles and a regular column in *Concrete*, and a contribution to the 75th anniversary issue of *World Cement*. His history of Concrete Publications Ltd was published in the academic journal, *Construction History*, in 2005.

Abbreviations and conventions

Author arrangement

Biographical entries are arranged in order of the date of first publication of the successive authors' earliest books to be included in the survey, any co-authors immediately following, and are otherwise generally in alphabetical order within each year. The qualifying titles are listed at the foot of each entry. At the head, above the author's name, is the date and jurisdiction (UK or USA) of publication, with the subject author's post-nominal initials following on the third line.

Affiliations

The first initial usually indicates the status of membership – licenciate, associate and/ or member, or fellow (sometimes hon. or honorary) of the following professional institutions:

AAAS	American Association for the Advancement of Science
AcerS	American Ceramic Society
ACI	American Concrete Institute
AIA	American Institute of Architects
AIME	American Institute of Mining Engineers
AIM&EE	American Institute of Mining & Electrical Engineers
ASAE	American Society of Agricultural Engineers
ASCE	American Society of Civil Engineers
ASEC	American Society Engineering Contractor
ASTM	American Society for Testing Materials
CanSCE	Canadian Society of Civil Engineers
CGI	City & Guilds Institute
CI	Concrete Institute
FGS	Geological Society
IOB	Institute of Building
IQ	Institute of Quarrying
ITP	Institute of Town Planning

ICE	Institution of Civil Engineers
IMechE	Institution of Mechanical Engineers
IStructE	Institution of Structural Engineers
RE	Royal Engineers
RIBA	Royal Institute of British Architects
RIC	Royal Institute of Chemistry
RSanI	Royal Sanitary Institute
RS	Royal Society
RSA	Royal Society of Arts
SA	Society of Architects
SCE (France)	Society of Civil Engineers (France)
SPEE	Society for Promoting Engineering Education
WSE	Western Society of Engineers

Qualifications

Other, more widely applicable post-nominal initials relate to academic, or social status:

BCE	Bachelor of Civil Engineering
BSc / BS	Bachelor of Science (British / American usage)
CB	Commander of the Bath
CBE	Commander of the British Empire
CE	Chartered Engineer
Dr Ing	Doctor of Engineering
DSc	Doctor of Science
JP	Justice of the Peace
KBE	Knight commander of the British Empire
MA	Master of Arts
MC	Military Cross
MEng	Master of Engineering
MSc / MS	Master of Science (British / American usage)
OBE	Order of the British Empire
PhD	Doctor of Philosophy

Secondary authors

Names highlighted in bold type within the descriptive text indicate the presence of a secondary author for whom there is no separate biographical entry. Typically these relate to contributors in a collective or encyclopaedic work.

Cross references

References to other authors included elsewhere in the survey consist of the surname followed, in square brackets, by the date of entry.

Bibliographical summaries

Bibliographical references at the foot of each biographical entry identify the location of each title in a major British library or contemporary listing, the source of any review quoted from and the collection from which physical examples of the book were examined.

Catalogues and bibliographies consulted

BL	British Library
C&CA 1959	Cement & Concrete Association's printed catalogue of 1959
Ch58	C&CA Library Bibliography, Ch58: *Historical and rare books* (1969)
ICE	Library of the Institution of Civil Engineers
IStructE	Library of the Institution of Structural Engineers
Moisseiff	*A review of the literature of reinforced concrete*. L.S. Moisseiff (1909)
Peddie	*Engineering and Metallurgical Books, 1907–1911*. R.A. Peddie (1912)
RIBA	British Architectural Library at the Royal Institute of British Architects
TCS	Library of The Concrete Society
Ykbk 1934	*Concrete Yearbook*, 1934 edition

Sources of reviews

Contemporary reviews are quoted mainly from the following journals:

CA	*Cement Age* (USA)
C&CE	*Cement & Constructional Engineering* (UK)
C-CA	*Concrete-Cement Age* (USA)
EN	*Engineering News* (USA)
SE	*Structural Engineer* (UK)
Trans CI	*Transactions of the Concrete Institute* (UK)

Journals cited

Other periodicals cited in the review include the Journals (*J.*), Proceedings (*Proc.*) and Transactions (*Trans.*) of the professional institutions.

Industrial and research organisations

Industrial and research organisations mentioned in the text include:

AAPCM	Association of American Portland Cement Manufacturers (USA)
APCM	Associated Portland Cement Manufacturers (UK)
BPCA	British Portland Cement Association (UK)
BRE(S)	Building Research Establishment (originally Station, UK)

C&CA	Cement & Concrete Association (UK)
CPL	Concrete Publications Ltd (UK)
DSIR	Department of Scientific & Industrial Research (UK)
PCA	Portland Cement Association (USA)
RRL	Road Research Laboratory (now TRL – Transport Research Laboratory, UK)

Introduction

Concrete in its modern form (that is to say, aggregate bound with Portland cement rather than the limes and ash of previous practice) emerged as a structural material in the middle of the 19th Century (initially in harbour works) and with it arose a literature to expound and explain its use. Much would have been incorporated in general books on building, but a number of authors chose to specialise in concrete. By the 1890s, just before the introduction of reinforced concrete construction to the UK, four men had made their name as authors in this field. Revised editions by John Newman and Thomas Potter were both issued in 1894, with Sutcliffe's first and Dobson's sixth edition a year earlier. They reflected the nature of concrete's use at the time, for civil engineering work and in the production of individual elements of buildings, such as floors and architectural dressings:

- Dobson, E. (rev. Dodd, George). *Foundations and concrete works*. Sixth edition. London: Crosby Lockwood & Co., 1893.
- Sutcliffe, George L. *Concrete, its nature and uses: a book for architects, builders, contractors and clerks of works*. London: Technical Press, 1893.
- Newman, John. *Notes on concrete and works in concrete: especially written to assist those engaged upon public works*. Second edition. London: E. & F. N. Spon, 1894.
- Potter, Thomas. *Concrete: its use in building and the construction of concrete walls, floors, etc*. New edition. London: Batsford, 1894.

Potter's pioneering work first appeared in 1877 and when it was revised for re-publication in 1894 it was still something of a novelty. In his preface (dated 1893), Potter claimed precedence in the literature:

> Although technical books on almost every conceivable subject have been written, it is somewhat singular that no practical work of Concrete – besides this – so far as its application to building is concerned, has hitherto appeared, except reports of papers read before professional societies, – pamphlets, – and two or three volumes dealing principally with its use in engineering works, and with Concrete in the abstract.

Henry Reid: author of A practical treatise on concrete and how to make it, *1869*
Courtesy of The Concrete Society

PRACTICAL

TREATISE ON CONCRETE,

AND

HOW TO MAKE IT;

WITH

OBSERVATIONS ON THE USES OF CEMENTS, LIMES AND MORTARS.

BY HENRY REID, C.E.,

AUTHOR OF "A PRACTICAL TREATISE ON THE MANUFACTURE OF PORTLAND CEMENT."

LONDON:
E. & F. N. SPON, 48 CHARING CROSS.
1869.

Actually, he was eight years later than Henry Reid, whose *A practical treatise on concrete and how to make it* had been brought out by E. & F. N. Spon in 1869 (and rather more so than George R. Burnell's *Rudimentary treatise on limes, cements, mortars, concretes, mastics, plastering, etc.*, and Dobson's *Rudimentary treatise on foundations and concrete works* of 1850, though the latter author considered concrete 'not fit to be used above ground'). But more importantly for our purposes, Potter also observed that 'very little is known, comparatively speaking, about the general application of Concrete to structural purposes'. Such an application, to the construction of whole buildings, was made practicable by the development of reinforced concrete that was soon to make an impact on the British market.

Though in 1854 W. B. Wilkinson had taken out an early patent in this country for the combination of steel with concrete, the commercial development of this new composite material took place in continental Europe (particularly France), and to a lesser degree in the USA (by Thaddeus Hyatt in 1877; E. L. Ransome in 1884; and later [in 1903], Julius Kahn). Lambot and Francois Coignet's French patents of 1855 attracted little attention at the time, but Joseph Monier's system – licensed by the German G. A. Wayss in 1879, and by Rudolph Schuster in Austria – enjoyed great popularity in central Europe. Other innovations were developed by the Frenchmen Edmond Coignet and Armand Considere, each of which took out British patents during the early 1890s for the reinforcement of concrete elements – pipes, beams, piles (Coignet in 1890, 1892 and 1894 respectively) – and for helical secondary reinforcement (Considere, 1895). But the most significant was that devised by Francois Hennebique. It was his complete system, with shear reinforcement patented in 1892, which was brought to Britain through the agency of Louis Gustave Mouchel. As Hennebique's agent from 1897, Mouchel was responsible for overseeing the construction of Weaver's Mill in Swansea – the UK's first building erected in reinforced concrete. Hennebique's rapid rise to dominance in the market for reinforced concrete during the 1890s was based as much on his commercial acumen as on the technical merits of his system, and for his brilliance at publicity. He relied upon a network of like-minded agents, trained and supported by Hennebique's own specialist staff, operating autonomously with licensed contractors, but tied in by payment of royalties. The house magazine *Le Beton Armee*, promoted the system in the wider market. It was this technical and commercial combination that dominated reinforced concrete throughout its early years in Britain and is described in *The first decade of reinforced concrete in the UK* (Gueritte, T. J. *C&CE* v.21, pp.92 et seq). During this time Mouchel expanded the business, moved to London in 1900 and founded L. G. Mouchel & Partners before his death in 1908. By this date the firm, now led by his partners – the French engineers T. J. Gueritte and J.S.E de Vesian – had been responsible for 130 significant buildings. Furthermore, the Hennebique system

Weaver's Mill
Courtesy of The Concrete Society

was employed in hundreds of civil engineering and construction projects as listed in Mouchel's published catalogue – as many as 1073 by 1911.

In 1905 most reinforced concrete work and all concrete-framed buildings in Britain employed Hennebique's system, though rivals were emerging. Coignet opened a British branch in 1904 and built his first reinforced concrete warehouse in 1905. Considere was represented in the British market too, as was Kahn (by the Trussed Concrete Steel Company, or 'Truscon' as it was known in the trade). The first building in each case though, was erected some while later; in 1908 (Kahn) and 1809 (Considere). British enterprise included E. P. Wells and the British Reinforced Concrete Co. (BRC) in 1905. In 1904 Charles Marsh (see biographical entry) identified as many as 50 different systems, though not all were integrated building systems or indeed, commercially available. This period in which these proprietary systems dominated, is described in detail by Cusack and Bussell (see Appendix A).

It was during these years that criticism of the prominence of Hennebique's system, of the 'system specialist', and of Mouchel personally, began to be voiced. Sir Henry Tanner, chairman of the RIBA Committee on Reinforced Concrete for instance, maintained that widespread adoption of reinforced concrete in Britain was being hampered because of the control exercised by specialists and patentees. Taking a step further, Marsh challenged the dependence on proprietary systems *per se*: 'it could not reasonably be expected that reinforced concrete would remain as a multitude of various more or less rational systems, relying on more or less valid patents.' And though such patents were protected by vigorous litigation at the time, Marsh was right in the longer term: reinforced concrete did not remain as a multitude of systems.

For our purposes in tracing the birth of the literature in Great Britain, we should recognise that the commercial confidentiality that surrounded these largely foreign systems acted as a spur to independent research and publication by the professions. This impulse – as we shall explore below – manifested itself early on in the founding of the new journal *Concrete & Constructional Engineering*); the drawing up of a set of recommendations by the Royal Institution of British Architects; and the establishment of a scientific body – the Concrete Institute.

Concrete & Constructional Engineering, 1906

Announcing the intentions of his new journal, Edwin Sachs aspired to meet 'the growing demand for reliable technical and economic information on concrete' and reinforced concrete especially – developments in which were then 'being accorded the closest possible attention by all interested in building and engineering work'. Describing this vogue for reinforced concrete, he noted that 'the foremost technical writers of books and treatises are presenting valuable information on the subject'. Significantly though, it was 'abroad – on the Continent of Europe and in the United States of America', that 'the literature in all languages on concrete in particular has become extraordinarily extensive'. Consequently, with the launch of *C&CE*, there would be 'reliable digest of the world's latest information on concrete and constructional engineering as applicable to the British reader'. The journal looked to the best expert assistance possible, promising 'no parsimony where expenditure is necessary for the presentation of the latest scientific data from pens of undoubted authority'.

Reviewing the technical journals already in existence, Sachs considered it a sign of 'great progress in the application of concrete and reinforced concrete on the Continent and in the United States that so extensive a contemporary specialist literature should already exist'. Beside *C&CE*, the other specialist journals in English were American. In 1906 these included *Cement*, *Cement Era*, *Cement Age*, *Cement & Engineering News*, *Concrete*, and *Rock Products*. The range was gradually extended in Great Britain by the addition of *Ferro-Concrete* in 1909 (in print to 1931, edited by the writer Noble Twelvetrees); the Concrete Institute's *Transactions & Notes* in 1909 (later [from 1923], the *Structural Engineer*); *Concrete for the Builder & Concrete Products* in 1926; *The Concrete Age* (Industrial Constructions Ltd, London, 1929–1931); and the *Reinforced Concrete Review* in 1933. Such journals nurtured the publication of books, by reviewing, serialising or reprinting the more substantial work of writers, and their editors preparing themed compilations from multiple contributors.

According to Barton the first extended work on reinforced concrete was published in Paris in 1902. Written in 1899 its title was *Le beton arme et ses applications* and the author was Paul Christophe; a Belgian employed by Hennebique. *La construction en ciment arme* by Berger and Guillerme came out the same year, also in Paris. Dependence on continental leadership was acknowledged not only by Sachs, but also repeatedly by other writers of the day and tacitly by the publication of translations. To judge simply by the names of several authors published in USA – Considere, Morsch and Melan – it would seem that the latter often found their way into English in America first, before being issued in Great Britain. From 1906 *C&CE* provided an alternative route.

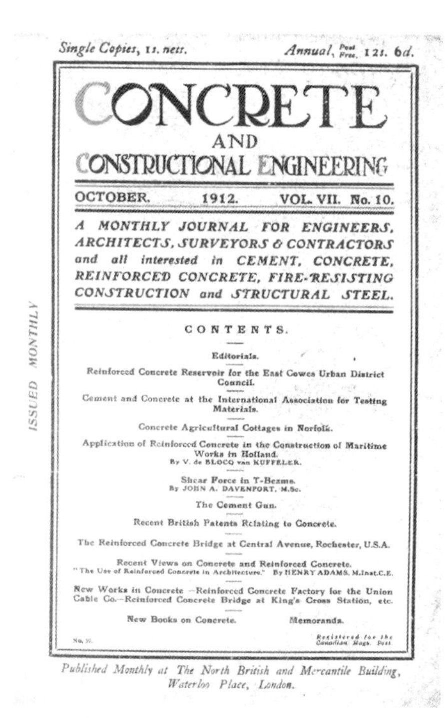

Single Copies, 1s. nett. *Annual, Post Free, 12s. 6d.*

CONCRETE
AND
CONSTRUCTIONAL ENGINEERING

OCTOBER. 1912. VOL. VII. No. 10.

A MONTHLY JOURNAL FOR ENGINEERS, ARCHITECTS, SURVEYORS & CONTRACTORS and all interested in CEMENT, CONCRETE, REINFORCED CONCRETE, FIRE-RESISTING CONSTRUCTION and STRUCTURAL STEEL.

ISSUED MONTHLY

CONTENTS.

Editorials.

Reinforced Concrete Reservoir for the East Cowes Urban District Council.

Cement and Concrete at the International Association for Testing Materials.

Concrete Agricultural Cottages in Norfolk.

Application of Reinforced Concrete in the Construction of Maritime Works in Holland.
By V. de BLOCQ van KUFFELER.

Shear Force in T-Beams.
By JOHN A. DAVENPORT, M.Sc.

The Cement Gun.

Recent British Patents Relating to Concrete.

The Reinforced Concrete Bridge at Central Avenue, Rochester, U.S.A.

Recent Views on Concrete and Reinforced Concrete.
"The Use of Reinforced Concrete in Architecture." By HENRY ADAMS, M.Inst.C.E.

New Works in Concrete—Reinforced Concrete Factory for the Union Cable Co.—Reinforced Concrete Bridge at King's Cross Station, etc.

New Books on Concrete. Memoranda.

Registered for the Canadian Mags. Post

Published Monthly at The North British and Mercantile Building, Waterloo Place, London.

An early issue of Concrete and Constructional Engineering
Courtesy of The Concrete Society

Nonetheless, there were a few pioneering British books available on the market – treatises by authors such as Marsh and Dunn, Twelvetrees,

Coleman and others; several of whom incidentally, had close personal connections with the continent. The number noticeably increased as the market expanded, the RIBA undertook its report and Sachs's other enterprise – the Concrete Institute – was established.

The RIBA Report, 1907

A committee was formed on the recommendation of the RIBA's Science Standing Committee. Its terms of reference were 'to draw up rules for the guidance of architects for the use of reinforced concrete', obviating the dependence on system specialists.

Sir Henry Tanner, I. S. O. – who had been appointed chief architect in the Office of Works in 1898 and was an advocate of concrete – was selected to be the chairman. Wisely, the institute decided to involve other parties with an interest in the outcome, so the membership comprised representatives of the Admiralty, War Office, Municipal and County Engineers' Association, District Surveyors' Association and the Institute of Builders, as well as the RIBA itself. William Dunn, Charles Marsh and Colonel J. Winn, RE, whom we shall meet later, were all members of the committee.

Seventeen concrete elements were supplied by the building firm Messrs Cubitt, as samples for a series of tests directed by Professor Unwin, but otherwise conclusions were drawn and theory developed from existing test results recorded by independent public and educational bodies. Dunn, Marsh and Winn had previously published their own reviews of the technology (based partly on foreign experience), and thus were significant contributors to the study. Marsh even went on to act as an expert consultant to the American A. L. Colby's private review of European practice the following year.

A provisional report was issued in 1907, with the aim of providing a 'good working guide, the laying down of the necessary conditions and setting safe rules'. The committee acknowledged the difficulty of determining the merits of rival systems, but claimed that the *Report* and *Rules* would 'enable an accurate judgement to be arrived at by the architect himself'. Likewise, the work was not

A presentation copy of the RIBA's *1907 report on reinforced concrete*
Courtesy of The Concrete Society

claimed to be final, 'but simply a reasonable guide in the present state of knowledge'. Consequently a second report was issued in 1911 to reflect increases in the use and knowledge of reinforced concrete. On the committee was the recently published Captain Fleming (replacing Colonel Winn), with the addition of E. Fiander Etchells, and representation from the newly formed Concrete Institute.

The Concrete Institute, 1908

The establishment of the Concrete Institute was closely linked to the above developments. The committee chairman, Sir Henry Tanner, was to become vice-president of the institute and Edwin Sachs – editor of *C&CE* and also a leading member of the British Fire Prevention Committee – became the first chairman of the executive (the Fire Prevention Committee incidentally, was responsible for conducting a series of fire tests on concrete at this time and publishing the results. Of the Committee's series of Red Books, for instance, numbers 173, 174 and 188 considered concrete. The FPC also set up a Special Commission on Concrete Aggregates whose interim report was published in December 1908). Sachs – as prime mover in establishing the institute – described the process in a memorandum published in volume one of *Transactions*:

> The formation of the Concrete Institute was primarily due to a number of members of the British Fire Prevention Committee, who having frequently met during 1906 at fire tests dealing with concrete and reinforced concrete floors, had found that there much that ought to be discussed and inquired into jointly by the various professions concerned, as also by those who on the one side had the control of building operations in a public capacity, and on the other had had experience as contractors or manufacturers.
>
> The first step in the direction of obtaining an opportunity for mutual discussion may be said to have comprised the appointment in December 1906, of a Special Commission on Concrete Aggregates, drawn entirely from the membership and subscribers of the Fire Prevention Committee, but representative both of the Government Departments, the Municipal Authorities, the various professions concerned and the industries affected.
>
> On the industrial side, too, a not dissimilar wish for a suitable opportunity for discussing technical questions had made itself felt amongst certain specialist designers and contractors connected with the execution of reinforced concrete work.
>
> A circular issued by a member of one of the reinforced concrete firms, during the autumn of 1907, may, in fact, be deemed to have been a direct, even if a negative cause of the formation of the Institute in its present form, inasmuch as the possibility of some society being instituted to represent only the trade interests, was so displeasing to many of the members of the technical professions concerned, that the writer of this memorandum was induced by some of them to take up the matter, and see if a suitable basis could not be found for the formation of a scientific society that, whilst fully appreciating and welcoming the co-operation and advice of those commercially concerned in concrete, would have

Left: Edwin O. Sachs
Courtesy of The Concrete Society
Right: Educator and writer on concrete, Prof. F.E. Turneaure, at the Engineering Department, University of Wisconsin
Courtesy of the University of Wisconsin, Madison Archives

> the standing of a technical institution of the first order and the active guidance of the leaders of the technical professions affected.
>
> In a manner it might thus be almost stated that the Institute was, on the one hand, formed to fulfil a very general demand, and on the other to prevent anything being constituted at the time on the commercial side that would be detrimental to the free development of the more modern applications of concrete.

The history of the Concrete Institute has been outlined elsewhere by Witten and others, and so the following remarks will relate only to its publishing activity. It might also be worth noting that the council, executive and standing committees included such published authorities as Henry Adams, Bertram Blount, William Dunn, E. Fiander Etchells, Charles Marsh, and E. R. Matthews; while commercial interests included Moritz Kahn and E. P. Wells (referred to above). Moreover, with the further inclusion of Sir Henry Tanner, A. E. Collins, C. H. Colson, E. D. Drury, W. G. Kirkaldy, and H. D. Searles-Wood, there was a considerable overlap with the members of the RIBA committee.

The Concrete Institute contributed to the literature in its own way: indirectly through its encouragement of expertise, formally through its official recommendations, and regularly through publication of its *Transactions & Notes*. The latter provided a transcript of papers presented at the institute's general meetings, together with a verbatim account of the ensuing discussions. At the first of these (on 19 November 1908), the paper was entitled *The composition and uses of plain and reinforced concrete* and was given by Charles Marsh. The speaker was introduced as 'one of our earliest English writers on the subject, and the author of a recognised text-book'. His co-author William Dunn, spoke at the second meeting on *The examination of designs for reinforced concrete work*, and so the institute went on to became an important forum for writers on concrete. Several authors of the day soon appeared on the pages of *Transactions*.

The first significant institutional contribution to the literature was its work on notation. Speaking to the institute's annual general meeting of 17 February 1910, Mr Wentworth-Shields explained the need for standardising literary conventions:

> We then found that the present state of things was intolerable. One turns to various English text-books on reinforced concrete, and one finds that in one C will stand for concrete, in another the same will stand for compression, and in another the same C will stand for a constant, and in all of them C stands for confusion. [*Trans.* vol.2]

The task of rationalising notation (at least for British use), fell to E. Fiander Etchells. Etchells was an important contributor to the promotion of reinforced concrete, a member of the RIBA committee, an energetic participant in the life of the Concrete Institute – he was the president that oversaw the transformation of the institute into the Institution of Structural Engineers in 1922–23 – and a writer on developments in structural engineering. Besides his papers and articles on concrete, and his report on notation however, his principal publication was on structural steelwork and so lies outside the scope of this review.

Much of the early work of the institute – besides coordinating the practice of notation – was to encourage the agreement on and publication of regulations that would permit building with reinforced concrete in London.

Recommendations and regulations

The senior engineering institution – the Institution of Civil Engineers – also looked at reinforced concrete and in 1910 its Committee on Reinforced Concrete issued a *Preliminary and interim report*. Though this was followed up with a second report in 1913, the institution considered that introducing regulations or a code of practice would be premature when there was still much about the new material to understand fully.

The RIBA's second report however, formed the basis of the regulations issued by the London County Council in 1915, following the advocacy of the Concrete Institute and *C&CE* and the amendment of drafts in 1911 and 1913. These *Regulations … with respect to the construction of buildings wholly or partly of reinforced concrete* – introduced under powers granted by the *London County Council (General Powers) Act* of 1909 – were important in being the first in Britain to have the force of law, even if only applicable locally. The American cities of New York and Chicago had introduced similar local regulations in 1903, 1911 and 1906. Both were examples followed by authorities elsewhere, and published local regulations form a distinct niche in the literature of the period.

Also based partly on the RIBA's earlier work was the Concrete Institute's own code of 1918: *Recommendations to clerks of works and foremen concerning the execution of reinforced concrete works*, and its *Standard specification for reinforced concrete work* of 1920.

The developing literature in Britain and America

The period from 1897 until the end of the First World War and 1919 was (for the UK) one of naturalising reinforced concrete: accepting foreign models, developing practical guidance, understanding the theoretical foundation of reinforced concrete design (in which Oscar Faber was an early specialist), establishing institutions, and drawing up recommendations and regulations. Publications included translations from French and German; treatises to introduce the new material; textbooks (mainly by American authors) to consolidate understanding; manuals and diagrams as empirical guidance from practical experience. But throughout these years a distinctive British literature of concrete was emerging. As Ewart Andrews said in a book review written in 1916:

> As Englishmen we admire the great amount of work that American engineers have done in developing the use of reinforced concrete and of attempting to place its design upon a scientific footing. We hope that the authors will not take amiss our suggestion that British engineers have also done something, and that some of their results and opinions might be given greater prominence in a book that has already established a strong position upon the bookshelves of British engineers.

Andrews was right to point out British progress, just as he was right to acknowledge American precedence. A similar pattern to that of the UK pertained in the USA, only starting a few years earlier. Ransome introduced reinforced concrete to the American market in the 1880s and America had responded with journals such as *Concrete* and *Cement Age*, and commercial book publishing by Myron C. Clark, John Wiley, McGraw-Hill and others. The early output was not without contemporary criticism:

> One of the faults of reinforced concrete literature has been the monumental treatises filled with abstruse and involved analysis of the problems of design in which one must wade through a slough of mathematics in order to separate the essentials from the non-essentials; and a few hand books based upon promises given in a separate volume and limited in their value to those particular values of stresses and ratios of Es to Ec which were favored by their authors. (*C-CA*, v.2, May 1913, p.284)

And as in Britain, efforts to improve practice and use led to the establishment of trade and technical bodies such as the Portland Cement Association (1902) and American Concrete Institute (1904).

The ACI started out as the National Association of Cement Users. An exploratory meeting (promoted through the pages of *Municipal Engineering*) was called on October 1904, initially with the idea of marketing good concrete practice on behalf of the manufacturers of concrete block-making machinery. It was clear from the enthusiastic response that interest was more general and at the hurriedly organised inaugural convention in Indianapolis (held in January 1905), the NACU was launched. Its objectives included 'the reading and discussion of papers' presented at conventions and the circulation 'among its members, by means of publications, the information so obtained'. In the early days this was principally in the form of *Proceedings* published annually after the convention, which from 1906 included committee reports and from 1907, ACI standards. On 2 July 1913 the association's name was changed to the American Concrete Institute, and at the end of its first ten years (in 1915), it could claim to have produced more than 18 standards and recommended practices.

A year later the Portland Cement Association was launched to represent the makers, rather than users, of cement. It evolved from the Association of American Portland Cement Manufacturers which had been in existence since 1902. Despite its name, the AAPCM – and then the PCA – was interested in promoting the use of concrete, as testified by an early publication: Maurice Sloan's *The concrete house and its construction* (1912). In 1916 the newly embodied PCA set up a research laboratory in conjunction with the Lewis Institute in Chicago.

Some American universities were also highly influential in deepening and extending an understanding of concrete technology, not only through research and teaching, but the publishing of research monographs and textbooks for correspondence courses. Of the latter, the series of blue-bound pocket textbooks compiled by the staff of the International Correspondence Schools in the 1930s provide a perfect example. Of the former, mention must be made of

A wartime advertisement in Concrete & Constructional Engineering *for BSP text books*
Courtesy of The Concrete Society

the University of Illinois Bulletins, derived from work at the university's Engineering Experiment Station. The work at Illinois in the years just before the First World War was dominated by A. N. Talbot, who established a tradition of research whose contributors (in many cases) went on become the leading lights of the American concrete establishment: Duff Abrams; Hardy Cross; Solomon Hollister; Arthur Lord; Willis Slater; C. A. P. Turner; Harald Westergaard; and Wilbur Wilson. To a lesser extent too, Columbia, Cornell, Kansas, Michigan, Purdue and Wisconsin each made a contribution to the literature.

The First World War marked the end of the proprietary systems' ascendancy. The systems did not disappear for several more years, but existed in a more open climate in which engineers and architects had much greater understanding of reinforced concrete design and construction. Not that this prevented companies from promoting their own peculiar approaches through the publication of manuals, as this contemporary advertisement for piling indicates (see p.xxix).

In the post-war period, many individual engineers who had been trained by the reinforcement specialists – Bowie, Faber, Geen and Turner, for instance, at Indented Bar – established themselves in private practice as consultants, making their expertise available for a fee, and disseminating it in print.

The 1920s: post-war expansion

After the First World War (as in so many aspects of Western society), attitudes to concrete changed. New markets were opened up, prompted largely by the example of America. In the immediate post-war period concrete was heavily promoted for highway construction, 'concrete cottages' and even shipbuilding – all subjects for new titles in the specialist literature, though concrete had already been widely extolled for house-building by American authors such as Hering and Radford. In Great Britain, H. P. Boulnois of the Concrete Institute prepared a report on *Concrete roads* for the Road Improvement Association in 1917; Ballard and Hilton's books on concrete houses came out in 1919; and all three themes were addressed by Concrete Publications Ltd.

At this time Edwin Sachs – for years one of the driving forces of the British concrete establishment – died in 1919. He was only 49. His passing propelled the young Oscar Faber to a position of prominence as technical consultant to *C&CE* and allowed the arrival of H. L. Childe as managing editor of Concrete Publications Ltd – the journal's publisher. These two men were to play a significant role in British publishing over the following years. Faber's writing and lecturing had a symbiotic relationship with *C&CE* and even after he left to concentrate on his own consultancy he still found time to act as editor of the new *Concrete Yearbook*, jointly with Childe, every year from 1924. Childe – having inherited *Concrete cottages* by Albert Lakeman (1918) and the anonymous *Concrete roads and their construction* (1920, see Smith for comment) – also launched the Concrete Series of monographs in 1924, with *A Hundred Years of Portland Cement* by A. C. Davis. Others followed throughout the 1920s, before the series really took off in the 1930s.

1925. *Reinforced Concrete Beams in Bending and Shear*, by Faber.

1925. *Elementary Guide to Reinforced Concrete*, by Lakeman

1926. *Design and Construction of Formwork for Concrete Structures*, by Wynn

1926. *Modern Methods of Concrete Making*, by Wynn

1928. *Concrete Products and Cast Stone*, by Childe

1929. *Precast Concrete Factory Operation*, by Childe

1929. *Concrete Construction Made Easy*, by Turner & Lakeman

Wynn and Lakeman were also closely associated with *C&CE* and this trend continued into the 1930s with the publication of authors such as Adams, Gray,

Above left: Harvey Whipple's 1917 compilation of American writing on concrete houses Courtesy of The Concrete Society

Above right: Concrete roads and their construction *Courtesy of The Concrete Society*

Right: Colophon identifying the 'Concrete Series' of Concrete Publications Ltd Courtesy of The Concrete Society

Reynolds and Scott. During these years – from 1924 to 1934 – the *Concrete Yearbook* usefully contained a bibliography that listed not only the Concrete Series, but also all other books on concrete available in the British market. It served as a sales catalogue, with Concrete Publications Ltd offering to supply copies of any title on receipt of the cover price and postage charge, but also provides a snapshot of the growing canon of published matter.

Much later (in 1956), Childe looked back at this period and observed how the Concrete Series made a significant contribution to the literature of concrete in this country:

> In the early 1920s some of the most popular textbooks in use in this country were of American origin. There were, of course, some excellent books by British authors, but more were needed if British students and British engineers were to have a sufficiently wide selection of British books. It was therefore decided to start the 'Concrete Series' books, and the guiding principle was to make available good books at low prices in the confident expectation that such a policy could not fail to be a financial success ... It is a source of great gratification that some of these books, of which nearly fifty have been published, are now used as textbooks in American universities.

Barry Barton (*op cit*, p.75) has described the 1920s as having 'something of an obsession with concrete, as if its practitioners had to demonstrate to the world that there was nothing that concrete could not do'. Hints of the 'exuberant and all-embracing use of concrete was occasionally seen before the First World War', but structural design tended to be cautious. This enthusiasm and correspondingly 'bold' designs of the 1920s was conspicuous in the work of the continental Modernists, and in this country, concrete construction was championed notably by Sir Owen Williams in a series of high-profile and prestigious projects from the Empire Exhibition of 1924 onwards.

Barton makes another observation: 'yet at the same time the 1920s are regarded as decade of indifferent construction standards.' Both the freedom to experiment and the resulting irregularity of work are attributed 'undoubtedly' to 'the lack of constraints imposed on engineers by formal Codes of Practice'.

It is not surprising that firms specialising in reinforced concrete design and construction should embody the highest levels of competence. What may be surprising is the extent to which some of them were prepared to share in-house expertise with the open market through branded textbooks. While the presentation and choice of examples may have had a promotional benefit to the company, the content of several (those by Sir William Arrol & Co. and Ketton Cement, for instance) was objective guidance – a genuine sharing of experience.

Very early in the period, Stuarts (the flooring company that had experimented with reinforced concrete floors), published its own contribution to the emerging literature, and the New Expanded Metal Co. followed with *Expanded steel in armoured concrete and plaster construction* in 1903. Similarly the American Steel & Wire Co. promoted its *Triangle*

Illustration of proprietary equipment published in O.N Rikof's Concrete making with up-to-date machinery, *1918*
Courtesy of The Concrete Society

mesh wire reinforced concrete pavements and roadways in 1912. In 1911 the British contracting firm William Moss & Sons Ltd placed its 'standard tables for girder reinforcement before the Engineering and Architectural professions'.

After the war a machinery manufacturer from London's Lee valley called Olaf Rikof (who, given recent upheavals on the continent, felt it necessary to clarify his nationality as Swedish) published a substantial booklet on developments in mixer technology. This may be regarded as typical of an emerging trade literature where marketing and technical progress became entwined.

In the 1920s the firms L. G. Mouchel & Partners and the British Reinforced Concrete Engineering Co., Ltd each published two clearly self-serving titles that mirrored each other. One explained the use of the respective company's own system and the other constituted a catalogue of work done:

- Mouchel. *List of works executed in the United Kingdom 1897–1919*, 1920
- Mouchel. *Hennebique ferro-concrete: theory and practice*, Fourth edition, 1921 (first published, 1909)
- BRC. *BRC Structures: a photographic record of the use of reinforce concrete in modern building construction*, 1923
- BRC. *Specification to be carried out to the designs of The British Reinforced Concrete Engineering Co., Ltd*, 1934

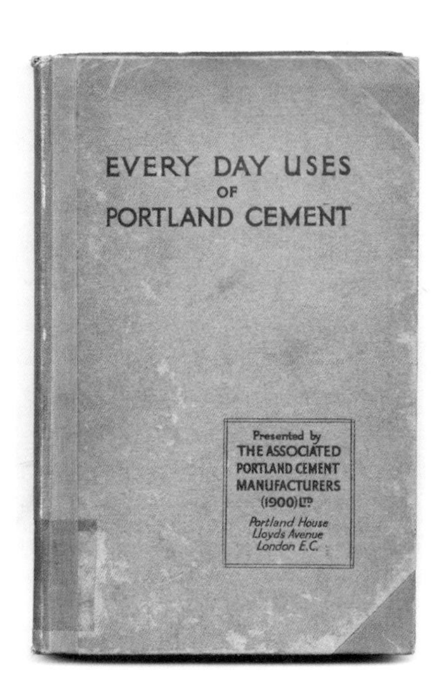

Every day uses of Portland cement,
first published by APCM in 1909
Courtesy of The Concrete Society

The cement manufactures had a vested interest in the success of concrete as a construction material, though their involvement in its use was one step removed. Their distance as a third party supplier allowed their publishing to be more impartial. In North America there are examples of marketing books published by cement makers before and shortly after the formation of the Portland Cement Association. The Canada Cement Company identified its market with *What the farmer can do with concrete* (1910); Atlas Portland Cement addressed *Concrete country residences*; and the Lehigh Portland Cement Company published *Concrete for town and country* in 1922. In the UK marketing was more centralised. After the formation of a national company – Associated Portland Cement Manufactures – the firm (and its later subsidiary, the Cement Marketing Co.) issued a hardback entitled *Every day uses of Portland cement* in successive editions from 1909, besides its more ephemeral pamphlets. Toward the end of our survey (in 1933) APCM published its more specialised *Swimming baths*. Wouldham Cement Co., Ltd. published *Portland cement and concrete* in 1914, and G & T Earle, an expensively produced volume – *The making and testing of Portland cement and concrete* in 1925. Ketton Cement's *Notes on cement and reinforced concrete* of 1934, published by Sir W. C. Leng & Co., attracted the comment that:

> This book, produced at the instance of a cement company, contains no advertisement beyond the mention of their own particular brand in the test results, and the quality of the text bears out their statement that the anonymous author is an eminent structural engineer [*SE*, May 1934, p.274].

This elegant volume was produced after the acceptance of the national Code of Practice in 1933, but *Arrol's reinforced concrete reference book* of 1930 was all the more remarkable for publishing internal company guidance before standardisation and without any direct commercial benefit expected. This important book will be reviewed below under the name of its author, Ernest A. Scott. Other promotional books at this time included rather more specialist titles: *Gunite: a handbook on cement gun work* from Concrete Proofing Co. in 1933; and *The McAlpine system of reinforced concrete tunnel lining*, from Sir Robert McAlpine in 1935.

As well as an age of experimentation, the decade was a time of technological development. With universities such as Illinois, Kansas, Wisconsin and Yale undertaking research, and the professional institutions formulating recommendations, much of this new work was published in limited-circulation reports rather than in book form. Perhaps Professor Duff Abrams – who discovered the importance of the water/cement ratio – may provide the best example of pioneering research published this way, but Wilbur Wilson, Inge Lyse, Timoshenko and Westergaard in the USA were similar academic specialists.

The focus for British research into concrete was provided by the Building Research Station, set up at Watford by the Department of Industrial and Scientific Research in 1926, after a brief spell at East Acton in the early 1920s. The initial output of reports and short guidance notes on concrete was written by a small group of specialists, listed here with their dates of publication:

F. L. Brady, 1927, '27, '30
A. D. Cowper, 1927,'27
R. E. Stradling, 1927, '28
E. Griffiths, 1928
F. M. Lea, 1928, '29
N. Davey, 1929, '30, '30, '32, '33, '34
W. N. Thomas, 1929, '29
G. E. Bessey, 1930, '31
W. H. Glanville, 1930, '30, '30, '31
B. Bakewell, 1931
H. M. Llewellyn, 1931
E. N. Fox, 1934

Of these Norman Davey was the most prolific, tackling topics such as bond and the effects of temperature on strength, but Glanville and Lea the most distinguished to judge from their roles in industry, later reputations and the weight of their published output. Glanville rose to be director of the RRL while an edition of Lea's classic 1935 book is still in print.

The 1920s saw a preoccupation among architects of how to respond to the newly fashionable material. Sir Owen Williams and Maxwell Ayrton's work on the Empire Exhibition in 1924 finally established concrete's presence in architecture (Wembley was even promoted by the Concrete Utilities Bureau in a publication of 1924 as *The first city of concrete*). The two men developed their ideas over the following few years (explored in depth by Yeomans and Cotton),

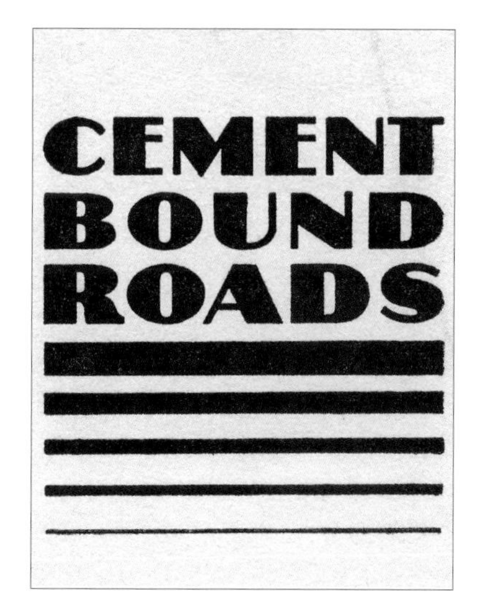

One of the many pamphlets by the Concrete Utilities Bureau, a precursor to the BPCA
Courtesy of The Concrete Society

but their various writings were in the form of papers presented at professional meetings or articles in the architectural press. The bibliography of works by Owen Williams for instance, comprises articles in *Engineering*, *Concrete & Constructional Engineering*, *British Engineers' Export Journal*, *Journal of the RIBA*, *Journal of the Institute of Municipal and County Engineers*, *Selected Engineering Papers* (ICE), *Aberdeen Press*, *The Studio*, *Architects' Journal*, *Architectural Review*, *Engineering News Record*, and *Architectural Association Journal*. Not one was a separately published book, booklet or even pamphlet. Despite the prominence of Ayrton and Williams at this time, it was Bennett and Yerbury's volume of 1927 that was the milestone in British architectural book publishing, and Onderdonk's classic *Ferro-concrete style* a year later fulfilled a similar role in America.

Not only were architects preoccupied by concrete. David Yeomans in his 'Engineering design books' argues convincingly that the explosion of textbook publishing during these years was indicative not just of the relative novelty and complexity of concrete compared with steel (the other 'new' structural material), but of the striving for and emergence of, a new professional identity – that of the consulting structural engineer. Perhaps it is no coincidence that the Concrete Institute transformed itself into the Institution of Structural Engineers in 1923.

The re-formed institution was to issue a new journal – the *Structural Engineer*, which, besides its own contribution to publishing in the form of original papers, noticed new book titles from the commercial and scientific press. Luminaries such as Oscar Faber, Albert Lakeman and Ewart S. Andrews were among the reviewers. Additionally, the institution went on to publish a series of reports, though these were not attributed to named authors:

1925 *Aluminous, rapid hardening Portland and other cements of a special character*
1927 *Loads and stresses*
1928 *Materials and workmanship for reinforced concrete*
1933 *Prevention of dusting of concrete floor surfaces*
1934 *Gravity retaining walls and concrete walls*
1934 *Water retaining concrete structures*

Likewise, the Institution of Civil Engineers (in association with the DSIR) commenced publication of a series of 22 reports in 1920 on the *Deterioration of structures of timber, metal and concrete exposed to the action of seawater*. Work continued on this long-running project until 1951. The ICE also published a series of 'Selected Engineering Papers', some of which – like Nos.107 and 108 on road slabs, by Reginald G. C. Batson – concerned the use of concrete. In similar fashion, the Institute of Water Engineers published a monograph by H. J. F. Gourley on *The use of grout in cut-off trenches, and concrete core walls for earthen embankments* (1922).

In the USA there were similar developments. The ACI – now with permanent premises, salaried staff and a systematic structure of committees to develop new

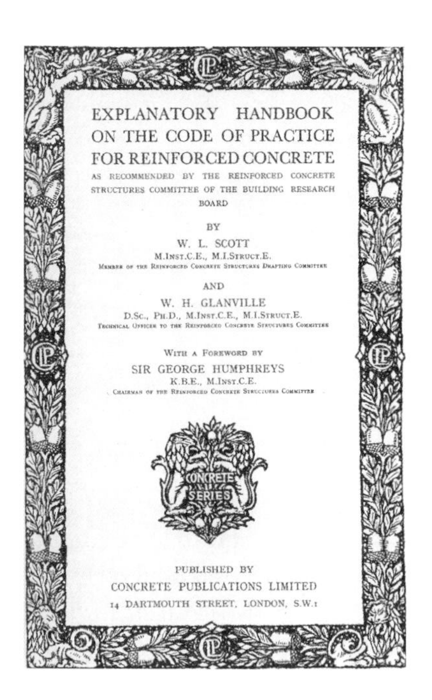
EXPLANATORY HANDBOOK
ON THE CODE OF PRACTICE
FOR REINFORCED CONCRETE
AS RECOMMENDED BY THE REINFORCED CONCRETE STRUCTURES COMMITTEE OF THE BUILDING RESEARCH BOARD

BY
W. L. SCOTT
M.INST.C.E., M.I.STRUCT.E.
MEMBER OF THE REINFORCED CONCRETE STRUCTURES DRAFTING COMMITTEE

AND
W. H. GLANVILLE
D.SC., PH.D., M.INST.C.E., M.I.STRUCT.E.
TECHNICAL OFFICER TO THE REINFORCED CONCRETE STRUCTURES COMMITTEE

WITH A FOREWORD BY
SIR GEORGE HUMPHREYS
K.B.E., M.INST.C.E.
CHAIRMAN OF THE REINFORCED CONCRETE STRUCTURES COMMITTEE

CONCRETE SERIES

PUBLISHED BY
CONCRETE PUBLICATIONS LIMITED
14 DARTMOUTH STREET, LONDON, S.W.1

The first edition of Scott & Glanville's Handbook on the Code of Practice, *1934*
Courtesy of The Concrete Society

publications – issued a series of 'Institute Codes', with editions published in 1925, 1927 and 1928. The year 1928 also saw the start of the 'Special Publications series' (starting with McMillan's *Concrete Primer*), and in 1929 the *ACI Journal* was launched. The PCA's output however, was more research-based, with a series of papers issued in the late 1920s and early 1930s by a team of materials specialists comprising R. H. Bogue, William Lerch, W. C. Hansen, L. T. Brownmiller, F. W. Ashton, W. C. Taylor and W. Dyckerhoff.

The 1930s: maturity, codification and institutionalisation

The distinctive atmosphere of the 1920s changed with the Great Depression, and the accelerated building programme designed to help economic recovery coincided with a more prescriptive trend by the authorities. In 1930 a new London Building Act was passed. It was applicable only in the capital, but became a benchmark for local government elsewhere – the Manchester Architects' and Builders' Consultative Board for instance, issued a *Specification of cement concrete* in 1931. Likewise the Institution of Structural Engineers developed its own draft – *Regulations concerning the design of flat slab floors in reinforced concrete* in 1932. But not long after (in 1933), the Government's Building Research Station issued its landmark *Report of the Reinforced Concrete Structures Committee of the Building Research Board*, with recommendations for a code of practice for the use of reinforced concrete in buildings. This Code of Practice was adopted and a supporting handbook by Scott and Glanville was issued the following year. It was the first formal and nationally applicable Code of Practice in the UK – the progenitor of a long and increasingly complex series of codes down to the present day. The Institution of Structural Engineers' long-standing *Recommendations to clerks of works* were immediately revised and enlarged in line with the new code.

By this date British Standard Specifications had become mainstream too. 'BS 12' – the standard for Portland cement dating from 1904 – was one of the first issued by the new Engineering Standards Committee after its establishment in 1901. It sold in considerable numbers. The publishing of standards grew apace, with 64 in print by 1914 and over 850 by 1939. At the end of our period (in 1935), those relating explicitly to concrete and its constituents included the following titles:

- Portland-Blastfurnace Cement
- Hard Drawn Steel Wire for concrete Reinforcement
- Concrete Kerbs, Channels and Quadrants
- Concrete Flags
- Precast Concrete Partition Slabs
- Concrete Roofing Tiles

The construction industry now seemed ready for a greater use of concrete. The material was now objectively validated. New supply techniques made concrete more readily available: ready-mixed concrete was introduced to Britain in 1931 and pre-cast concrete manufacture was experiencing an explosive period of growth, from the late 1920s and on through the 1930s. Concrete trade interests combined to promote the material in a way the institutions were either unwilling or unable.

Several associations had published work on concrete in the 1920s – the Hollow Building Tile Association for instance, published *Combination hollow tile and concrete floor and roof construction* in 1925, and the Association of Engineering & Shipbuilding Draughtsmen its *Elements of reinforced concrete design* by A. V. Crabtree. The Association of Floor Constructors issued *Coal residues in concrete* three years later. But these associations were not specifically representing concrete interests. The specialist Cast Concrete

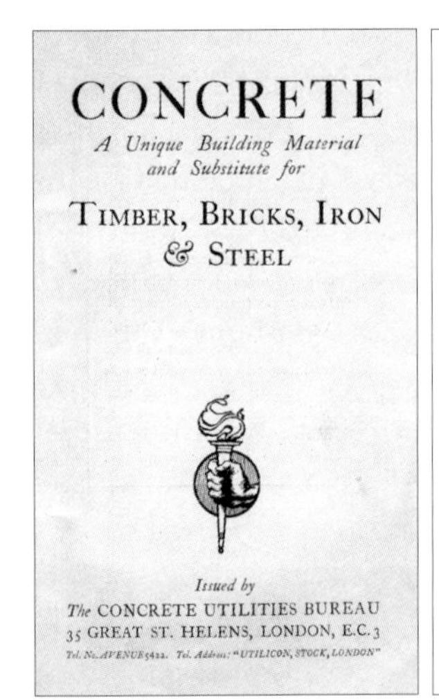

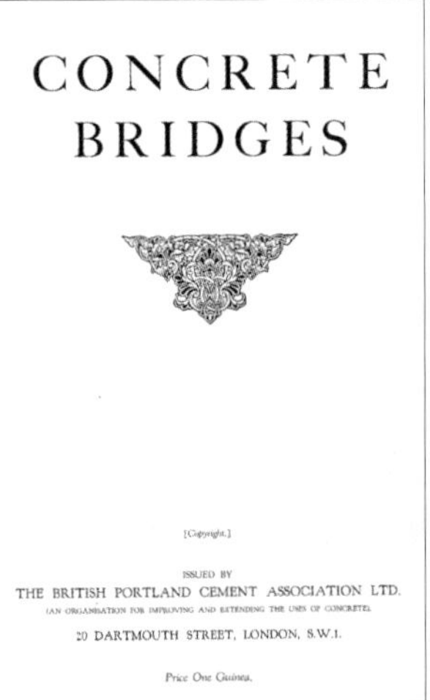

Left: One of the many pamphlets by the British Portland Cement Association
Courtesy of The Concrete Society

Middle: British Portland Cement Association's Concrete bridges, *1928*
Courtesy of The Concrete Society

Right: British Portland Cement Association's The artistic side of concrete
Courtesy of The Concrete Society

Products Association however, was formed in 1927, and it wasn't long before more generic concrete associations were founded. The Concrete Association of India was in existence and publishing by 1929, with Alan Moncrieff editing a handbook and directory of the Indian concrete industry that year, and Robert Mears a comprehensive *Students' text book on reinforced concrete theory and practice* in 1931. The year 1931 saw the National Ready Mixed Concrete Association in the USA publish its *Estimating quantities for concrete* by Stanton Walker. First in the UK was the Reinforced Concrete Association in 1933, which promptly added to the literature by issuing a monthly magazine – *Reinforced Concrete Review*. Then in 1935, building on the work of the earlier British Portland Cement Association, came the body that was to dominate concrete research, publishing, training and information for the following 50 years – the Cement & Concrete Association.

A *Report on proposed cement association*, which was published in May 1934, specifically looked to the example of the Portland Cement Association in the USA, which despite the apparent limitation of the name, was charged with promoting concrete as the market in which cement was used. The American cement industry was enjoying considerable success and it was observed in the *Report* that the PCA was spending £400,000 per year on publishing. Much would have been in the form of promotional material and slender guidance documents, but the PCA's monograph titles current in the mid 1930s included:

- *Concrete in architecture*, 1927
- *A handbook of reinforced concrete building design in accordance with the 1928 Joint Standard Building Code*, by A.R. Lord, 1928
- *Concrete bridges*, 1929
- *Design and control of concrete mixtures* (Sixth edition), 1934
- *Tests of the fire resistance and strength of walls of concrete masonry units*, by Carl A. Menzel, 1934
- *Analysis of rigid frame concrete bridges without higher mathematics* (Third edition), 1935
- *Concrete road design: simplified and correlated with traffic* (Third edition), 1935
- *Simplified design of concrete floor systems with design tables*, 1935
- *Vibration: a better method of placing concrete*, 1935
- *Continuity in concrete building frames*, 1935
- *Concrete bridge details*, 1935

The *Report* also recommended incorporating the assets of the British Portland Cement Association (itself the publisher of a series of 17 booklets), along with the handsome *Concrete bridges* of 1928 (see Faber, Tanner and W. L. Scott herein) and *The artistic side of concrete*. These included titles such as *Concrete: a unique building material*; *Concrete: how it is made*;

Concrete: its artistic possibilities; and several addressing paving and concrete for domestic and agricultural applications. These were duly taken over by the new association to help launch its publishing programme – a programme that would in time see the C&CA rank alongside PCA as a world-class publisher of research and guidance on concrete. As H. L. Childe observed of his own enterprise, British publishing had finally caught up.

C&CA's publication number A1, Cement-bound macadam, *issued in 1935*
Courtesy of The Concrete Society

1902–1911: Novelty and adoption

1902/USA

Cain, Professor William (1847–1930)

AMCE, MASCE

William Cain is an interesting figure with which to start this survey, having adopted reinforced concrete in his later years (aged 55) after an already long and distinguished career in engineering education. His position in the literature might be thought of as similar to Professor Adams [1911] in the UK, Professor Eddy [1913] in the USA, or Professor Warren [1921] in Australia. To take the parallel with Warren further, his first venture into print on the subject of concrete came as a revised edition to an earlier work, and bore a new title to reflect the new material. *Theory of steel-concrete arches and of vaulted arches* (of 1902) was a rewrite of his *Voussoir arches applied to stone bridges, tunnels, domes and groined arches*, published a year earlier – the proximity of the two editions indicating the rapid advancement of the use of reinforced concrete.

Cain was an engineer by profession, and a mathematician. In this second capacity he served for many years as Kenan professor of mathematics at the University of North Carolina. His professional interests may have been wider but a glance at his bibliography suggests an ongoing fascination with arches, bridges and retaining walls. As a young man of 27 he wrote *A practical theory of voussoir arches*, which was published in 1874 by the New York publisher, Van Nostrand. This publishing house issued most of Cain's output over the following years including titles such as: *Maximum stresses in framed bridges* (1878); *Symbolic algebra* (1884); *Practical designing of retaining walls* (1888); and *A brief course in the calculus* (1905).

By the start of the 20th Century Cain's book on reinforced concrete – or 'steel-concrete' to use the terminology of the day – had been published. The 1902 edition was a major rewrite, half of the text being new. His debt to continental sources of expertise was acknowledged, especially to Dr. Hermann Scheffler and experiments conducted by the Austrian Society of Civil Engineers & Architects in 1890–95, and also to his compatriot Professor Eddy's work on graphical statics. The book may have been limited in addressing the application of reinforced concrete to arches, but it was a pioneering work and made Cain's name in the English-speaking world as an early authority on concrete theory.

This interest was not sustained, however, and after a couple of further revisions his prominence was supplanted by the specialists that follow in this survey. He continued to teach at the University of North Carolina and to write. In 1914 he updated his *Practical designing of retaining walls* with appendices on stresses in masonry dams, and in 1916 Wiley published his *Earth pressure, retaining walls and bins*.

For many years Cain was a member of the American Society of Civil Engineers (ASCE) and submitted papers that were published in its *Transactions*. For instance, *Theory of the ideal column* appeared in June 1898 (v.39). In 1922 Cain was finally honoured by the ASCE for his writing and was awarded the J. James R. Croes Medal – an annual commendation for a published paper. He died eight years later, in 1930.

Cain, W. *Theory of steel-concrete arches and of vaulted arches*. Second edition. New York: Van Nostrand, 1902 [Fourth ed. Rev: *CA*, v.6, p.429]

1903/USA

Considere, Armand (1841–1914)

It is appropriate that the first general book specifically on reinforced concrete to be published in the English language – *Experimental researches on reinforced concrete* – was written by one of the pioneers of commercial reinforcement systems (though it was as a translation from the original French by **Leon Moisseiff** [who incidentally was to achieve notoriety years later as the designer of the Tacoma Narrows Bridge, known as 'Galloping Gertie', which collapsed spectacularly in 1940]). The market leader in the early years of the century – Hennebique – had to wait until after Mouchel's death in 1908 to make his way into print in Britain, at least in book form. Perhaps this was a reflection of the relative commercial positions of the two men and the consequent need for publicity. However, Considere's first title was not openly commercial, even if it was self-promoting. Published as a translation in 1903 the original would have consolidated Considere's reputation, following on from his patent of 1895. As chief engineer of roads and bridges in France, he had been appointed in 1900 to head a committee set up by the French Ministry of Public Works. The committee's purpose was to develop a specification for reinforced concrete, which it eventually published in 1906 and whose work was rendered into English by Nathaniel Martin [1912]. Considere became a focus for opposition to Hennebique's hegemony. Whatever his fortunes in his native France – where he became head of a laboratory for testing materials – or the America of his 1903 translation, in the United Kingdom his firm was in practice by 1908.

Armand Considere
Courtesy of The Concrete Society

In the year before his death, Considere's company issued *The Considere system of reinforced concrete*, the first in a series of booklets designed to display the merits of this proprietary system, particularly as applied to columns and piles. It was overtly an advertising vehicle, 'put forward to show the system to its fullest advantage' with photographs of the company's work, but considered at the time to be 'interesting and prepared with much good taste'. It also contained a biography of Considere, to which attention is drawn for more detail.

Cement Age, in its obituary of November 1914, described him as 'a voluminous and thorough writer upon the subject'.

Considere, Armand (trans. Leon S. Moisseiff). *Experimental researches on reinforced concrete*. New York: McGraw Hill, 1903. Second edition, 1906. [C&CA 1959, BL, ICE, IStructE]

The Considere system of reinforced concrete. London: Considere Construction Co., Ltd, 1913. [reviewed *C&CE*, v.8, p.662–663]

1903/UK

Winn, Lt. Col. John (1860–1928)

RE

Winn deserves recognition as the first British writer to produce a book on reinforced concrete, even if it was to a limited audience. Born in Stepney in early 1860, he had been gazetted as a Lieutenant in the Royal Engineers by 1881, and as a young man of 21 was on the active list, based at the School of Military Engineering, Brompton Barracks, in Gillingham, Kent. In 1888 Winn was made assistant instructor at the school, in which post he was later promoted to captain. After five years he was appointed to command 29th Company at Chatham. Four years later, in September 1897, he was again promoted, this time to major – and in this rank appointed to be an instructor in December 1898. In his new role Winn succeeded Major G. K. Scott Moncrieff, RE, as instructor in estimating and construction. He married in 1900, aged 40.

By now a senior officer in the Royal Engineers, Major Winn was responsible for teaching the rising generation of officers at the School of Military Engineering. One of his specialities was the use of recently developed methods of reinforcing concrete and his *Notes on steel concrete construction* – the first British book on the subject – was published in 1903. In 1904 Winn was retained as instructor, and in November he was promoted to Lieutenant Colonel, just a month before retiring in December 1904.

In civilian life Winn served on the Royal Institute of British Architects (RIBA) Committee on Reinforced Concrete, which was constituted in 1907, and was a member of the Special Commission on Concrete Aggregates appointed by the British Fire Prevention Committee, whose interim report was presented on 15 December 1908. He produced three articles for *Concrete & Constructional Engineering* (*C&CE*) in 1906, 1907 and 1908 and was pleased to write the foreword in Captain Fleming's book of 1910, commending the efforts of his brother officer and former student.

He also played a part in the early activities of the Concrete Institute, for which he was asked to read a paper at the first general meeting. In the event that honour went to Charles Marsh, but Winn was graciously described by the latter as 'well known as an authority on all matters relating to concrete and reinforced concrete'. Winn was a member of the institute's inaugural council and its Test Standing Committee, and was still in post in 1913. He died in late 1928, aged 77.

Winn, J. *Notes on steel concrete construction*. Chatham: Royal Engineers Institute, 1903. [IStructE]

1904/USA

Buel, Albert Wells (b.1861)

Buel was born on 20 November 1861 in Keokuk, Iowa. In middle age he wrote *Reinforced concrete construction*, issued by the Engineering News Publishing Company, and by Constable in London. In a preface dated 1 September 1904 in New York City, Buel and his collaborator C. S. Hill stated that: 'in preparing this book the authors have had in mind a treatise for designing and constructing engineers following American practice and governed by the conditions which prevail in America.' Overt theory was eschewed and in its place were practical working formulae, examples and a record of actual practice. In the search for material the authors made extensive reference to the proceedings of the civil engineering institutes on either side of the Atlantic and of French, German and American journals, including *Engineering News* in which the copyright of this book was vested. Part I – the working formulae – was by Buel, and Parts II and III – containing illustrations of representative structures and examples of materials and methods from practice – by Hill. A review by Professor A. N. Talbot [1917] in the *Engineering News* of 15 November 1906 referred to the second edition's 'excellent features'. Buel followed it up with a *General specification of steel railroad bridges and structures* (1906). He was still a consulting engineer in 1912 (based in New York) when he wrote to *Cement Age* to propose improvements to the design of reinforced concrete skeleton construction, declaring himself to be 'desirous that whatever merit they may possibly have shall be open to the free use of the engineering and architectural professions' [*CA* v.14, p.307].

Buel, A. W. and Hill, C. S. *Reinforced concrete construction*. New York: Engineering News Publishing, and London: Archibald Constable & Co., 1904 [ICE]. Second ed., 1906 [C&CA 1959, BL, ICE, IStructE. Copy: TCS. Rev: *EN*, 15 November 1906, *C&CE* May 1907, p.166]

1904/USA

Hill, Charles Shattuck (b.1868)

CE

C. S. Hill was the co-author of *Reinforced concrete construction* in 1904 (see Buel above), and with H. P. Gillette, of *Concrete construction: methods and cost* in 1908. At the latter date he was employed as associate editor of the periodical *Engineering-Contracting*, of which Gillette was managing editor.

His own paper *Comment on the advantages and limitations of reinforced concrete* was printed by the Association of American Portland Cement Manufacturers in 1906, and by 1909 *Cement Age* (v.9, p.444) was able to describe him as the 'author of several standard engineering works'. His work of that year was *Concrete inspection* (published by Myron C. Clark) – a 'convenient little volume' of 179 pages, comprising 'a series of rules and directions so presented that they may readily be understood by those entrusted with inspection'.

Nearly twenty years later in 1928, McGraw Hill brought out his *Winter construction methods* – a theme which had been trailed in an article of the same title in *Engineering News-Record* (v.99 [15] 1927, pp.597–9).

Buel, A. W. and Hill, C. S. *Reinforced concrete construction*. New York: Engineering News Publishing, and London: Archibald Constable & Co., 1904 [ICE]. Second ed., 1906 [C&CA 1959, BL, ICE, IStructE. Copy: TCS. Rev: *EN*, 15 November 1906, *C&CE* May 1907, p.166]

Gillette, H. P. and Hill, C. S. *Concrete construction: methods and cost*. New York: Myron Clark Publishing Co, 1908 [C&CA 1959, ICE. Rev: *EN*, 11 June 1908] and London: Spon, 1908 [BL]

Hill, C. S. *Concrete inspection*. Chicago and New York: Myron C. Clark [PCA. Rev: *CA*, v.9 1909, p.444]

Hill, C. S. *Winter construction methods*. McGraw Hill, 1928 [Yrbk 1934]

1904/UK

Marsh, Charles Fleming (1870–1940)

MICE, MASCE, MIMechE

Marsh, with his collaborator William Dunn, was the first important British writer on reinforced concrete and was acknowledged as such at the time. When about to present the paper at the Concrete Institute's inaugural meeting on 19 November 1908, Marsh was introduced as 'one of our earliest English writers on the subject, and the author of a recognised text-book'. Later, in 1912, he was described in *C&CE* as 'the well known author of various books on reinforced concrete' (v.7, p.71). These included 'two of the best known and best received books on the subject' in English.

Charles Fleming Marsh was born on 6 October 1869 to British parents in Belgium – a place of birth reflected by his middle name. He was the ninth of ten children of the Rev. Henry Augustus and Eliza Marsh and was baptized in Bruges on 19 December. When he was old enough, he was sent back to England for his education, boarding at Woking College in Surrey. In 1894 at the age of 24, he married Ethel Maud Reynolds of Totteridge House, High Wycombe, his senior by two years. They set up home in Twickenham.

Having chosen engineering as a career, Marsh was employed by the Metropolitan Water Board where he was to become chief assistant engineer. Membership of the ICE (and later the ASCE) was a natural accompaniment to such an occupation and, according to the *Minutes of proceedings*, the young man participated in several discussions on hydraulic engineering at ICE meetings between 1900 and the outbreak of the First World War, specifically:

- 'The Burator Works for the water-supply of Plymouth', discussion, session 1900–1901
- 'Coolgardie water supply', discussion, session 1904–1905
- 'Stresses in masonry dams', discussion, session 1907–1908
- 'Glasgow main drainage', correspondence, session 1912–1913

Marsh saw the possibilities of reinforced concrete in his branch of civil engineering and so took an interest in the new technology in the early stages of its emergence in this country. His paper *Construction in concrete and reinforced concrete* given at an ICE meeting during the 1901–1902 session, was presented to Hennebique himself. Marsh contributed to the discussion of 'Reinforced concrete structures' at the engineering conference of June 1907, which was reported in the ICE's transactions on *Harbours, Docks and Canals*. Similarly his contribution to 'Reinforced concrete wharves, warehouses and columns', during the session 1911–1912 combined his interests. His paper of 1916, *Reinforced concrete as applied to waterworks construction* (*C&CE*, v.11, p.142), makes the application explicit.

Long before then however, his newly found expertise was given expression in *Reinforced Concrete*, issued by Archibald Constable in 1904 as the first commercially published textbook on the subject by a British author. In his review of the subject, Marsh (assisted by the architect William Dunn) identified 50 proprietary systems. The book sold out and was promptly reissued in 1905. Further revision followed, drawing on the results of official investigations here and abroad, and an enlarged third edition (this time crediting Dunn as co-author) was published in 1906. With 650 pages it was described as 'encyclopaedic' in its coverage, 'systematic' in its arrangement, and the frequent mathematical data was thought to 'speak highly for the assiduity of the author'. In all, the reviewer in *C&CE* (v.1, p.405) congratulated Marsh on the 'general excellence of this splendid volume'. An American edition by Van Nostrand followed.

Given his status as an author, it is no surprise that when *Cement & Constructional Engineering* was launched in 1906 Marsh should appear in the inaugural issue. His topic – unrelated to his formal discipline of hydraulic engineering – was *Reinforced concrete for foundations of buildings*. Other articles in *C&CE* followed this, namely:

- *Fatigue of concrete and reinforced concrete*, 1907 (v.2, p.127)
- *The laws of proportioning concrete*, 1907 (v.2, p.291)
- *Experiments with reinforced concrete beams and columns*, 1908 (v.2, p.473)
- *Summary of the rules, regulations and recommendations for the use of reinforced concrete as published in various countries*, 1908 (v.3, p.181)

During 1907 Marsh served on the RIBA Committee on Reinforced Concrete, as did William Dunn, and it may be that the latter article was derived from investigative work undertaken on behalf of the committee.

It was a natural progression from involvement with *C&CE* and the RIBA committee to joining the Concrete Institute at its establishment in 1908. Indeed Marsh was the speaker at the Institute's first general meeting and consequently featured in the first issue of *Transactions*. His paper was entitled *The composition and uses of plain and reinforced concrete*. In it he commented on the nature of contemporary writing on concrete: 'It is unfortunate that so many systems of symbols should be used by different authors, and I am afraid that some may consider that I am one of the chief sinners in this respect.' Harmonising this disparate usage was to become one of the major early activities of the Concrete Institute. The other was to contribute to the formulation of the London County Council's regulations for reinforced concrete, and Marsh was able refer in his paper to his recent comparison of the official rules in force abroad. He followed this up a year later with a study of the *Recent Prussian regulations for the use of reinforced concrete*.

Having joined the Concrete Institute, he became a very active participant in its early proceedings: he was a member of the council and the six-man executive from its inception in 1908 and of the Science Standing Committee for several years.

William Dunn was the chairman of the Science Standing Committee (and contributing to ICE meetings at this date), so the two men would have found plenty of opportunity to collaborate on their next title for Constable: *Manual of reinforced concrete and concrete block construction*. This was published in 1909 and ran to four editions by 1922, though the reference to block construction was later dropped from the title. Marsh followed this up in the same year with a companion volume: *A concise treatise on reinforced concrete*. Both books were re-workings of the first, modified for different audiences.

The *Manual*'s object was 'to give in a concise and handy form for everyday use the methods employed for the solution of everyday problems, with the information most frequently required in as condensed a form as possible consistent with a clear presentation of the subject.' It was practical book with worked examples, omitting passages of theoretical reasoning. Consequently the content was briefer than *Reinforced Concrete* and so published in pocket book size. Written in 1908, the revision allowed incorporation of the recommendations of the French Commission's report and that of the RIBA's Committee on Reinforced Concrete, along with new sections from other sources. The material on block construction, for instance, was taken from the American journal *Cement Age*. The manual met with market approval and when the second edition was reviewed in 1911 (v.6, p.715), *C&CE* considered it likely to be familiar to 'most persons engaged in the practice of reinforced concrete construction'. It was considered to be 'strongest on the engineering side' and despite minor quibbles the book – and particularly the worked examples – was thought to be of 'great value' to students and 'practical men'. Its compilation was acknowledged to have involved a 'vast amount of labour and could only have been undertaken by men with a thorough knowledge of their subject'.

Marsh's companion volume was intended as an updated and condensed statement of the subjects covered by the original *Reinforced Concrete*, recognising that with the development of knowledge over the intervening years there was less justification in retaining detailed descriptions of experiments, the results of which were by then accepted (indeed, he reported the results of *Tests on full size reinforced concrete slabs* separately in *C&CE*, v.4, 1909, p.183). This re-working, emphasising the theoretical side ignored in the *Manual*, was based on a series of lectures given during the winter of 1908–1909 at the City and Guilds of London Technical College. A review in 1910 (*C&CE*, v.5, p.145) seems a little ambivalent, but accepted the book as an introductory work and as such considered it 'praiseworthy'.

His final publication, *Reinforced concrete compression member diagram*, was issued by Constable in 1912. It consisted of graphs to facilitate the rapid calculation of hooped reinforced concrete columns according to the rules set out by the RIBA committee. It was described at the time as 'useful', though *C&CE* hoped a still easier design might yet be devised (v.7, p.71). *C&CE* was to publish a technical article by Marsh two years later – *Shearing or diagonal tension reinforcement in beam* (*C&CE* 1914, v.9, p.307) – but much of his remaining output concentrated on regulations.

Having surveyed the various national regulations in force in 1908 and contributed to the RIBA committee's work in 1907–1911, Marsh was well qualified to comment on the London County Council Regulations of 1915. He did so at a meeting of the Concrete Institute in a paper entitled *Criticism of the London County Council Regulations relating to reinforced concrete*, which was summarised at the time in *C&CE* (v.13, pp.207–214) and reported at length in the Concrete Institute's *Transactions* of 1921 (v.9, pp.1–30). The ensuing discussion revealed differences of opinion between the exponents of theoretical and practical approaches, and between the requirements of design and construction. At this time, Marsh (in a continuation of his earlier survey) also reported on the regulations drawn up in Sydney, New South Wales (*C&CE*, v.13, p.232).

In 1918 Marsh was vice president of the Concrete Institute and still acting for the Water Board, which presumably he continued to do so for the rest of his career. He was to publish nothing further, other than a final revision of his earlier *Manual* and died in Cambridge in 1940.

Marsh, Charles F. (and Dunn, William in the third edition). *Reinforced Concrete*. London: Archibald Constable, 1904. Second ed, 1905, third ed, 1906. [C&CA 1959, BL, ICE, IStructE. Copy: TCS third ed. Rev: *C&CE*, v.1, p.405]

Marsh, Charles F. and Dunn, William. *Manual of reinforced concrete and concrete block construction*. London: Archibald Constable and New York: Van Nostrand, 1909. [Rev: *CA*, v.7, p.461]. Fourth ed, 1922. [C&CA 1959, BL, TCS, Yrbk 1934. Copies: TCS. Rev: *C&CE*, v.3, p.178; v.6, p.715–16]

Marsh, Charles F. *A concise treatise on reinforced concrete: a companion to 'the reinforced concrete manual'*. London: Constable, 1909 and New York: Van Nostrand, 1909. Second ed, 1911, third ed, 1920. [C&CA 1959, BL. Copy: TCS. Rev. *C&CE*, v.5, pp.144–5]

Marsh, Charles F. *Reinforced concrete compression member diagram*. Archibald Constable, 1911 [Peddie, Yrbk 1934. Rev: *C&CE*, v.7, p.71]

1904/UK

Dunn, William Newton (1859–1934)

FRIBA

Like his co-author Charles Marsh (above), Dunn enjoyed a high regard in the early days of reinforced concrete in Britain. In 1908 he was described as 'a leading exponent on all questions relating to reinforced concrete', and as we shall see, his experiences in this field of work overlapped with Marsh's for several years.

Dunn was a Scotsman and as a youth in 1876 was articled to the architect Duncan McNaughtan of Glasgow. Five years later he secured a position in the office of William Flockhart, followed by turns with James Marjoribanks MacLaren and Thomas Chatfield Clark. Though he was largely self-taught (according to an obituary by William Curtis Green), concentrating on maths and building construction, he qualified in 1886. He moved to London and established himself in independent practice at 21 King William Street. Throughout his childhood years he was troubled with bandy legs and as his new circumstances enabled him to afford treatment, he had his legs surgically broken and straightened.

His relationship with MacLaren continued – Dunn undertaking work on a fee-paying basis – and after MacLaren's death he entered a partnership with the latter's assistant Robert Watson. This arrangement was underpinned by guaranteed support from a major client that enabled the new partnership to support MacLaren's widow and children. Dunn & Watson operated from 35 Lincoln's Inn Fields. As the practice expanded, Archibald Campbell Dickie was taken on as senior assistant, and then in 1900 William Curtis Green joined.

In 1904 both Dunn and Wilson were elected as Fellows of the RIBA. Besides an expertise in dome construction – he advised on the repair of St Paul's Cathedral – Dunn's skills lay in mathematics, and his interests developed in engineering and the emerging technology of reinforced concrete. He increasingly left architecture to his associates.

Although he was not credited in Marsh's *Reinforced Concrete* when first published in 1904, his assistance was acknowledged and he was named as co-author in the third edition of 1906. Part of Dunn's contribution may have derived from tests he carried out for the well-known building contractor W. Cubitt & Co, in 1903. These tests were on a column using the Considere system – then competing with the Hennebique system for market presence.

The years immediately following publication of this third edition were a period of intense activity. Combining his interest with professional affiliation, Dunn was a natural choice for involvement in the RIBA Committee on Reinforced Concrete of 1907 – indeed he was described by Sachs as 'one of the primary movers'.

Dunn's article on 'graphical statics of reinforced concrete structures' was published the following year in *Engineering*, but 1908 was more notable for the founding of the Concrete Institute. Dunn was a founder member and like Marsh, Sachs and Tanner, appointed to the council in its first year. Like Marsh, he was also on the Science Standing Committee during 1908–1913 – with Dunn in the capacity of chairman. It was at only the second general meeting of the Institute on 17 December 1908 that he read his paper *The examination of designs for reinforced concrete work*, following on from Marsh as speaker at the first. In introducing the speaker, Edwin Sachs had this to say of Dunn:

> Whenever there has been good work to be done, either in the advancement of reinforced concrete or its reasonable control, Mr. Dunn has been to the fore, and only in the past weeks several of us have seen him actively engaged on the question of the regulation of the load periods as granted by the Local Government Board on reinforced concrete structures, and also on the proposed new enactments for the regulation of buildings in the Metropolis under the London Building Act.

Dunn was chief advisor to the Office of Works on reinforced concrete, and maybe this is what Sachs alluded to when he referred to Dunn's 'experience of designs on behalf of a certain government department'. Subsequently, in 1910, he was elected to the council with 239 votes – a large share of the ballot.

Besides lecturing at the Concrete Institute, Dunn was invited by the University of London to be the guest speaker at a series of advanced lectures on reinforced concrete given at the Institution of Civil Engineers. The university press subsequently published these lectures and the 'handsome and excellently printed volume' was subject to a detailed but positive review in *C&CE* the following year (v.6, pp.716–17). In 1911 London University also published Dunn's *Diagrams for the solution of T beams in reinforced concrete*.

Dunn's involvement in architecture – rather than specifically in concrete – was probably quite limited by this time and so to expand, the practice welcomed Curtis Green as a partner in 1912. By 1914 the firm had offices in Cape Town and Durban, as well as London, and Curtis Green went on to make a name for himself, styling for instance, the new Dorchester hotel designed in 1930 by the great concrete-specialist Owen Williams.

In 1915 Dunn planned to retire and move to Kenya, but the progress of the war delayed his departure. On eventual arrival in East Africa in 1919 he built himself bachelor quarters adjoining his brother's premises and took up farming. He died there on 7 February 1934.

An obituary by Curtis Green described Dunn as a 'rugged and uncompromising' man who 'always chose the difficult path', yet had 'the heart of a child and the clear outlook that abhorred all that was mean and second rate: fortunately [life's] rigours were sweetened to him by an unusually keen sense of humour, and a love of young people. He was an extraordinarily good talker and the best of good company.' (*RIBA Journal*, v.41, 24 February 1934, p.418).

A fuller account of his professional life and a list of his architectural commissions can be found in the *Dictionary of Scottish Architects*.

Marsh, Charles F. (and Dunn, William in the third edition). *Reinforced Concrete*. London: Archibald Constable, 1904. Second ed, 1905, third ed, 1906. [C&CA 1959, BL, ICE, IStructE. Copy: TCS third ed. Rev: *C&CE*, v.1, p.405]

Marsh, Charles F. and Dunn, William. *Manual of reinforced concrete and concrete block construction*. London: Archibald Constable, and New York: Van Nostrand, 1909. [Rev: *CA*, v.7, p.461]. Fourth ed,

1922. [C&CA 1959, BL, TCS. Copies: TCS. Rev: *C&CE*, v.6, p.715–16]

Dunn, William. *Lectures on reinforced concrete delivered at the Institution of Civil engineers in November 1910*. London: University of London Press and Hodder & Stoughton, 1911 [ICE, IStructE, PCA, Peddie]

Dunn, William. *Diagrams for the solution of T beams in reinforced concrete*. London: University of London Press and Hodder & Stoughon, 1911 [Peddie]

1904/USA

Falk, Myron S.

PhD

Dr Falk was Instructor in Civil Engineering at Columbia College (later Columbia University), New York. Between early 1903 and June 1904 he supervised John Hawkesworth [1906] and Walter T. Derlath in a programme of tests on concrete for tensile strength and tensile efficiency of elasticity. During 1904 his book *Cements, mortars and concretes: their physical properties* was published by the Myron C. Clark Co. A notice published in the August issue of *Cement & Engineering News* makes the following claims for the author:

> The author has made a careful study of the physical properties of concrete and cement mortars...The data has been gathered from numerous scattered articles and books in various languages... classified and arranged under proper heads and subjects...The wide variety of tests covered by the data submitted must be highly appreciated by the supervising architect and engineer who desire the latest information.

'The purpose of this book', paraphrased Moisseiff subsequently, was 'to present in compact form the best of the numerous scattered results of tests and studies of the physical properties of cements, mortars and concrete'. Moisseiff considered it 'compiled with a good knowledge of what is best in the field'.

Falk, M. S. *Cements, mortars and concrete: their physical properties*. New York: Myron C. Clark, 1904. [ICE. Rev: *EN*, 13 Oct 1904]

1904/USA

Heidenreich, Eyvind Lee

Member AIME and Western Society of Engineers

Having had about ten years experience of working on silos in both North and South America, E. Lee Heidenreich was invited by the ASCE to present a paper on American grain elevators at the 1893 International Congress of Engineers in Chicago. With the advent of reinforced concrete he revised his treatment of the theme as *The design of concrete grain elevators*, which was published in the ACI's *Journal Proceedings* of 1 March 1912. By that date he had written both of his contributions to the corpus of books on concrete.

His first title was one of the earliest to appear in America. In anticipation of the book a series of monthly articles, commencing in 1903, appeared in *Cement & Engineering News* under the title 'Armored concrete construction'. The intention, as announced in 1904, was to issue the serialised work as a treatise:

> The *Cement & Engineering News* will publish the above named treatise on

Armored Concrete Constructions. It will cover from 500 to 700 pages with over 400 illustrations tables and diagrams. Abstracts from the text will appear monthly up to the date of publication. The price of this treatise has not been definitely fixed, but orders will be entered for the present at $7 C.O.D. on delivery of the book.

The full title of the book was, *Treatise on armored concrete construction: with general applications under the various systems in use in the United States and Europe with numerous illustrations and calculations*. The synopsis advertised, and the chapters published individually, show the first half of the work to consist of descriptions of structural type or element (bridges, chimneys, docks, domes, floors, stairs, tanks – and Heidenreich's speciality, silos, are included), with the second half setting out calculations, costs and practical details.

For his second title, in 1909, Heidenreich was based in New York in 1909, and though the *Pocketbook* was published by Chicago-based Myron C. Clark (the publisher had New York offices too). It was well received: 'This is the most satisfactory of the handbooks yet published for reinforced concrete work', declared Moisseiff at the time; 'compact, well arranged and clear, and cognoscent of rapidly evolving standards. The author's experience in pipes, culverts and sewers illuminates those subjects, whereas the treatment of bridges is somewhat perfunctory.'

Heidenreich, E.L. *Treatise on armored concrete construction*. Chicago: Cement & Engineering News, 1904 [C&CN, v.15, 1903/04]

Heidenreich, E.L. *Engineers' pocketbook of reinforced concrete*. Chicago: Myron C. Clark Publishing Co., 1909 [IStructE, PCA. Rev: EN, 14 Jan 1909], and London: Spon, 1909 [Peddie]. Second edition, 1915 [BL]

1904/USA

Mensch, L. J.

MASCE

Little is known of this author, save the books attributed to him and that he wrote five articles for publication in the American Concrete Institute's *Journal Proceedings*. These appeared in 1911, 1914, 1917, 1930 and 1932, on the subject of columns and the deflection of beams and slabs. With reference to these articles, Mensch was described by *Cement Age* as an 'eminent authority' [v.13, 1911, p.284].

As a civil engineer and contractor Mensch had practised in various parts of America, but it was appropriately in Chicago that his first book was published. *Architects' and Engineers' Handbook* was serialised in *Cement & Engineering News* before being issued in book form toward the end of 1904. This was a format following Heidenreich's *Armored concrete construction* that was published earlier in the year, and with which it shared many illustrations supplied by William Seafert, the publisher and copyright holder. It drew on Mensch's practical experience and consisted of the various standard answers he had developed in response to questions from clients and other professionals. He hoped to redress the 'absence of suitable literature in the English language on the subject'. His contemporary Leon Moisseiff described it as 'a handsomely printed and much illustrated book of more than 200 pages', but as it promoted the Hennebique system, it was 'rather a trade publication than an engineering book'. Indeed Mensch announced from the outset that he would distribute copies 'gratis to clients or responsible persons intending to become his clients'. More dismissive was Moisseiff's comment: 'it is of the propaganda type, demonstrating what reinforced concrete can do, but contains no engineering or scientific information which could be made use of.' Similarly, but perhaps more fairly, he wrote: 'Judged as a trade publication, the work is a good one, but as a technical work for engineers, it can hardly be considered so favourable, as the treatment of the subject is partisan and incomplete.'

In 1909 Mensch followed his earlier promotional work with *The reinforced concrete pocketbook: a practical manual of tables, rules and illustrations to help in design and estimating*. At 216 pages, *Cement Age* described this as 'a convenient volume for field or office, but comprehensive in character … not a textbook, but designed to be of practical value'. The reviewer judged that it 'should appeal to those qualified to comprehend it, for it has given space for information sometimes crowded out of books of an elementary character', and declares 'the chief claim of the author is the presentation in convenient form of established facts'.

Mensch, L. J. *Architects' and engineers' handbook or reinforced concrete constructions*. Chicago: Cement & Engineering News, 1904 [PCA. Copy: TCS]

Mensch, L. J. *The reinforced concrete pocket book*. San Francisco: Mensch, 1909 [PCA. Rev: *CA*, v.9] and New York: *Engineering News*, 1909 [Peddie]

1905/USA

Sabin, Louis Carlton (b.1867)

BS, MASCE

The one book on concrete by Louis Sabin was derived from a period in early-married life (his wife was Mary Ruth Elizabeth Doucette) when he was employed as assistant engineer in the American army. In the preface to *Cement and Concrete* he explained that the 'original investigations forming the basis of the work were made in connection with the construction of the Poe Lock at St Mary's Falls Canal, Michigan, under the Direction of the Corps of Engineers, US Army'. Completed in January 1905 the book was published by McGraw later that year and by Constable in Great Britain. It was well received. According to *The Surveyor*, 'Mr Sabin's notable book is as comprehensive as it is original in its investigations, and clear and concise in its arrangement'. *Cement Age* (v.2, p.210) was almost strident: 'this important volume of 500 pages is a most valuable contribution to the field of cement literature.' Its review concluded with the assertion that 'no one who is dealing with cement, concrete, or reinforced concrete can fail to find what he is seeking in Mr. Sabin's book, presented in a clear, compact and logical manner'. It ran to a second edition, revised and enlarged in 1907, and the *Engineering News* approved: 'The book as it now stands is an admirable treatise on concrete as material, but must be taken in connection with some reference book of design and construction to make a complete survey.' Commentators at the time noticed the similarity of subject matter treated by Sabin, and by Taylor and Thompson, with Sabin leading in his coverage of physical testing.

Sabin, L. C. *Cement and concrete*. New York: McGraw Publishing Co., and London: Constable, 1905 [BL, ICE, IStructE. Rev: *CA*, v.2, p.210; *C&CE*, v.2 May 1907, p.165]; and Boston: Stanhope Press, 1907 [*EN*, 18 July 1907]

1905/USA

Taylor, Frederick Winslow (1856–1915)

MEScD

Frederick Winslow Taylor
Courtesy of the Stevens Institute

Known as 'the father of scientific management', Taylor's involvement in concrete appears as a sideline to his life's main work, yet the textbook *Concrete plain and reinforced* (that he co-wrote with Sanford Thompson) was one of the most widely circulated of its day.

He was born in Germantown, near Philadelphia, Pennsylvania, the youngest child of Franklin Taylor and Emily Annette (née Winslow) on 20 March 1856. After an early education with his mother and two years of school in France and Germany, he entered Phillips Exeter Academy in New Hampshire to prepare for Harvard Law School and follow his father's profession. He graduated two years later but with such poor eyesight that he abandoned further study. Between 1874 and 1878 he worked for a local pump manufacturer, learning the trades of pattern maker and machinist at the Enterprise Hydraulic Works. He then joined the Midvale Steel Company as a common labourer, but after studying at night for a degree in mechanical engineering, rose to be chief engineer in 1884. On 3 May 1884 he married Louise M. Spooner of Philadelphia. During his 12 years with the steel company he introduced many improvements to machinery and manufacturing methods, patenting the largest successful steam hammer ever built in the USA. After three years (1890–1893) as general manager of a paper mill business, he began a consulting practice in Philadelphia. His card read: 'Systematizing Shop Management and Manufacturing Costs a Speciality.'

From years of observation Taylor had developed a theory that by scientific study of every process in a factory, data could be gained to gauge the fair and reasonable capacities of man and machine, and application of such intelligence would improve efficiency and industrial harmony. For five years he made a living applying his ideas in a variety of establishments until in 1898 he was retained by the Bethlehem Steel Company. There his work resulted in increased cutting capacities of tool steel by 200–300 per cent and his Taylor-White process became accepted in machine shops worldwide. For treating high-speed tool steels he was awarded a gold medal at the Paris Exposition in 1900 and the Elliott Cresson gold medal by Philadelphia's Franklin Institute the same year.

Convinced that his scientific management methods could be more widely adopted, he left Bethlehem in 1901 and devoted the rest of his life to promulgating his theories. Active in the American Society of Mechanical Engineers, his work on management appeared in *Transactions* in 1895 and 1903, and after two years as vice president he served as president in 1906. In 1911 a society of his own – the Society to Promote the Science of Management (posthumously the Taylor Society) – was established to extend his work. That year saw publication of his book *The principles of scientific management*, with *Scientific management* (edited by C. B. Thompson) following in 1914.

More significantly for us, he was associated with another Thompson during his autumn years. It was in this period of consultancy that he and Sanford Thompson turned their attention to the new technology of reinforced concrete and applied their modernising principles to this emerging industry. The application of steel reinforcement to concrete may have provided the link to Taylor's long experience of the steel industry. Maybe it was the construction background of Taylor's fellow advocate of scientific management? Perhaps concrete was seen as a means to enhance the productivity of construction. Whatever the cause, Taylor and Thompson were engaged on collaborative work in the building trades from 1892 until Taylor's death in 1915. Their correspondence seems to have been particularly intense in the three years before publication of their *Treatise on concrete: plain and reinforced* and led to joint articles in the periodical press, such as 'Quantities of water to use in gauging mortars' in *C&EN* (1903/4, p.112–113). Indeed both men corresponded with a third prophet of scientific management, Frank Gilbreth [1908] between 1904–1906.

Concrete: plain and reinforced was initially published by Wiley in 1905, and announced in the trade press as 'full of valuable data ... The entire literature on the subjected treated has been studied, condensed and arranged under appropriate headings and incorporated into the text ... A compilation bristling with valuable data.' (*C&EN*, v.17, p.216). It was well received, as evident in this detailed review in *Cement Age*:

> This noteworthy contribution to cement literature is especially welcome at this time when the field for the use of cement and concrete has so largely expanded. The book itself is prepared in the most accurate and pains-taking way, embraces some 600 pages and is practically all 'meat'.

The reviewer described its authors as 'two men well-known in scientific research', and considered their collaboration as representing 'the tendency of the period'.

It was successively revised and reissued: Chapman & Hall published British editions, where the book was recognised as extensive in coverage, but not too advanced as to hinder its role as a basic working tool for the engineer:

> The treatment in this book is very full, and the summary is not short. Just the information needed for design is given, though, of course, other special books provide other information that is really

> required for the finished designer; but, still, Taylor & Thompson are a sufficient guide to enable many engineers to deal efficiently with the small problems of reinforced concrete design. (*C&CE*, 1910, v.5, pp.855–56)

The second edition represented a 'distinct improvement upon the first', incorporating over 200 pages of new material, including the conclusions of Taylor and Thompson's testing activity in 1909. A review in *C&CE* (1910, v.5, pp.855–56) considered the book 'most praiseworthy. It should be in the possession of every one interested in concrete'.

An extended review of the third edition (1919) was written for *C&CE* by Ewart S. Andrews [1912] – a notable author in his own right. Though it comments in detail upon the particular edition it gives an indication of how Taylor and Thompson had come to be regarded by peers. First, Andrews recognised that by its third edition the work 'holds a very high place in the literature of concrete engineering'. He considered that 'the book as a whole is excellent and is almost indispensable to all who deal literally in concrete matters'. Identifying its niche, he said it is 'not a student's book ... it is a book for men actually engaged in the design and execution of concrete work'. Moving onto a detailed assessment, Andrews thought 'it bears evidence in several places of thoughtful presentation which helps to make its very extensive subject matter easily accessible.' Yet, if anything, the referencing and indexing were too thorough to be helpful in ordinary use. Andrews challenged the book's treatment of shear, where its understanding lagged behind recent developments in Britain. Likewise, he took issue with Edward Smulski – a contributor to the book on double reinforcement. A feature of the book was the participation of contributors for particular chapters: Fuller and Feret appeared as well as Smulski.

Reverting to more general comments, Andrews raised the relative positions of British and American literature:

> As Englishmen we admire the great amount of work that American engineers have done in developing the use of reinforced concrete and of attempting to place its design upon a scientific footing. We hope that the authors will not take amiss our suggestion that British engineers have also done something, and that some of their results and opinions might be given greater prominence in a book that has already established a strong position upon the bookshelves of British engineers.

The fourth and final edition was rewritten and issued in three volumes long after Taylor's death: 'a complete departure has been made from the old Taylor and Thompson we have known'. Thompson provided the continuity and Smulski – who had contributed to the third edition – was cited as a co-author. The American emphasis on flat-slab or 'mushroom' floors was commented on in review as an example to follow in Britain.

The second book by Taylor and Thompson was one that sits more naturally with Taylor's expertise in scientific management. Entitled *Concrete costs* and published both by Wiley in America and Chapman & Hall in Britain, it expounded the benefits of motion study, effective estimating, job organisation and bonus payments. It was favourably reviewed but did not run to a second edition. It did however (along with *Concrete: plain and reinforced*) appear in French as *Practique de la construction en beton et mortier de ciment, armes on non armes*. The combined work was translated and adapted by M. Darras in 1914.

So, with these and other works on steel and scientific management, Taylor achieved a literary immortality and left a legacy in his eponymous society. In a lifetime of achievement he filed about 100 patents, gained two gold medals and (in his younger days) was the tennis doubles champion of the USA for 1881. A biography of this outstanding man was prepared by Ordway Tead in the 1930s.

Taylor, F. W. and Thompson, S. E. *(A treatise on) concrete plain and reinforced*. New York: John Wiley, 1905 [BL, C&CA 1959, ICE. Rev: CA, v.2, p.610]. Second ed, 1910 [BL. Rev: *CA*, v.10; *C&CE*, 1910, p.855]. Third ed, 1919 [BL, ICE]. Fourth ed, 1925 [BL, C&CA 1959, ICE, IStructE, TCS, Yrbk 1934. Copy: TCS. Rev: *SE*, 1926, p.67]

Taylor, F. W. *Concrete costs*. New York: John Wiley & Sons, 1912. [BL, ICE, Yrbk 1934. Rev. *C&CE*, 1912, p.553]

1905/USA

Thompson, Sanford Eleazer (1867–1949)

SB, MASCE

A collaborator and follower of Taylor (above), Thompson made his name applying the latter's principles to a range of industries in the years up to and after his co-author's death in 1915. He had a particular interest in concrete and was described in his middle years as 'the well known engineering expert' (*CA* v.6, p.411).

He was born on 13 February 1867 at Ogdensburg in New York. The education of his early years culminated a degree in chemistry and civil engineering from Massachusetts Institute of Technology. Thereafter he became an engineer and superintendent of construction at various plants, particularly paper mills, until 1896. He turned to consultancy next and in this period, although at an earlier stage in his career than the older man, he began his collaboration with Taylor.

The two men shared various career experiences, both working in the paper and iron industries (Midvale Steel, 1895; Bethlehem Steel, 1901–1903), but our interest is in their joint work in 'the building trades' during the early years of the century, when Thompson was based at Newton Highlands, Massachusetts. Thompson's papers include: summaries of concrete and soils studies from 1892–1914; a contract of 1895 signed by both men for work measurement in construction; letters from 1903–1905; and further papers on concrete from the year of Taylor's death.

During the period 1897–1905 Thompson maintained a correspondence with Frank Gilbreth [1908] on concrete, construction and measurement studies. Perhaps most importantly was correspondence on experiments in reinforced concrete dating from 1904–1906 and 1908.

The outcome of this shared work (as we have noted above) was the *Treatise on concrete plain and reinforced* of 1905, co-authored by Taylor and Thompson. It was revised shortly after the series of experiments that were the subject of correspondence with Gilbreth, and was followed in due course by *Reinforced concrete in factory construction*, which Thompson wrote in 1907 for the Atlas Portland Cement Co. It provided for a market demand, as by 1913 it had reached its seventh edition.

His reputation enhanced by these publications, Thompson enjoyed the acquaintance of Robert Lesley [1907] – a prominent cement industry figure and prime mover at the ASTM – who also edited *Cement Age*. Thompson appeared regularly in the magazine between 1906–1908, for example:

- *Determining quantities in concrete mixing* (v.2, p.866)
- *Collapse of a filter roof* (v.6, p.411)
- *Concrete in the Chelsea fire* (v.6, p.569)
- *Permeability tests of concrete* (v.7, p.47)

Also at this time, Thompson was commissioned to write a report for the Association of American Portland Cement Manufacturers (AAPCM) on the 400 concrete chimneys then being, or lately, constructed in the USA. He gave detailed consideration to the failures and defects of 120 of these, but concluded that concrete, given adequate design and workmanship, was an eminently suitable material for such construction. His report was presented to the association in December 1907 and summarised in *Cement Age* as *Reinforced concrete chimneys* early in the following year. Moreover he was a judge for an AAPCM competition in 1907, presumably a consequence of his consultancy work in the cement industry. Further work for Atlas Cement followed, in which he edited the company's 1909 volume *Concrete in railroad construction*.

His involvement as an occasional writer for *Cement Age* continued after its merger with *Concrete*, and during the years 1911 to 1914 he was a regular contributor to the discussion columns as an expert respondent.

In 1917 – a couple of years after Taylor's death – Thompson founded the Thompson & Lichtner Co: an engineering and management consultancy that was to bear his name for the rest of his life. He continued (with increasing help from Edward Smulski) to revise and update 'Taylor & Thompson' in a further two editions. During his twilight years, he and Smulski released another book, *Reinforced concrete bridges*, published by Wiley in 1939.

These years, particularly from 1921 to 1943, were notable for appointments to various government posts, including work for the American Institute for Economic Research on national economic policies in 1934–1935 – a far cry from concrete! Still he was made an honorary member of the American Concrete Institute (ACI) in 1932 and kept up his involvement with the Boston Society of Civil Engineers from 1894 until the year before his death. He died at a grand old age in 1949.

Thompson's papers – reports and correspondence with Taylor, Gilbreth and others – are held at the Kheel Center for

Labor-Management Documentation and Archives at Cornell University Library.

Taylor, F. W. and Thompson, S. E. *(A treatise on) concrete plain and reinforced*. New York: John Wiley, 1905 [BL, C&CA 1959, ICE. Rev: *CA*, v.2, p.610]. Second ed, 1910 [BL. Rev: *CA*, v.10; *C&CE*, 1910, p.855]. Third ed, 1919 [BL, ICE]. Fourth ed, 1925 [BL, C&CA 1959, ICE, IStructE, TCS, Yrbk 1934. Copy: TCS. Rev: *SE*, 1926, p.67]

1905/UK

Twelvetrees, Walter Noble (1853–1941)

MIMechE, MSCE (France)

Walter Noble Twelvetrees Courtesy of the IStructE

One of the most prolific of early British writers on concrete, W. Noble Twelvetrees was among the first to write on the subject and sustained his output until 1932 – almost the entire span of our review.

He was born in Boxton, Lincolnshire, and moved to London in the course of his career in engineering. His earlier years were spent in mechanical and electrical engineering, before developing interests in structural steelwork and finally, reinforced concrete. In these he reflected his occupation in membership of the appropriate professional associations: the Institution of Mechanical Engineers and the Institution of Electrical Engineers, as well as the engineering societies.

After a spell as Vice President of the Civil and Mechanical Engineers' Society, Twelvetrees served as President for the years 1907 and 1908. His presidential address for 1908 on 1 May at Caxton Hall, Westminster, was a farsighted proposal for 'improved control of vehicular traffic at important street junctions'. This was later published in *Progress: the Scientific New Zealander* [v.3(7), p.228] under the title *Gyratory traffic regulator*. In this capacity he worked hard to bring about the amalgamation of the society with the Society of Engineers. This was achieved and in 1910 the new Society of Engineers (Incorporated) was launched. He became a Fellow and later (in 1919) served as President. In addition to this society he was a member of the Manchester University Engineering Society, becoming a Life Member at some point between 1905 and 1911.

In 1900 he wrote a treatise entitled *Structural iron and steel*. By the following year he was described in the census of 1901 as a civil engineer, resident in Streatham. He wrote *Electricity for small hospitals* for the Royal Society for Promotion of Health in 1903 (*J.R.S.P.H.*, v.24, pp.362–369), but by this date his attention had been captured by the emergence of a newer technology. His long-standing friend, T. J. Gueritte – one of the original partners in the firm of L. G. Mouchel – described him as 'one of the first in this country to realise the possibilities of reinforced concrete'. He started to visit important examples of concrete construction and his first book on the subject – *Concrete-Steel* – was published by Whittaker in 1905.

Twelvetrees' preface makes his purpose clear, and indicates the precedence and importance of this British book on the subject (notwithstanding Marsh's work of the previous year):

> The object of this work is to present definite and reliable information relative to concrete-steel construction, a subject whose literature is scattered over articles in various technical journals, chiefly of foreign origin, and upon which no treatise of convenient form and dimensions has hitherto been published in the English language for the guidance of engineers, architects, and others who desire to make use of the new material.

With an established expertise in steel, he was able to give weight to this component of the new material – an aspect of the book picked up by an otherwise purely descriptive review on *Cement Age* (v.2, p.686).

This was followed by the accompanying *Concrete-steel buildings* from the same publisher a couple of years later. With no claim to originality, it was a compilation of case studies of recent reinforced concrete buildings both in Great Britain and America, carefully described and illustrated with numerous drawings and photographs. While several of these had previously appeared in the Periodical Press, others were published for the first time. It may be noted that the expression 'reinforced concrete' had not yet become fully established – 'concrete-steel', 'ferro-concrete' and 'beton-arme' were also competing for acceptance.

By this date *Cement & Constructional Engineering* had been launched and Twelvetrees was to become a regular contributor, with a series of articles on reinforced concrete bridges in volume 1 and whaves and quays in volume 2. His involvement in the periodical press intensified in 1909 when he became editor of *Ferro-concrete*: the new house journal of L.G. Mouchel & Partners; a monthly review promoting reinforced concrete and the Hennebique system in particular. Despite his advancing age he retained this post until 1932 and it would have been through this role that his friendship with Gueritte developed. In his obituary of Twelvetrees (*C&CE*, Sept 1941, p.378), Gueritte wrote: 'During this period he described hundreds of concrete works and gave the results of many tests and other useful information. Very thorough in his work, he would rarely describe what he had not inspected personally. For instance, Gueritte and he were guests at the launch of the first 1000-ton reinforced concrete barge at Poole Harbour in 1918. Later, in June 1928, he attended the opening of the landmark structure the Royal Tweed Bridge, about which he wrote a souvenir number.

He also wrote *Francoise Hennebique: a biographical memoire* as a tribute to his journal's inspiration [*Ferro-concrete*, v.12 (1921), pp.119–144]. Perhaps too, his Francophile employment led him to membership of the *Societe des Ingenieurs Civils de France* the year after taking on *Ferro-concrete*. With Gueritte he was instrumental in setting up a British Section in 1919, becoming President in 1922. Interestingly he succeeded Gueritte as President of the British Section within a year or two, while the latter followed him as President of the Society of Engineers. The two men collaborated on the translation into English of B. de Fontviolant's *The design of circular bridges* in 1921. His work in this area was recognised and he was decorated as an *Officier de l'Instruction Publique* by the French Government.

Back in 1909, at the start of his term with *Ferro-concrete*, he changed publisher for his next book. *Simplified methods of calculating reinforced concrete beams* was issued by Sir Isaac Pitman & Sons – the first of three titles by this publishing house. *Concrete and reinforced concrete* followed, as did *A treatise on reinforced concrete* in 1920.

Back with Whittaker in 1911, Twelvetrees wrote *The practical design of reinforced concrete beams and columns*. This was intended to develop the general principles outlined in *Concrete-steel* and *Concrete-steel buildings*, and now that reinforced concrete construction was becoming more widespread, present engineers with a complete series of formulae for practical design, along with simple working rules and labour-saving diagrams. In doing so, its author 'endeavoured to render the work a thoroughly practical guide'. However, by concentrating just on beams and columns he restricted the value of the book and its reviewer in *C&CE* (v.11, pp.799–800) anticipated that it was 'hardly likely to be adopted either as a text-book for students or as an essential reference book by the majority of concrete engineers'. Criticism was also levelled at Twelvetrees for failing to adopt the Concrete Institute's recently-published standard notation, despite him deploring the lack of agreement in this matter; indeed he was a close friend of E. Fiander Etchells, who had led the Concrete Institute's work in this direction. Nonetheless the diagrams and their explanation were praised for their clarity and cited as sufficient recommendation for the book.

In the years up to and during the Great War, Twelvetrees – now resident in Bedford Hill, southwest London – turned his attention to more general construction and took up the challenge of editing the Rivington's *Notes on Building Construction*. Parts I and II were issued by Longmans, Green & Co in 1916. This was a long-established title brought 'thoroughly up to date' under Twelvetrees's editorship. Chapters – by different contributors, all distinguished in their field – were however, of an uneven standard. Twelvetrees contributed those on gas, electricity, roofing materials, scaffold, but apparently not concrete. The chapter on reinforced concrete was dismissed by its reviewer as possessing 'little merit, being merely a description of systems employed by various firms which are found in almost every book on the market, together with a few general notes as to the principles governing the use of the material'. Despite these reservations the book was regarded as 'still quite above the average volume dealing with construction' (*C&CE* v.11 1916, pp.105–106).

The early 1920s saw a brief flurry of publishing activity. His work for Pitman continued with *A treatise on reinforced concrete* (noted above) in 1920, which this time featured standard notation – indeed it claimed to be 'the first book in the English language embodying the improved notation' – and a foreword by Etchells. Supportively, Etchells identified his friend's achievement, arguing that it 'probably contains a more complete set of formulae for the resistance movements of various types of reinforced concrete beams than any book hitherto published in this country'. It was followed by a second edition of *Simplified methods* a year later. In 1922 his *Reinforced concrete piers and marine work* appeared in the Concrete Institute's journal (v.12). This period culminated in 1925 with the appearance of his last book – *Concrete making machinery* – from yet another publisher, Greenwood & Son.

He retired from *Ferro-concrete* in 1932 and was able to devote himself to his private interests. These were described in the obituary penned by Gueritte in 1941:

> He made a hobby of the study of the origin of family names and his knowledge in this matter was considerable as well as entertaining. Another hobby was music, and he constructed with his own hands a very fine organ, from the playing of which he derived and gave great pleasure. To the end he retained a really beautiful handwriting. He was a very loyal friend, and Fiander Etchells characterised him in a very neat way when he said that he 'combined the knowledge of a savant with the punctuality of a man of business.'

Gueritte described him as 'a most attractive and interesting personality' and recognised him as an authority on concrete:

> 'Few men in this country had such an all-embracing acquaintance with so many aspects of reinforced concrete work, and, with his methodical ways, he was a mine of information on the subject … Although nearing 89 years of age, Twelvetrees retained to the very last his intellectual keenness and his thirst for information concerning the latest developments in engineering; less than a week before his heart suddenly failed he had written for elucidation on a point concerning pre-stressed concrete.'

Twelvetrees, W. N. *Concrete-steel: a treatise on the theory and practice of reinforced concrete construction*. London: Whittaker & Co, 1905 [BL, C&CA 1959, IStructE. Copy: TCS. Rev: *EN*, 14 Sep 1905; *CA*, v.2, p.686]

Twelvetrees, W. N. *Concrete-steel buildings*. London: Whittaker & Co, 1907 [ICE. Rev: *EN*, 15 Aug 1907; *C&CE*, v.2 Sep 1907, p.320], and New York: Macmillan, 1907 [Peddie]

Twelvetrees, W. N. *Simplified methods of calculating reinforced concrete beams*. London: Sir Isaac Pitman, 1909 [ICE], and New York: Macmillan, 1909 [Peddie]. Second ed., 1921 [C&CA 1959, Yrbk 1934. Copy: TCS]

Twelvetrees, W. N. *The practical design of reinforced concrete beams and columns*. London: Whittaker & Co, 1911 [C&CA 1959, ICE. Copy: TCS. Rev: *C&CE*, v.6, p.799], and New York: Macmillan, 1911 [*CA*, v.14, p.164]

Twelvetrees, W. N. *A treatise on reinforced concrete*. London: Sir Isaac Pitman, 1920 [C&CA 1959, Yrbk 1934]

Twelvetrees, W. N. *Concrete and reinforced concrete*. London: Sir Isaac Pitman & Sons, 1922 [C&CA 1959, Yrbk 1934]

Twelvetress, W. N. *Concrete making machinery*. London: Scott, Greenwood & Son, 1925 [C&CA 1959, ICE]

1906/USA

Brayton, Louis F.

Louis Brayton, a consulting engineer from Minneapolis, filed for and was awarded a number of American and Canadian patents relating to reinforced concrete in 1904 and 1905. On 25 April 1906 he took the unusual but magnanimous step of legally dedicating these patent rights to the public and accompanying this gesture with an explanatory book, *Brayton standards*.

In a refrain familiar from the writing of other early authors, Brayton referred to the stranglehold of the reinforced concrete specialist: 'the principal difficulty at the present time lies in the fact that the specialists doing this class of work are largely in the employ of companies who are exploiting some particular feature in the line of reinforcement.' As the designs for reinforced concrete work were the responsibility of the contractor at this date, contractors had to rely on such specialists in order to tender for the work, and the proprietary nature of reinforcement meant that no two bids were based on comparable designs. Brayton's book attempted to make generic information more widely available, 'to enable architects and engineers who have not made a specialty of this class of work to show the complete drawings required to properly illustrate a structure in reinforced concrete, so that all contractors bidding upon the work will bid on a uniform basis'. The author also hoped that it would 'render the design of reinforced concrete as easy for architects, engineers and builders as is the design of steel structures'. In this regard *Cement Age* (v.3, p.129) noted that the book had 'been likened to the *Carnegie handbook on structural steel*'.

Brayton's book, the author claimed, was 'a compilation of information acquired from actual experience, coupled with the necessary theory. The methods of construction shown are not merely theoretical, but have been put into practice and found highly efficient and economical'. The 110-page volume took the form of a handbook (convenient for office use) and included illustrations and about 20 tables. A review in *Engineering News* of 16 August 1906 commented on the defects of its typographical appearance, but a second edition came out in 1907.

Brayton, L. F. *Brayton standards: a pocket companion for the uniform design of reinforced concrete*. Minneapolis, 1906 [Moisseiff, Peddie. Rev: *CA* v.3, p.129]; second ed., 1907

1906/USA

Hawkesworth, John

Hawkesworth studied civil engineering at Columbia University during the first years of the century and between early 1903 and June 1904. While 'graduating students of the fourth class', he and Walter T. Derlath undertook a programme of tests on concrete for tensile strength and tensile efficiency of elasticity. In this he was supervised by Myron Falk, the author of *Cements, mortars and concretes: their physical properties* (1904). His own book *Graphical handbook for reinforced concrete design* followed in 1906. A quarto volume, it contained 15 large folding plates of diagrams of sufficient size and clarity for office use.

Hawkesworth, John. *Graphical handbook for reinforced concrete design*. New York: Van Nostrand, 1906 and London: Crosby, Lockwood & Son, 1907 [C&CA 1959, IStructE, Peddie. Rev: *EN*, 16 May 1907]

1906/USA

Hodgson, Fred T.

Editor of *The Builder & Woodworker*, Hodgson had been writing books on carpentry, steel and construction subjects since at least 1890. His *Steel squares* of that year sold more than 150,000 copies at a time when 50,000 to 60,000 constituted a best seller. In 1906 his *Concrete, cements, mortars, artificial marbles, plasters and stucco* continued this popular journalistic vein (it was reissued in 1916). A reprint is still available in the 21st Century.

Hodgson, F.T. *Concrete, cements, mortars, artificial marbles, plasters and stucco: how to use and how to prepare them*. Chicago: Drake, 1906

1906/USA

Howe, Malverd A. (b.1863)

CE

Howe was an American author about whom little is known other than his work specialised in arches. *Symmetrical masonry arches* of 1906 specifically included coverage of reinforced concrete, and his theme was extended with a later article entitled *The analytic calculation of a concrete arch* in *Engineering News* (12 May 1910). His book merited a revised and enlarged second edition in 1914 – a year that saw two other titles by Howe: *Influence diagrams for the determination of maximum moments in trusses and beams* (whose 'subject is carried far enough to cover swing spans, arches, etc'); and *Foundations: a short text-book on ordinary foundations, including a brief description of the methods used for difficult foundations*. Both were published by Wiley and a thousand copies of the latter were printed. *Concrete-Cement Age* described Howe's writing as of 'a clear, short style, which makes it easy to read and understand' (v.5, p.259).

Howe, M. A. *Symmetrical masonry arches: including natural stone, plain concrete and reinforced-concrete arches*. New York: Wiley, 1906 [IStructE]. Second ed. 1914 [ICE]

1906/USA

Rice, Harmon Howard (b.1870)

Harmon Rice appears to have specialised in concrete blocks – probably as a manufacturer, as his books on the subject suggest. In 1906 he was the secretary of the American Hydraulic Stone Co. of Denver, Colorado, when his first book *The manufacture of concrete blocks and their use in building construction* was published in New York by the Engineering News Publishing Company (and by Constable in London for the British market). Both he and his co-author William Torrance (below), had won prizes for papers submitted to a competition run by *Engineering News* and *Cement Age*, and these, with excerpts from nine others, formed the basis of the book. It is clearly related to Rice's articles in *Cement Age*: *The manufacture of concrete blocks and their use in building construction* (v.2, pp.307–321); and *The waterproofing of concrete blocks and walls* (v.2, pp.491 *et seq*). At the same time, his own slightly longer *Concrete block manufacture* was published by John Wiley & Sons and their British counterpart Chapman & Hall in London. It was described as 'nicely illustrated throughout', with a 'timely' chapter on causes of failure (*CA*, v.3, p.58).

Rice next appeared in print on 1 January 1907 as the author of *Concrete blocks* in the *Proceedings* of the American Concrete Institute. The ACI had recently been established (in 1904), largely by representatives of block-making manufacturers. It is worth quoting Rice's own abstract as a lofty statement of his belief in value of this industry:

> Because I believe that concrete block construction presents a representative style of architecture, typical of that spirit of independent enterprise characteristic of American progress, and because Europe, Asia, Africa and the islands of the seas are looking to the States for light upon this industry, I characterize the conduct of our business, or, if you please the practice of our profession, as the National Game, believing that in the play of the Diamond we shall find an interesting and instructive parallel to the opportunities and responsibilities of co-workers in the domain of concrete blocks.

Further articles on concrete blocks appeared in the journal, *Concrete*, namely: *The evolution of the concrete block in residence construction* (v.6(6), pp.17–20, 43–44); and '*The utility of concrete block construction*' (v.7(11), pp.21–23).

Rice, H. H. and Torrance, W. M. *The manufacture of concrete blocks and their use in building construction*. New York: Engineering News Publishing Co., 1906 and London: A. Constable & Co, 1906 [C&CA 1959. Rev: *EN*, 16 Aug 1906]

Rice, H. H. *Concrete block manufacture: processes and machines*. New York: John Wiley & Sons, 1906 [Rev: *EN*, 16 Aug 1906; *CA*, v.3, p.3] and London: Chapman & Hall, 1906 [Rev: *C&CE* v.2 May 1907, p.166]

1906/USA

Torrance, W. M.

MWSE

An alumnus of Cornell University – where he was a member of the Society of Civil Engineers – Torrance became assistant engineer for the Hudson Co. of New York. Like his co-author Rice (above), he was a prize-winner in the *Engineering News* competition of 1906.

1906/USA

Warren, Frank Dinsmore (1879–1930)

A graduate of the Massachusetts Institute Technology in 1900, Frank Warren consciously chose to write a reference handbook (in preference to a textbook) for the benefit of designers whose 'business methods and competition demand an economy of time'. Like Brayton (above), he was keen to treat general forms of design rather than those that were dependant on patented systems. His approach was to base his formulae on the theory of elasticity, modified by 'the usual assumptions', supported by test results, and not derived from empirical experimentation. Indeed, in his preface to his 268-page *A handbook on reinforced concrete*, Warrant declared that: 'if this volume will tend to do away with the use of some of the 'empirical formulae' and 'rule of thumb' methods of designing reinforced concrete structures, and tend to concentrate all toward a standard and universal system … it will have accomplished its purpose.' It was welcomed by *Cement Age*, which considered the tables setting out comparable costs, spans and safe loads to be 'an especially valuable and convenient feature of the work'. It also noted that a quarter of the book was devoted to the design of trussed roofs.

Warren, F. D. *A handbook on reinforced concrete: for architects engineers and contractors*. New York: D. Van Nostrand Co., 1906 [Rev: *CA*, v.2, p.888]. Second edition, 1907 [Moisseiff]

1907/USA

Douglas, Walter J. (b.1872)

Douglas served in the Army Corps of Engineers, but it was his work as engineer of bridges in Maryland and Washington DC, for which he is principally remembered today.

In 1902 he designed the notable Boulder Bridge in Rock Creek Park, Washington DC, which is now on the National Register of Historic Places. It was an early reinforced concrete arch bridge, built according to the Melan system at a cost of $17,636.

He developed his formulae and a number of design aids in the years following and at the end of 1906 shared his experience in two papers for *Engineering News*. These papers (dated 20 December 1906 and 4 January 1907) were reprinted in book form by the Engineering News Publishing Company, as *Practical hints for concrete constructors*. Douglas explained his aim in the preface:

> The purpose of this Book of Charts is to furnish to the Designing Engineer, the Draftsman, the Estimator, and the Student a concise and complete apparatus for Proportioning the Parts and Estimating the Cost of Concrete Structures ... The formulas for Moments of Resistance were derived in 1904 ... They are here given algebraically and are plotted on charts for convenience in computation. In this form they can be used to determine the sizes, shapes, composition and cost of concrete beams, slabs, columns, etc. with the minimum expenditure of time and risk of error. ... Careful directions accompany these charts ... greatly expediting the process of design.

His ideas apparently stood him in good stead, as he was engaged on several concrete bridges in the years after publication.

Douglas was the engineer in charge of construction of the Connecticut Avenue Bridge in Washington DC. This structure consisted of nine 150-foot spans and two 82-foot spans, all arches of mass concrete, trimmed with tool-dressed concrete blocks. At 52 feet wide and 1,341 feet long it was one of the largest concrete arch bridges in the world, and containing 80,000 cubic yards of concrete it cost $850,000 to complete. An account of the methods and cost of construction was prepared by Gillette and Hill [1908], based on information obtained from Douglas and personal visits to the site. The account was published in their book *Concrete construction: methods and cost* (1908).

After the success of its first reinforced concrete bridge in 1904, the authorities in neighbouring city of Baltimore appear to have made a commitment to the arch design too. In 1908 – according to the *Annual Report of the City Engineer, 1909* – the construction of three reinforced concrete bridges began at Hollins Street, University Parkway, and Edmondson Avenue. The plans for the Edmondson Avenue Bridge were prepared by W. J. Douglas, described at the time as 'a bridge engineer from the District of Columbia'. The Baltimore Ferro-Concrete Co. constructed the resulting multiple-span, 540-foot-long bridge between 1908 and 1910.

Further experience (but this time abroad) was written up (with his collaborator Eugene Klapp) in December

1911 as *Reinforced concrete bridge across the Almendares River, Havana, Cuba*, in *Transations* of the ASCE v.74(4), pp.216–240. This paper attracted the attention of the 'eminent authority' L. J. Mensch, who contributed to the ensuing published discussion.

Douglas, W. J. *Practical hints for concrete constructors*. (Reprint No.12). New York: Engineering News Publishing Co., 1907 [Peddie. Copy: Internet Archive]

1907/USA

Leffler, Burton R. (b.1871)

While we cannot be certain it was the author in question, a Burton R. Leffler was born in Illinois during 1871, the son of a house carpenter called Jeremiah Leffler (b.1846, Pennsylvania). A lifetime later the social pages of a 1948 newspaper record: 'Mr & Mrs Burton R. Leffler of Cleveland, Ohio came over last week from Clearwater where they are spending the season, to visit several days with Mrs Edwin D. Buell of Sylvan Drive.' At least for our purposes however, Leffler was the author of *The elastic arch*, published by Holt of New York in 1907.

Leffler, B. R. *The elastic arch: with special reference to the reinforced concrete arch*. New York: Holt, 1907 [BL]

1907/USA

Lesley, Robert Whitman (b.1853)

AASCE

Although a civil engineer by professional qualification, and for a while a journalist, writer and editor, Robert Lesley was first and foremost a cement manufacturer who achieved national distinction in the field of cement-testing and standardisation. This was largely through his involvement in establishing both the American Society for Testing Materials (ASTM) and Association of American Portland Cement Manufacturers (AAPCM) – forerunner of the Portland Cement Association (PCA).

Born in Philadelphia on 3 July 1853, the son of James Lesley Jr and Elizabeth Thomson, he attended the University of Pennsylvania in 1871. Foreshadowing his subsequent career, that year saw the start of Portland cement production in the USA, by David O. Saylor's Coplay Cement Co. in Pennsylvania's Lehigh Valley. On graduation Lesley joined the emerging cement industry and founded the cement broking firm of Lesley & Trinkle in 1874. Nine years later he established his own manufacturing firm – the American Improved Cements Co., in Egypt, Pennsylvania. There he introduced improvements in manufacturing processes and equipment. He was the first cement maker to install an iron mill of the Griffin type and one of the first to adopt the rotary kiln. The combination of efficient new production methods with rapidly increasing domestic demand in the 1890s fuelled a dramatic rise in output by America's burgeoning cement industry. New entrants to the industry pursued their own manufacturing and marketing methods with a resulting dilution of quality tolerated by an expanding market. Lesley (one of the early manufacturers) drew attention to the proliferation of specifications for American-made cement, when in a paper to the Engineers Club of Philadelphia in 1898 he noted more than

200 different types, including some issued by different engineers employed in the same government department!

The new century saw moves to bring order to the industry – to harmonise the production of cement as an essential prerequisite for its use in the emerging market for reinforced concrete and concrete products. Both the ASTM and AAPCM were founded in 1902, partly on the basis of a common interest in cement sacks and standards. Lesley became the AAPCM's first president and by 1906 was vice-president of the ASTM. With the American Society of Civil Engineers and the American Railway Engineering & Maintenance of Way Association, these organisations collaborated as the Joint Committee on Concrete and Reinforced Concrete.

It was as an offshoot of his involvement in this work on industry collaboration that he wrote his *Concrete factories* in 1907. The subtitle explains the book's contents: 'An illustrated review of the principles of construction of reinforced concrete buildings, including reports of the Sub-Committee on Tests, the US Geological Survey and the French Rules on Reinforced Concrete.' The latter were those in which Armand Considere [1903] had lately been involved.

In 1907 Lesley's involvement in cement and concrete was at its height. He was chairman of the Committee on Competitive Plans for the AAPCM's competition for Concrete Houses of Moderate Cost, held in June, for which Sanford Thompson [1905] was one of the judges. He had also recently funded the establishment of a cement-testing laboratory at the University of Pennsylvania – a facility that was named after him:

> With imposing ceremonies the new building for the Engineering Departments of the University of Pennsylvania, Philadelphia, was dedicated on Friday, October 19th. The importance of the occasion is indicated by the fact that in addition to the large number of visitors representing local and State interest in the institution, six foreign governments, including France and Germany, more than one hundred universities, engineering and scientific societies were represented by special delegates. Honorary degrees were conferred upon thirteen American and foreign engineers. The building is believed to take first rank in the English-speaking world in the completeness and excellence of its appointments and equipment ... A laboratory for the testing of cement and concrete and for research work in connection with these products is one of the most important features of the equipment of the new building. The fund for this equipment was generously provided by Robert W. Lesley, Esq., of Philadelphia, Assoc. M. Am. Soc. C. E., Vice-President of the American Society for Testing Materials, and of the class of '71 college, who has long been engaged in the manufacture of cement ... The University authorities have named the laboratory after the donor and have placed on its walls the bronze tablet shown in the accompanying illustration (*CA* v.3, pp.431–435).

Lesley was prominent in the professional press, not least as the founder and editor of *Cement Age*. This monthly journal came into being in 1904 on his return from a tour of the cement manufacturers of England, France and Germany. On its first anniversary, Lesley looked back over the first year of publication and was careful to thank Messrs. Herbert W. Anderson, Edouard Candlot and Rudolph Dyckerhoff, along with several American authors, for their 'instant, cheerful and unselfish response' in getting the 'best material' for the magazine.

It was also a vehicle – though a discreet one – for his own interests and featured several of his own articles – some of which had been published elsewhere or were presented as papers at meetings he had attended. The titles indicate the nature of his preoccupations at the time; in testing, standardisation, cooperation and organising American cement producers as an industry. His article for the ACI's *Proceedings* entitled *Cooperation – what it is and what it can accomplish* (v.4, 1908) was typical. Others of the period include:

- *British standard specifications for cement* (*CA* v.2, 1906, p.497)
- *Review of cement industry* [for the year 1905] (*CA* v.2, 1906, p.571)
- *What cement users owe to the public: a review of the present tendency on the part of manufacturers and users of cement to cooperate to their own advantage and to the interests of the cement industry* (CA v.2, 1906, p.651). This was read before the Convention of National Cement Users Association, 10 January 1906 and also published in the ACI's *Proceedings* (v.2, 1906)
- *American Society for Testing Materials: a review of the published proceedings issue to members for 1906, including extracts from … the address of Vice-President Robert W. Lesley* (CA v.3, p.354)
- *Hydraulic properties of reground cement mortars* with Henry S. Spackman (*CA* v.6, 1908). This was a paper read before the meeting of the AAPCM, New York, 9th, 10th and 11 Dec 1907, and referred to a research project of the previous two or three years.

Annual reviews of the American cement industry became a regular feature of the magazine, and from 1909 this summary was paralleled by a review of the German cement industry as presented in translation to the AAPCM at its annual meeting in New York each December. These reviews continued until 1915 when wartime conditions caused the German association to suspend its own meetings. Before then however, Lesley's role in bringing German achievements to an American audience was recognised:

> Robert W. Lesley, consulting editor of *Concrete-Cement Age*, was signally honored recently by the Association of German Portland Cement Manufactures. A special invitation was sent him to attend the annual meeting of the Association in Berlin, Feb. 10-12 last. Dr. Muller, who sends the invitation, adds a personal note that it would give the members pleasure to see Mr. Lesley personally and to thank him for the many courtesies extended to them at the recent Congress of the International Society for Testing Materials, held in New York. (*C-CA* v.2, p.182) April 1913.

By this date Lesley's involvement in publishing had receded. In January 1911 *Cement Age* took over its rival *Concrete En-*

gineer – established in 1907 in Cleveland, Ohio and edited by Allan Brett. With a slightly larger page size and a two-column format, the new magazine continued much the same. The business aspects of the enterprise were divided between east and west offices and continued by the staff of the former magazines under the presidency of F. F. Lincoln at 30 Church St., New York. Lesley continued as editor and Brett joined Edward Trego as joint associate editor.

Eighteen months later however, the combined business entered an agreement with a third magazine – *Concrete*, in Detroit, to merge the three titles on an equal footing. In July 1912 *Concrete-Cement Age* was launched with the business roles largely in the hands of former *Concrete* staff, but with Lincoln as eastern manager and *Concrete Engineering*'s Brett as managing editor, with Lesley as consulting editor. Trego and Harvey Whipple from *Cement Age* and *Concrete* respectively, continued as associate editors. Perhaps significantly, Lesley (who was nearly 60 years of age) had sold his American Cement Co. to Giant Cement in 1912, and had stepped back from such an active role in the industry. The tone of the editorial announcing his new role was valedictory:

> Robert W. Lesley, of Philadelphia, will give to the enterprise the benefit of his ripe and practical experience in the capacity of consulting editor. An experienced newspaperman, and one of the first to engage in the manufacture of Portland cement in this country, the services of Mr. Lesley, not only as a prominent manufacturer, but as an active member of the various scientific associations that have put the industry on this present high plane, will be of immediate and practical benefit.

Twelve years later, with the assistance of J. B. Lober and G. S. Bartlett, he wrote *History of the Portland Cement Industry in the United States* (Chicago: International Trade Press, 1924), which became the standard source on its subject. It also sheds light on Lesley's key role in the industry. This, and his work on ACI committees, was recognised and in 1932, at the age of 79, Lesley was awarded honorary membership of the American Concrete Institute.

Lesley, R.W. *Concrete factories: an illustrated review of the principles of construction of reinforced concrete buildings.* New York: Bruce & Banning (for the Cement Age Co.), 1907 [Peddie. Rev: *C&CE* May 1906, p.166]

1907/USA

Reid, Homer Austin (b.1869?)

AASCE

From Warren, Ohio, the young Homer Reid enrolled at Lehigh University to read Civil Engineering. He was in the senior class of 1896. His early career as a civil engineer coincided with an economic boom in the cement industry and the material's use in concrete construction. As he saw it, 'the marvellous growth of the cement industry during the past few years has led to the present time being spoken of as the cement age'. Reid was appointed assistant engineer at the Bureau of Buildings for the New York City authorities and became an associate member of the ASCE. In 1907 (while still a young man) he wrote his book *A complete treatise on the properties and uses of concrete and reinforced concrete, as applied to construction*. He intended it as a reference work for those professionally engaged in construction – and it is certainly a substantial work – but hoped his approach was sufficiently clear and simple to be accessible to the student and general reader. It was published in 1907 by the Myron C. Clark Company, and reissued a year later. *Engineering News* reviewed it: A. N. Talbot [1917] describing it as 'valuable'; the presentation 'commendable'; and a book that 'deserves to rank among the standard works on the subject' (v.57, p.301).

Reid, H. A. *Concrete and reinforced concrete construction*. New York: Myron C. Clark Publishing Co, 1907 [C&CA 1959, PCA. Copy: TCS], and London: Spon, 1907; 1908 [BL, ICE]

1907/UK

Tanner, Sir Henry (1849–1935)

CB, ISO, RVO, FRIBA, FSI

Sir Henry Tanner, 2nd President of the Concrete Institute, 1910–12 Courtesy of the IStructE

Sir Henry Tanner is a borderline case for inclusion in this review as the institutional press – rather than the commercial press – published his writing, even though it was intended for a wider circulation. However, his significance as a character in the story of concrete has been more than sufficient to swing the decision in his favour.

Born in London, Tanner studied architecture at the Architectural Association and trained with Anthony Salvin. He joined HM Office of Works in 1871 and carved a career designing public buildings. Specialising in post offices, his oeuvre included those at Birmingham, Bradford, Cardiff, East Grinstead, Leeds, Preston, Sunderland and Wolverhampton. He progressed from Assistant Surveyor, through Surveyor (Northern District) and Surveyor (London) to an appointment in 1884 as Principal Architect. He was elected a fellow of the RIBA, among his other professional affiliations, and was awarded the Imperial Service Order. In 1898 he became chief architect to HM Office of Works and from 1899 he was responsible for continuing the work of the late John McKean Brydon on the Government Offices, Great George Street (the Treasury Building).

His appointment coincided with the advent of reinforced concrete in Britain and Tanner duly became aware of the new material's economic advantages. He encouraged its use in official projects; because government buildings were crown property and thus exempt from local authority building control, he (unlike commercial architects) was free to do this. Such patronage did much to encourage others.

Tanner was thus a natural choice to chair the RIBA's Committee on Reinforced Concrete when it was convened in 1907. Though not published commercially, the committee's report has seminal importance, and Tanner (as chairman) wrote the introduction:

> There has not been hitherto in this country any authoritative pronouncement on the necessary rules to be ob-

served in such construction. In many ways this has prevented employment of reinforced concrete, such employment being practically prohibited for complete buildings under the ordinary building rules and regulations; and it is only those bodies who are free from these restriction, such as railway and dock companies, who have been able to avail themselves of so economical and space-saving a method of construction, and on these points I speak from experience. Other countries have been more lenient, and in consequence those countries are far in advance of this, both as to general knowledge of the material and skill in its use. However, we hope that, if the Meeting adopts the rules which have been prepared, this country may not for long occupy the backward position that it now does.

He again acted as chairman for the RIBA's second report in 1911, and by that time Tanner's stature in the concrete establishment had grown. His role as RIBA chairman was enough to suggest him as one of four (titled) vice-presidents of the Concrete Institute when was it was established in 1908, and in 1910 he succeeded Lord Plymouth as president. His presidential address – the first such address to the Institute – was published in volume four of *Transactions & Notes* (and reported in *C&CE* v.6, p.930–37).

The year 1910 saw completion of his sorting office at the GPO's King Edward Street building in St Martin's le Grand (1907–1910) in the City of London. Built in the Hennebique system it was the first of its type in central London and the largest for many years. He wrote it up for *C&CE* (v.6, 1911, p.132).

Now 62 years of age, Tanner started to fade from the record and in 1913 he retired from his position as chief architect to join his son in private practice, though this allowed him more time for his role in RIBA's Finance & General Purpose Committee. During these final years he worked on the design of several commercial buildings in Regent Street and Oxford Circus, including Dickins & Jones and Peter Robinson. In retirement (aged 77) he wrote a thoughtful piece – *Concrete: its present handicaps and its future possibilities* – for *Concrete & Constructional Engineering* (v.21, 1926, pp.15–18); and finally he contributed a chapter on Caversham Bridge in the British Portland Cement Association's handsome hardback, *Concrete bridges*.

He died in 1935 – the closing year of our review – and obituaries appeared in both *The Builder* (v.149, 6 September 1935, p.394) and *RIBA Journal* (v.42, 1935, p.1160).

1907/USA

Turneaure, Frederick Eugene (1866–1951)

Turneaure was an educationalist – appropriately enough for the author of one of the most widely circulated textbooks of the period – and a professor at the University of Wisconsin.

Professor Frederick Eugene Turneaure in 1935
Courtesy of the University of Wisconsin, Madison Archives

He was born 30 July 1866

on a farm near Freeport, Illinois. A successful education led him to Cornell University, from which he graduated in civil engineering in 1889. After a year employed on railway work he took up a post as instructor of civil engineering at Washington University, but in 1892 he transferred to the University of Wisconsin as professor of bridge and hydraulic engineering. In 1902 he was appointed acting dean and then (in 1904) dean, of the College of Mechanics and Engineering – one of the university's four constituent colleges. He remained in post for 33 years, until his retirement in 1937.

His expertise lay in bridge building and he developed an early interest in reinforced concrete construction, undertaking research on impact and stress loading. He co-authored three books: *The theory and practice of modern framed structures* with J. B. Johnson and C. W. Bryan in 1893; *Public water supply* in 1900; and *Principles of reinforced concrete construction* with E. R. Maurer in 1907.

The success of his second book was probably instrumental in his appointment as city engineer of Madison. During his tenure of this post (from 1902–1904) he was responsible for building the first modern sewage-disposal plant in the city.

During the early years of the new century, Turneaure turned his attention to researching reinforced concrete. In 1905 he was bestowed with an honorary doctorate of engineering from the University of Illinois – an institution becoming noted for its concrete research. H. A. Reid, writing in January 1907, acknowledged his use of observations from Turneaure's laboratory tests. Later that same year, Turneaure and his colleague Maurer published their *Principles of reinforced concrete,* based in part on these very tests.

The stated intention of this book was 'to present in a systematic manner those principles in mechanics underlying the design of reinforced concrete; to present the results of all available tests that will aid in establishing coefficients in working stresses; and to give such illustrative material from actual designs as will serve to make clear the principles involved'. The treatment of concrete's elastic properties was based on his experiments at Wisconsin and the section on failure of concrete beams was illustrated with photographs taken in the laboratory.

In review it was described as 'admirably clear' (*C&CE* v.3, p.258) and (circulated in Britain by Chapman & Hall) as 'one of the most useful' American books on the subject. *Engineering News* thought it 'should be in the library of every engineer'. The second edition of 1909 built on this success and the reviewers pressed their praise further: 'this is one of the best books on the subject that have been published', and 'one which ought to be purchased by every student of the subject' (*C&CE* v.5, p.612). 'A large amount of new material' had been added, with 'new experimental data … noted in nearly every division of the subject' (*CA* v.9). Particularly important were test results on bond, the strength of beams in shear, and of columns. A comprehensive chapter was added on chimneys.

Further professional recognition followed. In 1907 he was awarded the Octave Chanute Medal by the Western

Society of Engineers. A year later he was elected president of the American Society for the Promotion of Engineering Education, and then in 1909 by a term as president of the Wisconsin Engineering Society.

By the time the third edition was issued in 1919, 20,000 copies of 'Turneaure & Maurer' had been issued, and so it was with some justice that it was described as 'a well-known book'. More significantly, however, the process of revision drew attention to the changing state of knowledge: 'a good deal of fresh information is given, principally in regard to flat slab (or mushroom) construction.' That said, the reviewer for *C&CE* (v.15, p.650) took issue on its treatment of two topics: of moments in beams and slabs at mid span (where British practice differed); and of shear – the understanding of which was being transformed by Faber's work of that year.

Similar comments on the changing technology apply to the final edition:

The new edition shows how greatly the subject has advanced since 1919; it contains long descriptions of 'shrinkage and plastic flow' in their separate relations to concrete itself, and to reinforced concrete in the form of beams, columns and arches. The subject of flat slabs has been developed and much analysis has been added on the subject of interaction of beams and columns in framed structures. The chapter on arches has been greatly revised and enlarged.

An appendix contained the full text of the 1924 report prepared by the Joint Committee on Standard Specifications for Concrete and Reinforced Concrete.

Overlapping with his time at the university, Turneaure was also a member of the Wisconsin Highway Commission between 1911 and 1929, during which time 15,000 miles of road were laid out.

As an engineer – both academic and practicing – Turneaure was active in his profession. He was a member of ASTM, the American Railway Engineering Association and the American Association for the Advancement of Science. The American Society Civil Engineers made him an honorary member in 1933 and the American Society for the Promotion of Engineering Education awarded him the Benjamin Carter Lamme Medal in 1937 for excellence in engineering teaching. More germane to this study, the American Concrete Institute granted him the Henry C. Turner Gold Medal in 1930, for 'distinguished service in formulating sound principles of reinforced concrete design', and one of the first honorary members in 1932.

He died 31 March 1951 (aged 84) after a lengthy career of service to both the University of Wisconsin and the city of Madison, as well as (through his books) to engineers throughout the English-speaking world. An appreciation entitled *F. E. Turneaure: an everything engineer* by Julie Nickchen, appeared in the *Wisconsin Engineer* of May 1989 (v.93(4)).

Turneaure, F. E. and Maurer, E. R. *Principles of reinforced concrete construction*. New York: John Wiley & Sons, 1907 [Copy: TCS. Rev: C&CE, v.3, p.258]. Second ed, 1909 [BL, ICE. Rev: C&CE, v.5, p.612; CA, v.9]. Third ed, 1919 [C&CA 1959, BL, ICE. Copy: TCS. Rev: C&CE, v.15, p.650]. Fourth ed, 1932 (BL, ICE, Yrbk 1934. Rev: SE, v.11, p.102]

1907/USA

Maurer, Edward Rose (1869–1948)

Colleague and co-author of Turneaure, Maurer was born three years later on 18 February 1869. Unlike Turneaure, he was a graduate of Wisconsin – the university at which they were both to make their names. Maurer graduated in 1890, and after a short break to gain practical experience, returned to become an instructor at the university's College of Engineering. There he served for 47 years, most of which as professor and chairman of the department of mechanics.

His writing was concentrated in the early years of the new century: *Technical mechanics* in 1903; *Principles of reinforced concrete construction* with Turneaure in 1907 (discussed above); and then as co-author of *Strength of materials*. Added to these were numerous papers on engineering in practice and education.

An inspirational teacher, he introduced courses in aerodynamics and vibration and in 1934 became the seventh ever recipient of the Lamme Medal, presented by the Society for the Promotion of Engineering Education.

After his death on 1 May 1948, the *Wisconsin Engineer* paid his memory the following tribute:

> As an administrator he was revered and admired by his colleagues. His numerous committee assignments were always handled promptly, carefully, and with the considered application of good judgement. A thorough teacher and a strict disciplinarian, he was respected and loved by his students. His colleagues remember him particularly or the encouragement that he gave to the younger men on his staff, looking toward their professional development. His highly developed sense of honesty, justice and fairness will always be remembered by his friend and pupils ('In Memoriam', *Wisconsin Engineer*, v.52(8), May 1948).

1908/USA

Andrews, Hiram Bertrand

MASCE

A slim volume of 46 pages, Andrews' *Practical reinforced concrete standards for the designing of reinforced concrete building* aimed to be just that: a tool containing tables and charts of the standardized sizes for buildings and sets of specifications for concrete work. It was directed at architects, the majority of whom (at that date) had qualified long before there was a demand for buildings in reinforced concrete. Andrews hoped that as the early theories of concrete design were thought to be achieving consensus, the time was ripe for presenting ready-reference data to enable concrete to compete with steel or timber. Much of the guidance was based on the design of several buildings actually constructed in Boston – the city in which his book was published. *Engineering News* reviewed it on 13 August, and in September *Cement Age* described it as an 'attractively bound volume' whose 'especially valuable' feature was the set of tables giving mix proportions, strengths, working loads, etc.

Such guidance might now be thought of as tentative or preliminary, but Andrews was to gain wider experience of concrete. In 1910 he issued a pamphlet – *Design of reinforced concrete slabs, beams and columns, confirming to the recommendations of the Joint Committee on Concrete and Reinforced Concrete*. He also become active in the ACI, by which his paper *The design of reinforced concrete fuel-oil reservoirs*, was published in the *Proceedings* for 1919 (v.15).

Andrews, H. B. *Practical reinforced concrete standards for the designing of reinforced concrete buildings*. Boston: Simpson Bros., 1908 [Moisseiff. Rev: *CA*, v.7, September 1908]

Andrews, H. B. *Design of reinforced concrete slabs, beams and columns, confirming to the recommendations of the Joint Committee on Concrete and Reinforced Concrete*. Boston: Andrews, 1910 [Peddie]

1908/USA

Balet, Joseph W.

Balet was a New York-based bridge engineer, whose offices were at 261 Broadway. He was notable for his plans for the $29m New Jersey Bridge submitted in May 1895, which combined cantilever and suspension cable systems; and for the design of a 1,400 ft bridge over the St Lawrence at Montreal. His practical bridge-building was accompanied by original thinking on the theory and calculation of stresses. For example, *Engineering News* carried an article by him entitled *Graphic method for stresses in three-hinged arches* (1904, ii, p.356); and 'original communication' of 1910 (held at the ICE in London) discusses the *Computation of stresses in open webbed arches without linings*. However, his principal contribution to arch theory was his book of 1908: *Analysis of elastic arches: three-hinged, two-hinged and hingeless, of steel, masonry and reinforced concrete*.

Balet, J. W. *Analysis of elastic arches: three-hinged, two-hinged and hingeless, of steel, masonry and reinforced concrete*. New York: Engineering News Publishing, 1908 [ICE]

1908/UK

Coleman, George Stephen

G. S. Coleman wrote his *Reinforced concrete diagrams* (1908) early in his career and later wrote for *Cement & Constructional Engineering*, namely:

- *Uniform rolling loads on rigid arched ribs*. *C&CE*, v.21, September 1926, pp.625–629
- *The moment distribution method*. *C&CE*, v.27, April 1932, pp.212–217

His final work – on the more general field of civil engineering rather than concrete particularly – was in keeping with the trends of the early 1930s. *Civil engineering specifications and quantities* (Longman, Green & Co., 1932) was written jointly with G. M. Flood.

He was awarded the ICE's Telford Premium in 1924.

Goleman, G. S. *Reinforced concrete diagrams for the calculation of beams, slabs, and columns in reinforced concrete*. London: Crosby Lockwood, 1908 [ICE], and New York: Van Nostrand, 1908 [Peddie]

1908/USA

Gilbreth, Frank Bunker (1868–1924)

Like Taylor and Thompson [1905], Gilbreth combined an interest in concrete with a career as a pioneer of scientific management and motion study. He became known for his work on improving the efficiency of motion – initially in the construction and then other industries – to increase output and enhance the ease of operation. Throughout his career he was in turn, a bricklayer, building contractor, inventor, management engineer and university lecturer.

Born on 7 July 1868 in Fairfield, Maine, Frank Bunker Gilbreth was largely self-made, with no formal education beyond high school, despite passing the MIT entrance examination. He started work for a construction firm in Boston, initially as a bricklayer. A thoughtful young man, his experiences in the trade acted as a spur to developing ways of improving the building process and by 1895 he was the firm's general superintendent. In that year he started up his own construction company, relying on 'his own good name among dealers of supplies'. Early work on waterproofing cellars in the Boston area soon evolved to bigger projects and the firm grew rapidly. He moved to New York in 1904 and began to obtain contracts all along the east coast, throughout the south, as far north as Canada, and to the west in California and Washington. As well as traditional brick and stone, Gilbreth's firm adopted the new material reinforced concrete, and undertook major building projects such as factories, dams and power stations.

As the company expanded Gilbreth developed labour-saving machinery and methods of working to enhance efficiency in construction. Jane Morley in *Frank Bunker Gilbreth's Concrete System* (*Concrete International*, November 1990, pp.57–62) defines his approach:

> The Gilbreth System was intended to be a comprehensive method for completing construction work. Its components were interactive organisational, operational, and technical subsystems coordinated by various pre-computer record-keeping methods.

A contemporary explanation can be found in the pages of *Cement Age*:

> The code of roles and instruction to his foremen, and the printed forms for their reports, were bound in book form, and this book Mr. Gilbreth called his 'Field System'. He makes the interesting statement that when it became known that he had supplied his foremen and superintendents with printed instructions, copies of the book were bribe from his employees, and photographed page by page, from cover to cover. Other copies were 'lost', and thus made way with entirely. Huge salaries were used to hire away Mr. Gilbreth's superintendents, for the purpose of getting not only their trained services, but for their knowledge of the contents of the 'Field System'. Mr. Gilbreth finally decided to make his system public (1908, p.127).

He did so in 1908 and Myron C. Clark was the publisher.

Gilbreth was at this time a general and civil engineering contractor, but he also advertised his business as maintaining 'a permanent force of … concrete engineers and the largest equipment of concrete machinery'. Concrete was for the advocates of scientific management (as much as for contractors) just one speciality – though an important one – used to develop and display their theories. Gilbreth was active in this emerging industry at the same time as fellow theorists Taylor and Thompson.

Seeking to harness the power of gravity, Gilbreth invented the gravity concrete mixer, and then to develop the precast pile and a pile-driving apparatus (described in detail by Morley). These provided the technical base of the firm's concreting operations and added to Gilbreth's reputation. His paper *Concrete as piling* was read to the AAPCM in 1906 and published in *Cement Age* (v.3, p.72). In November 1907 he was a guest speaker at the Cornell Society of Civil Engineering. Then in 1908 his *Concrete system* was published by Engineering News Publishing and by Hive as volume 70 of the Hive Management History series. It was reviewed in *Cement Age* (November 1908 p.394), and described as 'a printed set of instructions prepared by Mr. Gilbreth for his concrete field forces. The review continued:

> [It] contains some especially valuable illustrations taken from the works of the author, comprising half-tone pictures of buildings in process of construction and completed and detailed drawings of form work, *etc.*, together with a number of organisational charts ... Throughout the volume are innumerable paragraphs setting forth in a brief and concise way the things to avoid in concrete construction as well as proper methods of work. In this respect alone, the book should be a great value to concrete workers, especially such as may not have had extended experience in this field.

A similar work entitled *Bricklaying System* followed in 1909 – an aspect of building that was also the subject of his *Motion study* (Van Nostrand) of 1911.

Gilbreth married Lillian Moller in 1904 and together they founded a consulting practice, whilst finding time to raise a family of 12. Frank and Lillian's employment of their children as efficiency 'guinea pigs' was endearingly described in the 1948 book *Cheaper by the Dozen*, written by Frank Jr and portrayed in subsequent film adaptations and sequels.

The Gilbreths' approach to efficiency concentrated on identifying and reducing the individual motions involved in a process, rather than on simply cutting time – the hallmark of Taylor's methods. They were also more concerned with workers' welfare than maximising profits, and although they corresponded with Taylor and Thompson, their philosophical differences led in time to a personal rift.

Gilbreth went on to become an occasional lecturer at Purdue University, but died of heart failure on 14 June 1924 aged 55. He was at a railway station in Montclair, NJ, and collapsed whilst on the telephone. His wife outlived him by 38 years and ensured his legacy. From 1962 the Institute of Industrial Engineers has presented an Industrial Engineering Award in their joint name.

Gilbreth, F. B. *Concrete system*. New York: Engineering News, 1908 [PCA. Rev: *CA*, November 1908 p.394], Easton: Hive Publishing Co, 1908 [BL], and London: Constable, 1908 [Peddie]

1908/USA

Gillette, Halbert Powers (1869–1958)

MASCE, MSEC, MAIME, MSPEE

As editor of a major engineering journal, Gillette combined his engineering expertise with a professional writing career and left a prolific legacy of engineering books to his name, including one specifically on concrete construction.

The son of Theodore Weld Gillette and his wife Laetitia Sofronia Powers, both of Lorain Co., Ohio, Halbert was born on 5 August 1869 at Waverley in Bremer Co., Iowa. Having graduated in 1886 from Hammond Hall Academy, Salt Lake City, he studied at the School of Mines at Columbia University until 1892. Having gained some early experience in civil engineering as Assistant State Engineer for New York, Gillette married on 28 April 1897 at the age of 28. His bride was Julia Washburn Scranton of Philadelphia, with whom he was to have a son and business heir, Edward Scranton Gillette. He worked as a contractor from 1898–1902, before turning to journalism – he was an early contributor to *Cement Age* – and publishing, ending up as president of the Gillette Publishing Company in 1906.

His first book – *Economics of road construction* – was published in 1901 by the Engineering News Publishing Co., which suggests at least a connection with the specialist periodical press with which he was to become associated. In 1904 his *Rock excavation* was published by the Myron C. Clark Publishing Co. as the start of a long series of titles and editions with that house. Published by Clark and – after 1918 – McGraw-Hill, his output included:

- *Economics of road construction*. Engineering News, 1901, 1906, 1908
- *Rock excavation*, Myron Clark, 1904
- *Handbook of cost data for contractors and engineers*. Myron Clark, 1905, 1907, 1910, 1918
- *Concrete construction* (with C. S. Hill). Myron Clark, 1908
- *Cost keeping and management engineering* (with R. T. Dana). Myron Clark, 1909
- *Earthwork and its cost*. McGraw-Hill, 1912, 1920
- *Handbook of rock excavation, methods and cost*. Myron Clark, 1916
- *Handbook of clearing and grubbing methods and cost*. Myron Clark, 1917
- *Handbook of mechanical and electrical cost data*. McGraw-Hill, 1918, 1920
- *Handbook of construction cost keeping and management* (with R. T. Dana). McGraw-Hill, 1922, 1927

With such an emphasis on costs, it may come as little surprise to find that the distribution of many of Gillette's titles was undertaken in Britain by E. & F. N. Spon – publisher of the long-running *Architects' and builders' price book*.

In the opening phase of his publishing career he was taken on by the McGraw Publishing Company in 1903 as associate editor of *Engineering News* and moved to New York. Then, in 1905 he joined forces with the McGraw book division's marketing director to establish the Engineering & Contracting Publishing Co., whose first title was the journal *Engineering-Contracting*.

In 1908, at the time of writing *Concrete construction*, the 39-year old Gillette was managing editor of *Engineering-Contracting* and based in Chicago, where, as it happened his publisher, Myron C. Clark, had offices. He was a member of the American Society of Civil Engineers, the American Society of Engineering Contractors, the American Institute of Mining Engineers and (like his contemporaries, Turneaure and Maurer) the Society for the Promotion of Engineering Education.

His 690-page book on concrete was written jointly with Charles Hill [1904] who, at this point in his career was Gillette's associate editor at *Engineering-Contracting*. The authors intended it to be a practical book for the builder of concrete structures and interestingly, in defining what it was *not*, they summarised the contemporary state of concrete book publishing in America to date. It was not a description of the physical properties of concrete (like the books by Falk and Sabin); it was not a consideration of reinforced concrete design (such as those by Turneaure & Maurer and Buel & Hill); and it was not a general treatise on concrete construction (as written by Reid and Taylor & Thompson).

In acknowledging sources from 'the volumes of *Engineering News*, *Engineering Record* and *Engineering-Contracting*, and from the *Transactions* of the American Society of Civil Engineers and the proceedings and papers of various other civil engineering societies and organizations of concrete workers', the authors declared 'the work done by these journals and societies in gathering and publishing information on concrete construction is of great and enduring value'. In his commentary Leon Moissieff considered that contents had been 'compiled with more care than is usual in compilations, and much original matter has been added'. *Cement & Engineering News* praised the book for its treatment of formwork and declared it 'should be in the hands of all persons engaged in concrete construction work.'

Gillette's first marriage failed and at the age of 48 he married Winifred Essery in 1918. Born in 1885 she was 16 years his junior, but they were to have no children. By 1927 the long outpouring of regularly revised practical construction guides had run its course, and Gillette's retirement was to be a long one. He spent much of his time in later years studying and writing on aspects of astronomy and meteorology, and he died at the age of 88 on 18 June 1958, at San Marino, Los Angeles.

Gillette, H. P. & Hill, C. S. *Concrete construction: methods and cost*. New York: Myron Clark Publishing Co., 1908 [C&CA 1959, ICE, PCA. Rev: *EN*, 11 June 1908 and *C&EN*, 1908, v.20, p.153] and London: Spon, 1908 [BL, Peddie]

1908/USA

Godfrey, Edward (b.1871)

MASCE

Godfrey's early publications were written while employed as structural engineer for Robert W. Hunt & Co. – a construction company with offices in Chicago, New York, Pittsburg and London – and presumably draw on his experiences in that capacity. However, although he used his employer's name to establish his credentials, the books were published privately rather than as promotional material for the firm.

As a structural engineer, and a member of the ASCE, his interests were not limited to concrete alone, and his first title (in 1905) was the more general *Structural engineering*. But it was his *Concrete* of 1908 – published 'by the author' in Pittsburg – that brought him into the specialist concrete literature. It comprised four parts: the first described concreting materials; whilst the second and third were derived from articles Godfrey had submitted the previous year to *Engineering News* and *Concrete Engineering* respectively, along with responses from other engineers. The fourth section consisted of illustrations from *Engineering News*, *Engineering Record*, *Railroad Gazette*, and the *American Railway Journal*. That much of the book was derived from articles submitted to the periodical press and then published privately is a clue to Godfrey's future *modus operandi*. His interests were vigorously pursued, and consequently he became widely known through the publicity of the press. Besides the journals above, he was a regular contributor to *Concrete-Cement Age* in the years before the First World War, writing about codes and building failures and acting as a resident commentator in the 'information and consultation' department.

A noncommittal review of the book in *Cement Age* (v.7, p.79) quotes passages by the author that convey an irritable tone: 'laying bare the falsity of much that passes for good engineering'; 'the author has found attempted standard nomenclature extremely annoying'; 'it is one of the greatest faults of books of reference', *etc*. They are an early hint that Godfrey was a man of strong opinions. Indeed *Cement & Engineering News* described him as 'an original' and 'often at variance with some of the accepted formulas used in designing reinforced concrete structures.' (*C&EN*, 1908, v.20, p.98).

As befitting the employee of a company than maintained offices in London, he appears to have spent time visiting Britain and became quite well known on this side of the Atlantic. At an early date in its proceedings, he addressed a meeting of the fledgling Concrete Institute (*Transactions* v.3). Back in America he featured in *Transactions* of the ASCE (v.80, December 1910) too.

His next books – *Godfrey's tables* and *Steel and reinforced concrete in buildings* – followed in 1911. Godfrey aimed to 'to supply a want in which where designing is done on a small scale that does not justify the employment of an engineer', and 'to lay down the principles of correct and consistent design as applied to buildings and to give simple rules and tables to be used in designing'.

By this date Godfrey had become driven by the perceived need to learn from mistakes and improve the quality of engineering design – an attitude that sometimes pitted him against the establishment and vested interests. Little surprise then that he took a keen interest in the American Concrete

Institute's move towards regulation. In December 1914 he wrote *A critical review of current practice in reinforced concrete design as embodied in building regulations and the Joint Committee report* (*Proceedings* 10(12), pp.10–27, 1 December 1914). The abstract – in his own words – is worth repeating as an expression of the man's approach:

> Mark Twain once said that he had nothing against the German language, he would just like to reform it. I have nothing against reinforced concrete, I would just like to reform it. At the outset I wish to make two statements that no sane engineer can contradict. The first is: There are entirely too many failures in reinforced concrete construction. The second is: there has been entirely too little attention paid to these failures.

He was not always in agreement with the official proposals to reform and the *Proceedings* of 1 June 1919 contained Godfrey's dissenting opinion to the *Proposed Standard Building Regulations for the use of Reinforced Concrete* – a preliminary draft of which was circulated to members in May 1919 (*Proceedings* 15(6), pp.402–403).

He continued to express his opinions throughout these years – years in which he entered practice as a consulting engineer and which led to publication of his *magnum opus*. Much later (in 1964), Jacob Feld and Kenneth Carper described this work in terms that convey a sense of the man and his mission:

> In 1924, Edward Godfrey, a consulting engineer well known for his frank and tireless criticism of improper design techniques, published a book on *Engineering failures and their lessons*, consisting principally of discussion and letters written in the years 1910–1923. Every phase of civil engineering is covered in the 20 chapters of Godfrey's book. The discussion of concrete failures is somewhat compromised by two of Godfrey's incorrect beliefs: that stirrups in concrete beams are of no value as shear resistance and that 'hoop-tied' column bars are a detriment to the performance of a concrete column. (*Construction Failure* by Jacob Feld & Kenneth L. Carper).

Perhaps resulting from his campaigning efforts and the reception of his book he continued to find an audience in Britain, especially at the Institution of Structural Engineers. The *Structural Engineer* published several of his papers at this time and in the succeeding years: on shear (1923); reinforced concrete columns (1925); dams (1928 and 1931); and strength of compression members (1932). They were not universally well received. His delivery, for instance, was described as in 'his typical sledge-hammer style' [*SE* v.2 p.62]. We get a hint of controversy in Godfrey's response to published comments on his paper of 1925: 'It is gratifying to have the courtesy of the Institution extended, and a vote of thanks tendered for my paper, even though it received so little approval of the points raised.'

Later in the year E. S. Andrews reflected on Godfrey's character and contribution to concrete engineering in a revealing commentary that is worth quoting at length:

Mr Edward Godfrey, who has been for many years persistently pointing out what he regards as fundamental errors in conventional engineering, has given use a new proverb: 'To err is human: to persist in known errors is devilish.' Our Institution may be proud of having accepted at least three thoroughly heterodox papers in recent years – one by Mr. Kempton Dyson and two by Mr. Godfrey, and I believe that in ten years time we shall be more inclined to accept the views expressed in these papers than we were when they were read. Mr. Godfrey has recently published a book entitled *Engineering failures and their lessons*, in which he has collected notes that he has made of engineering failures over a long period during which he has been busily engaged in structural engineering. It is a stimulating book – brimful of pep – and gives the reader an impression of a man bursting with energy, and always 'agin the government'. Mr. Godfrey claims to be an iconoclastic high priest of the cult of 'horse sense' and lashes out with fear or favour on mathematicians and theorists … He complains that the publication committees of engineering societies have refused to publish his contributions to the discussion of papers – this is always the fate of iconoclasts, but big men are not silenced by these means. Martyrdom has always been the most effective form of publicity … We have had two papers by Mr. Godfrey at our Institution; after reading his book I want more than ever to see in the flesh the man who, while engaged for so many years in a busy professional life, still has burning within him such an unquenchable fire of zeal. Iconoclasts may be a nuisance; but it is only by the breaking of images that we are emancipated from the shackles of ignorance (*Structural Engineer*, 1925, p.420).

Godfrey, E. *Concrete*. Pittsburg: Godfrey, 1908 [C&CA 1959. Rev: *CA*, v.7, *C&EN*, v.20, p.98 and *EN*, 14 May 1908]

Godfrey, E. *Steel and reinforced concrete in buildings*. Pittsburg: Godfrey, 1911 [Peddie, *CA* v.13, p.178]

1908/USA

Ketchum, Milo Smith (1872–1934)

MASCE

Milo S. Ketchum in 1904, from an unpublished autobiography dated 11 March 1930 Courtesy of Mark A. Ketchum, PhD, SE (Berkeley, California)

Ketchum, the son of a Primitive Baptist minister, was born on a farm in Illinois in 1872. On completing school he sat for his teaching certificate in 1889 and taught locally for a year. In 1890 he entered the University of Illinois to read civil engineering and whilst there, during the summer of his junior year, he became an instructor in surveying at the Michigan School of Mines. He graduated from Illinois in 1896 and spent the next two years as an assistant in civil engineering.

Leaving academia he worked for a steel fabrication and machinery company from 1897 to 1899, based at a branch office in Butte, Montana. There he designed buildings, bridges, coal tipplers and mining equipment. Proving his worth he transferred to head office in Minneapolis as a problem solver.

In 1899 he returned to the University of Illinois to take up the post of assistant professor on a salary of $1200 per year for the next three years. While there he wrote the *Surveying Manual* in collaboration with Professor W. D. Pence. This rapidly became a standard work of which 56,000 copies were issued. He combined his writing and university work with consultancy during the summers.

When he married in 1903, Ketchum changed employment, doubling his salary as contracting manager for the American Bridge Co. in Kansas City. This was to be a major source of information for many of his future books.

Months later however, he was offered the professorship of civil engineering at the University of Colorado. There, for $2,000 per year and with the help of a part-time assistant, he taught all the department's courses: lectures in the morning; laboratory in the afternoon. In 1904 he was promoted to dean, aged only 33. Student registrations rose from 171 in the first year to 225 in the second, and 350 by the time he left Colorado in 1919. During this defining period in his life he was an active member of the Society for the Promotion of Engineering Education (like Turneaure and Maurer [1907]) – national secretary for a while, then president in 1917. He also wrote a number of books:

- 1903 *Design of steel mill buildings, and the calculation of the stresses in framed structures*
- 1907 *Design of walls, bins and grain elevators*
- 1908 *Design of highway bridges of steel, timber and concrete*
- 1912 *Design of mine structures*
- 1914 *Structural engineers' handbook*

These titles were published by Hill (and then McGraw-Hill), but according to his son, Ketchum undertook the proof-reading and arranged for the printing and binding, over and above writing the text. Under this arrangement he received half of the retail price. To help him, 'he trained and kept busy many a student drafting the illustrations which were beautifully done'. The profusion of these illustrations were commented on in *C&CE*: 'much time and labour must have been expended by the author in its preparation, as it contains numerous drawings and tables, each of which represents a considerable amount of work.'

Of these books only *The design of highway bridges* mentions concrete in the title, but both the *Design of walls, bins and grain* and *Structural engineers' handbook* touched on the material. They were reviewed by *C&CE* and both unstintingly recommended. When *Highway bridges* was revised in 1920, 15,500 had been issued. It had – according to *Cement Age* (v.7, p.461) – opened 'practically a new field so far as literature upon the subject is concerned'.

In 1920 Ketchum was working at the University of Pennsylvania. Whilst at Colorado, in 1917 he had been approached by the War Department to take the role of assistant director of explosive plants and to take charge of the construction plant in Nitro, near Charleston in West Virginia. When he returned at the end of the war he felt detached from his former work and in 1919 became head of the department of civil engineering at the University of Pennsylvania in Philadelphia.

Later still he became dean of engineering at the University of Illinois and director of the Engineering Experiment Station there. Like Wisconsin and Colorado, Illinois was a major centre for concrete research at the time – if not the pre-eminent centre.

He wrote an autobiography and died in 1934. A biographical sketch by Milo Ketchum Jr was presented at the centenary of the department of civil engineering, University of Colorado. His son – perhaps partial, but as a writer on concrete himself, well-informed – described him as 'one of the outstanding engineering educators of the early 20th Century'.

Ketchum, M. S. *The design of highway bridges of steel, timber and concrete*. New York: Engineering News Publishing Co., 1908 [Rev: *CA*, v.7, 1908]. Second ed. New York: Hill, 1908 [C&CA 1959, BL, ICE, Yrbk 1934. Copy: TCS]

1908/USA

McCullough, Col. Ernest (b.1867)

MWSE, MASCE, MIInstructE

McCullough – 'a frequent contributor to the engineering press' (*CA*, v.7, p.79) – wrote widely on various subjects not limited to concrete, and had at least seven books to his name. His first was *Public works* (issued by Courier in 1894) and some years later wrote *Engineering work in towns and cities*. Myron C. Clark published this in 1908 as well as the first of his concrete books.

Reinforced concrete first appeared in serial form on the pages of the *Cement Era*, before being tidied up for publication by McCullough's son George Seymour. The preface declares its purpose to be as indicated by the subtitle *a manual of practice*: 'The intention in the sections on design has been to keep within the usual requirements of the ordinary conservative building ordinances of American cities ... So far as construction is concerned the principles stated herein as the result of personal experience, apply to all manner of work in reinforced concrete, and to this extent the book should be of some value to a large number of men.' *Cement Age* thought it a 'convenient little book' presented in 'clear and concise terms' for those without a technical education and *C&EN* considered 'the chapter on forms is alone well worth the price of the book'.

McCullough followed this with another practical book on concrete – *Practical structural design in timber, steel and concrete* – and other engineering works:

- *Engineering as a vocation* (David Williams Co., 1911)
- *Practical surveying for surveyors' assistants, vocational and high schools* (Van Nostrand, 1915)
- *Practical structural design* (U.P.C. Book Co., 1918)

Like his *Reinforced concrete*, *Practical structural design* was issued first as a series of journal articles – in this instance in *Building Age* between 1914 and 1916, 'especially adapted for the needs of the self-tutored man'. At this date McCullough was describing himself as a licensed structural engineer and architect in the State of Illinois. In 1924 *Practical structural design* was promoted in the UK and was favourably reviewed by Ewart S. Andrews in the *Structural Engineer*. Andrews welcomed it as the first American textbook aimed at 'those students who have not had the opportunity of university training'. He noted that McCullough, currently in Great Britain, was expected to read a paper at the Institution of Structural Engineers: 'if it is as good as his book we shall have an excellent evening.'

McCullough, E. *Reinforced concrete: a manual of practice*. Chicago: Cement Era / Myron C. Clark, 1908 [PCA. Rev: *CA*, v.7, p.79 and *C&EN* 1908, v.20, p.208]

McCullough, E. *Practical structural design in timber, steel and concrete*. New York: United Publishers, 1918 [PCA]. Third edition, New York: Scientific Book Corporation, 1926 [ICE] and 1927 [BL]

1908/USA

Palliser, Charles

It is not clear where Palliser's antecedents lie. The architectural firm of Palliser, Palliser & Co. had flourished in the American housing market, specialising in the standard pattern-book designs that were so much a feature of the expanding suburbs of the late 19th Century. The firm comprised the brothers George and Charles and was noted for a series of publications, including:

- *Palliser's model homes*, 1878
- *Palliser's American cottage homes*, 1878
- *Palliser's new cottage homes and details*, 1887
- *Palliser's American architecture*, 1888

George died in 1903 and perhaps his brother was responsible for a final publication in 1908 – harnessing the new material to the old firm's purposes?

By coincidence, however, there was also a Palliser building firm in Timaru, New Zealand, during the early years of the 20th Century that was known for its development of an early concrete block made using a distinctive mould. Such a specialism would be reflected in a book of this title, *Practical concrete block making*.

The book itself was a practical treatise covering materials, moulds and machines, and offering 'hints and suggestions'. It included the standard specifications of the National Association of Cement Users, later the American Concrete Institute.

Palliser, C. *Practical concrete block making*. New York: Industrial Publications, 1908 [C&CA 1959. Rev: *CA*, v.8]

1908/USA

Reuterdahl, Arvid (1876–1933)

We cannot separate contemporary writers' treatment of concrete from the wider contexts of the day. We have seen concrete drawn into the case for scientific management; with Arvid Reuterdahl concrete collides with the philosophy of science and religion.

Reuterdahl was born on 15 February 1876 in Karlstad, Sweden, and came to America as a boy when his parents emigrated in 1882. They settled in Providence, Rhode Island. He studied at Brown University from 1893, gaining a BSc in 1897 and an MA in 1899. For a while he taught mathematics and physics at the Technical High School in Providence and spent a year as professor of physics at Colby College in Waterville, Maine. Having developed a type of storage battery, he founded the Reuterdahl Electric Company to exploit his design. In 1902 he married Elinor Louise Morrison, with whom he had a son, Norman.

In 1905 he moved to Spokane, Washington, where he established a private engineering practice. Later he served as assistant city engineer, water commissioner, chief of the bridge department and president of the Board of Public Works. He also taught engineering at Spokane College. It was during these years at Spokane that he wrote his one contribution to the literature of concrete, although he was a prolific author on other subjects. His *Theory and design of reinforced concrete arches* of 1908 was addressed to both engineers and technical students, reflecting his twin occupations in Spokane. Its purpose was to explain in detail the theory of arch design, making sure there were 'no missing steps in the necessary mathematical analysis' and thus removing the obscurity of previous books on the subject.

Remaining in the Northwest for another three years or so, Reuterdahl transferred his consultancy to nearby Boise, Idaho in 1910. In 1913 he moved to Kansas City and between 1915 and 1917 he was professor of theoretical and applied mechanics at Kansas City Polytechnic Institute. In 1918 he joined the College of St. Thomas in St Paul, Minnesota as dean of the engineering and architecture department, and in 1922 he established the Ramsay Institute of Technology.

The 1920s saw Reuterdahl increasingly interested in the relationship of science to religion. In 1921 he, with Robert T. Browne, founded the Academy of Nations in a bid to reverse the increasing degree of specialisation and abstraction within scientific knowledge. Perhaps most controversially the academy attacked Einstein and his theory of relativity. In 1923 Reuterdahl was awarded a doctorate in science by the academy in recognition of his discovery of the physical basis of Planck's constant and the theory of 'Septad-Constants'. The movement however, though it had extended to Germany, Sweden and Yugoslavia, dissolved before the end of the decade. By that time Reuterdahl had become the editor of *Theistic Monthly* – a post held from 1926 – and wrote extensively on the relationship of science, philosophy and religion. He died in 1933.

He left behind him an autobiography entitled *My life: the first years*, and his collected papers with a portrait photograph are preserved at the University of St Thomas.

Besides his battery, Reuterdahl improved the design of culverts, theorized about the velocity of light and relativity and developed a world alphabet. He was a member of many organisations including the Mathematical Association of America, History of Science Society,

Swedish Physical Society and American Society of Mechanical Engineers – the range of which reflects his interests. It is fair to say his horizons were not bound by concrete design or construction.

Reuterdahl, A. *Theory and design of reinforced concrete arches: a treatise for engineers and technical students*. Chicago: Myron C. Clark, 1908 [BL, ICE. Rev: *EN*, 17 Dec 1908], and London: Spon, 1908 [Peddie]

1908/USA

Watson, Wilbur J. (1871–1939)

W. J. Watson was a civil engineer who specialised in bridges and made a name for himself in his native Ohio and beyond as an innovatory bridge designer concerned with aesthetics. His early written output consisted of practical guides to concrete work, while his mature writing concentrated on bridges.

Wilbur J. Watson
Courtesy of Cleveland State University

Wilbur was born in Berea on 5 April 1871, the son of David R. and Maria Watson. His education culminated with a BSc from the Case School of Engineering and he entered the engineering profession. He started to design bridges from 1898 whilst working for Osborn Engineering Company of Cleveland. In 1900 he married Harriett Martha Barnes, with whom he had two daughters, Sara and Emily. By 1907 he had established his own firm: Wilbur J. Watson & Associates. From this point on he made his mark in his chosen field. In 1908 he introduced precast beams for bridges; in 1910 pioneered a design of steel centring; in 1911 he undertook experiments with the new method of mushroom slab construction; and in 1912 (at Akron) he built the highest bridge of its kind. He also began to share his experience in print.

His first book – a pamphlet really, consisting of only 46 pages – was entitled *General specifications for concrete work: as applied to building construction* (1908). It was modelled on the lines of the regulations adopted by many American cities, though with a generally wider view. Watson drew on his own experience, articles in the engineering press, and experiments conducted by Professor Arthur Talbot [1917] at the University of Illinois. Two months later in March, he extended his coverage to bridges with a companion volume of nearly twice the length: *General specifications for concrete bridges* (1908). Combining first hand knowledge with guidance from the authorities, the second edition of 1911 was amended in line with the report of the Joint Committee on Concrete and Reinforced Concrete that had been issued in the interim. His personal experience as an innovator came to the fore in his next title: *Steel centring used in the construction of the Rocky River Bridge*. This referred directly to the bridge he erected in 1910. He continued to write – contributing to discussions in technical journals such as *Concrete-Cement Age*.

Despite earning the distinction of erecting the largest uninterrupted interior space in the world at the time –

the Akron Goodyear Zeppelin Airdock of 1929 – Watson's career was shaped by his specialisation in bridges. His *Bridge architecture* was published in 1927 and the 1930s saw *A decade of bridges: 1926–1936*; *Bridges in history and legend*; and *Great bridges: from ancient times to the twentieth century*.

Beside his legacy of concrete structures, his collection of rare books survives. In 1983 his daughter Sara presented the Watson Bridge Book Collection to Cleveland State University Library. An engineer herself, Dr. Sara Ruth Watson added to the collection and on her death in 1996 bequeathed a fund for its ongoing maintenance.

Watson, W. J. *General specifications for concrete work: as applied to building construction*. New York: Engineering News Publishing Co., 1908 [Rev: *Engineering News*, 20 February 1908] and New York: McGraw-Hill, 1915 [PCA]

Watson, W. J. *General specifications for concrete bridges*. New York: Engineering News Publishing Co., 1908 [Rev: *Engineering News*, 12 March 1908; *CA*, v.6, p.429]. Second edition, 1910 [Rev: *CA*, v.13]. Third edition, McGraw-Hill, 1916

1908/USA

Webb, Walter Loring (1863–1941)

BCE, CE, MASCE

Born in 1863, Webb studied civil engineering at Cornell University, graduating in 1884. Five years later he was chartered. He went on to gain a reputation in railway engineering that was reflected in the various long-lasting titles that he wrote on that subject:

- *Railroad construction: theory and practice* (Wiley, 1900–1932)
- *The economics of railroad construction* (Wiley, 1906, 1907 and 1912)
- *Railroad engineering* (American School of Correspondence, 1912)

He was also interested in instruments – the tools of his trade – and in fact his first book was entitled *Problems in the use and adjustment of engineering instruments* (Wiley, 1896). This he followed 21 years later with *Technic of surveying instruments and methods* (Wiley, 1917). He also compiled a volume of trigonometric tables for students.

Much of this work, writing and revising took place in the first decade or so of the century, with clusters of activity around 1906–1909 and 1912–13. This period of course, coincided with the emergence of a new body of literature on reinforced concrete, and Webb was not long in offering his contribution. With his collaborator, W. H. Gibson (below), he wrote two titles for the American School of Correspondence: *Reinforced concrete* (1908); and *Masonry and reinforced concrete* (1909). The school offered a technical education by (as its name indicates) correspondence. It was established in 1897 and located in a new building erected in Hyde Park, Chicago, in 1906/7. Brought to the British market as well (by Crosby Lockwood), Webb and Gibson's books would have been widely read. The latter (*Masonry*) was reviewed in *C&CE* (v.6, pp.154–155), where it was considered adequate, but suffering a little from a compromise between the needs of the designer and the practical superintendent. The treatment of masonry was dismissed as 'cursory' and the reviewer thought it 'a pity that this volume should not have been devoted to reinforced concrete alone'. That said, the book's coverage of reinforced concrete was 'dealt with very well', so far as it went, but should perhaps be thought of as a taster, prompting the student to further study.

The year 1909 (the date of the *Masonry* volume) also saw Webb attending the 25th anniversary reunion of his class from Cornell at the Ithica Hotel. A similar event was recorded by the *Alumni News* of 2 April 1912 when the by-now Professor Webb of the University of Pennsylvania visited the Association of Civil Engineers for its ninth Annual Banquet at the Ithica Hotel.

In 1916 the Chicago-based American Techincal Society published Webb and Gibson's third collaboration, *Concrete and reinforced concrete*. Not long afterwards however, the USA entered the First World War and Webb played his part in the armed forces. *Alumni News* of 14 Feb 1918 records him as a major in the Reserve Corps, on active duty with the American Expeditionary Force.

His final book, still with Gibson, was completed much later. The American Technical Society published *Concrete design and construction* in 1931. Like the book on masonry, it tried to tackle two subjects – design and construction – and in the view of the *Structural Engineer* (v.10, p.271), failed to accomplish either adequately. Too condensed, more detail was called for, though the authors' practical knowledge of the subjects was credited, and the book received a qualified welcome.

Webb, W. L. & Gibson, W. H. *Reinforced concrete: a treatise on cement, concrete and concrete steel and their applications to modern structural work*. Chicago: American School of Correspondence, 1908 and London: Crosby Lockwood, 1927 [Yrbk 1927]

Webb, W. L. and Gibson, W. H. *Masonry and reinforced concrete*. Chicago: American School of Correspondence and London: Crosby Lockwood, 1909 [Peddie. Rev: *C&CE* v.6, p.154]

Webb, W. L. and Gibson, W. H. *Concrete and reinforced concrete: a condensed practical treatise on the problems of concrete construction in cement mixtures, tests, beams an d slab design, construction work, retaining walls, etc*. Chicago: American Technical Society, 1916 [BL, PCA] and 1919 and 1925

Webb, W. L. and Gibson, W. H. *Concrete design and construction*. Chicago: American Technical Society, 1931 [BL, ICE. Rev: *SE* v.10, p271] 1940

1908/USA

Gibson, William Herbert (b.1877)

BSc, CE, MASCE

Consulting engineer and collaborator with Walter Loring Webb (above) for all four of Webb's concrete titles.

1909/USA

Ballinger, Walter Francis (1867–1924)

AAIA, MASCE

A self-made man, rising from humble origins, Ballinger was an early exponent of concrete architecture and co-inventor of the 'super-span sawtooth' roof.

Walter Ballinger was born on 31 August 1867 in the curiously named Petroleum Center, Venango County, Pennsylvania. His father, Jacob Howe Ballinger, ran a machine shop until his death in 1869, whereupon his mother Sarah Wolfenden moved the family to Woodstown, New Jersey. The family lived there for the next dozen years and by the age of 13 Walter was working on a local farm. Transferring to a factory, the boy showed promise and was promoted. He attended evening classes at the local grammar school, the YMCA and the Drexel Institute and finally enrolled in a business school for fulltime study. In 1889 he entered the service of an architectural and engineering firm – Geissinger & Hales of Philadelphia. He started as a bookkeeper and clerk, but by 1895 had an autonomous partnership at the same premises, with another member of the staff – William H. Brinkworth. This was short-lived and in the same year he replaced Geissinger as a principal in the older firm, which then traded as Hales & Ballinger. In July 1901 Edward M. Hales retired and Ballinger took the chief draughtsman – Emile G. Perrot (below) – into partnership, operating as Ballinger & Perrot.

An early commission – the renovation of Old St Joseph's Church, St Joseph's University in 1904 – gave a pointer to the firm's future specialisation in ecclesiastical architecture. The construction of the Crane Company Building on Girard Avenue a year later, was an example of a concentration on concrete construction. This industrial building comprised a cast concrete structure faced with brick. Though the firm never limited itself to concrete design – fire escapes were another speciality – it was clearly taking on an increasing importance. Ballinger's article in *The Manufacturer* entitled *Industrial plants, the design, construction and equipment* (15 June 1905), involved a favourable consideration of concrete. By 1909 the partners had identified a gap in the market, where the demand for inspectors or clerks of works with suitable expertise in reinforced concrete far exceeded supply, and addressed the problem with their book *Inspector's handbook of reinforced concrete*. They declared their hand in the preface:

> Those architects or engineers who have specialized in this construction have had applications from inspectors continually who wish to avail themselves of contact with this work under the direction of specialists. The authors' experience in selecting men to represent their firm in this capacity has led them to publish this little volume with the hope that it would fill a place in the literature on reinforced concrete not covered by existing books. Personal contact with the inspectors and their viewpoint of various details of the work convince us of the necessity of such a book of instruction. While not endeavouring to inform the inspector of all his duties, this volume is written to point out the essentials governing the construction of this class of buildings.

Cement Age (v.10) pronounced it a 'very complete and concise handbook' and observed that it was 'based on actual experience gained in the erection of several hundred reinforced concrete structures with which the authors were connected'. Indeed, some material was 'derived from tests on full-sized members made by the authors'.

Architectural work followed this essay into the literature, including the following buildings:

- 1911 Four Points, Philadelphia

- 1912 Isaac H. Goodman Building, Philadelphia
- 1914 134 Bay St, Jersey City
- 1916 The Victor, Camden, NJ
- 1920 Traylor Building, Philadelphia

Notably the firm undertook the planning and design of the industrial community 'Viscose Village' (discussed under Perrot below), and in 1914 worked on two contrasting church commissions: Villa Maria Hall – the central building of Immaculata University (along with Nazareth Hall on the same campus); and the Wesley Building – a six-storey office for the Methodist Church, built in the gothic style.

By 1917 Ballinger & Perrot had expanded beyond Philadelphia to maintain a New York office with a staff of 125. However in 1920, Ballinger bought out Perrot's share of the business and the firm continued into the 21st Century as the Ballinger Company. Besides the still ongoing business, architectural drawings from the years 1920–1927 are preserved at the University of Philadelphia.

Ballinger, W. F. & Perrot, E. G. *Inspector's handbook of reinforced concrete*. New York: Engineering News Publishing Co.,1909 [PCA], McGraw-Hill, 1909 [ICE. Rev: *CA*, v.10], and London: Constable, 1909 [Peddie]

1909/USA

Perrot, Emile G. (1872–1954)

AAIA, AASCE

Chief draughtsman for Hales & Ballinger, Perrot became Ballinger's partner in 1901 on the retirement of Mr Hales. As a principal of Ballinger & Perrot, he contributed much to that firm's expertise in reinforced concrete construction. We can surmise the balance of interest in the new material from Perrot's record of writing for the ACI's *Proceedings*, contrasted with the absence of a comparable output from Ballinger (other than the book they co-authored in 1909):

- *1908* *The unit versus the loose bar system of reinforced concrete construction (v.4)*
- 1909 *Comparative cost of reinforced concrete buildings* (v.5)
- 1910 *Method for long span, light floor, reinforced construction, with comparative cost* (v.6)
- 1911 *Analysis of results of load tests on panels of reinforced concrete buildings* (v.7)
- 1911 *Incident of the value of concrete reducing the cost of insurance* (v.7)
- 1918 *Architectural design of the concrete house* (v.14)

A recurring theme in the ACI papers is Perrot's interest in costs. Similarly he contributed to *Cement Age*, though with a more advocatory tone and reference to concrete's fire resistance qualities:

- 1908 *No building material offers the possibilities found in concrete* (v.6)
- 1908 *The unit versus the loose bar system of reinforced concrete construction* (v.6)

- 1908 *Status of reinforced concrete from the fireproof standpoint* (v.7)
- 1909 *A comparison of tests of steel and reinforced concrete beams* (v.9)

He also contributed to discussions conducted on the pages of *Concrete-Cement Age* in 1913.

Perhaps his greatest architectural achievement during these years (specifically 1912–1915) was 'Viscose Village', a planned community designed by Perrot for the newly established American Viscose Co. at Marcus Hook, Pennsylvania. The site comprised 261 model homes, two boarding houses and a shop on a 20-acre tract of land across the road from the fibre works, with a dining hall and dispensary on the factory grounds. According to a University of Pennsylvania dissertation by Christine Tate, Viscose Village was based on Bourneville, Port Sunlight and Hampstead Garden Suburb, and influenced by Bedford Park and Letchworth Garden City: 'English antecedents figured prominently in Perrot's plan and design' (Tate, C. *Viscose village: model industrial workers' housing in Marcus Hook, Delaware County, Pennsylvania*).

After the dissolution of the partnership Perrot wrote a further book in 1921 entitled *The groundwork of architecture: or, the study of how architectural styles are affected by structural engineering*. Interplay between the sister disciplines was an increasingly topical issue in the 1920s on both sides of the Atlantic, which in the UK was explored at the time by Maxwell Ayrton and Owen Williams.

More church-inspired building came Perrot's way in the twenties – in 1929 he added a transept to the gothic-styled St John's Church, New York, later known as the University Church.

Perrot lived to a ripe old age, dying in 1954, with grandchildren to mourn his passing. His name and achievements are recorded on a plaque at Viscose Village.

1909/USA

Brett, Allen (b.1884)

A journalist, Brett was the founding editor of *Concrete Engineering* – based in Cleveland, Ohio throughout is publication life from 1907 to 1911. Doubtless drawing on the material appearing in *Concrete Engineering*, Brett compiled a book entitled *Reinforced concrete field handbook*, which was issued in 1909 by the magazine's publisher, Technical Publishing Co.

As a consequence of the proliferation of cement and concrete magazines in the new century's first decade, *Concrete Engineering* was merged with Robert Lesley's New York title *Cement Age* in January 1911, in which the name was recalled by the strapline, 'with which is combined *Concrete Engineering*'. Brett was appointed associate editor of the new venture. In July 1912 the combined magazine merged with *Concrete* of Detroit, and Brett became managing editor of *Concrete-Cement Age*, with Lesley [1907] taking a more distant role as consulting editor.

Brett, A. (ed). *Reinforced concrete field handbook*. Cleveland: Technical Publishing Co., 1909 [Peddie]

1909/USA

Colby, Albert Ladd (1860–1924)

MASCE, MASME, MIATM, MASTM, MISI, MAIME

Colby was born in New York City on 26 June 1860 and baptised on 13 July. In 1870 he was recorded as resident in the city with his parents – John Ladd Colby, MD and Mary Ann Stuart Tannatt. He was educated at the local public schools and the College of the City of New York, and in 1880 he was still living with his father. A year later (in 1881), he was awarded his degree by the Columbia School of Mines. He continued at Columbia until 1883, acting as assistant to Professor Chandler. The next three years were spent as assistant professor in chemistry at Lehigh University in Bethlehem, Pennsylvania.

Transferring from university life he took a position as head chemist – and then metallurgical engineer – for the Bethlehem Steel Co. in 1886 and remained with this firm until 1903. By coincidence the author Frederick Taylor [1905] was engaged by this company too, from 1898 to 1901. As an engineer 'he had entire charge of the inspection, assignment and economic use of all metallurgical materials, and he early gave his attention to the formulation of specifications ... an activity in which he long held a distinguished part in this country and as a representative of his own country in England and Europe'. He also acted as an expert for his company in patent suits and between 1897 and 1905 he was secretary of the Association of American Steel Manufacturers.

During his long service with Bethlehem Steel he married and raised a family. His bride – whom he wed on 20 June 1894 – was Agnes Wilson Lee. They had two children: Richard and Mary; and dwelt at Fountain Hill, Lehigh County at least until 1920.

Leaving Bethlehem Steel in 1903 Colby worked for the International Nickel Co. for two years, before setting up as a private consultant. As a consulting engineer Colby continued in a familiar vein, as an expert in court and a representative of American practice in Britain and the Continent. In 1908 he returned to the USA from Cuxhaven, sailing on the *Amerika*. The nature of his business is revealed by the publication a year later, of his *Reinforced concrete in Europe* by the Chemical Publishing Company of Pennsylvania – a publisher whose name and location closely parallel the author's background. The book – the subject of which combined Colby's interests in chemistry and steel – was an outcome of a personal visit abroad:

> The Report is a compilation of information, on current practice in Reinforced Concrete Construction in Great Britain and on the Continent, collected during 1908, chiefly by personal interviews with the leading authorities in each Country. The writer visited England, France, Germany, Austria, Hungary, Switzerland and Italy and desires to here record his appreciation of the courteous attention to his inquiries, according to him in each Country.

It was a detailed compendium of current official and commercial practice, supported by numerous sources of further information. There were addresses of experts, companies and associations; references to building regulations, standard specifications for steel and cement, and committee recommendations on reinforced concrete; and an extensive bibliography. Besides technical data on reinforcing steel and descriptions of the systems available in each country – as supplied by steel producers and reinforcement agents – the report reproduced statements from consulting engineers and national experts such as Charles Meik (GB), Gerard Lavergne (F), O. Kohlmoren (G), R. Janesch (A) and Professor F. Schule (S). Notably there was an interview with Charles Marsh [1904], who with Dunn and Twelvetrees remained the most widely published authority in Britain at that date.

A private edition of 50 copies was printed May 1909 for distribution to subscribers. This evidently made an impact: 'In response to numerous requests, permission has been given to the Printer to reprint the Report for sale by him as a Book.' Further trips followed, including one in 1913 from which he returned to the USA on the *Kaiserin Auguste Victoria* from Cherbourg.

Various appearances in litigation are exemplified by his reports for the 1919 case of the Boston molasses tank explosion, for which he was called by the defendant – Purity Distilling Co. – during a hearing that was the longest in the history of the Massachusetts courts. With 21 deaths and 50 injuries, Purity and its parent company paid out over $1m as a response to the 125 lawsuits filed against them.

The American Institute of Mining Engineers recognised Colby's talents as an expert in such cases:

> 'His mind was ingenious, thorough, clear, and keen, both in the preparation of evidence and its presentation to the Court. He very quickly appraised the value of evidence, or of an answer to a question, and knew how to secure the information, either from the available literature of any language, or by special researches or plant studies. In the executive control of literature research for patent purposes, his long training, infinite capacity for digging through long series of articles in English, German or French, and clear formulation of the vital subject matter for the guidance of assistance made him invaluable. He was also an adept in collecting information on special branches of his profession.' (AIME *Transactions*).

Colby died on 30 April 1924 of influenza. Characteristically he was abroad, in Torquay. He was buried in his native New York City.

Colby, A. L. *Reinforced concrete in Europe*. Easton, PA: Chemical Publishing, 1909 [BL, PCA, Peddie. Copy: Internet Archive]

1909/USA

Hanson, Edward Smith (b.1871)

Editors of several of the specialist trade journals – Hodgson (*Builder & Woodworker*), Gillette (*Engineering Contracting*) and Whipple (*Concrete-Cement Age*) in the USA; and Jones (*Building World*) and Childe (*C&CE*) in Great Britain – turned their writing skills to the production of books as spin-offs from their main business. Hanson – editor of the *Cement Era* – was another such journalist turned author. In 1909 he produced his *Cement pipe and tile*: a 100-pager on a topical, and at the time much discussed, subject in relation to rival clay. Besides its discursive introduction, the book included information on costs and methods of manufacture, 'illustrated in a way to emphasize the text'. He followed it in 1913 with *Concrete roads and pavements*.

Hanson did not confine his efforts to Chicago and the *Cement Era*, but also participated in the American Concrete Institute at around the time it re-branded from the National Association of Cement Users. In December 1914 his article *The layout of concrete production plants* was published in the ACI's *Proceedings* (pp.485–505). We last hear of him with his *Concrete silos*, written in 1915 but published the following year. Despite his membership of the ACI and of the Western Society of Engineers, this book was composed in a manner that you would expect from a magazine editor, derived from an assortment of bulletins issued by the Portland Cement Association and its members, and various state agricultural research stations. Nonetheless, it fulfilled a need in a rapidly expanding farming market.

Hanson, E. S. *Cement pipe and tile*. Chicago: Cement Era Publishing Co., 1909 [Rev: *CA*, v.8]

Hanson, E. S. *Concrete roads and pavements*. Chicago: Cement Era Publishing Co., 1913 [BL, C&CA 1959]

Hanson, E. S. *Concrete silos: their advantages, different types, how to build them*. Chicago: Cement Era Publishing Co., 1916 [BL]

1909/USA

Johnson, Lewis Jerome (b.1867)

MASCE

Lewis Jerome Johnson, from The Biographical History of Massachusetts, *1909*
Courtesy of The Concrete Society

Born on 24 Sep 1867 in Milford, Massachusetts, Lewis was the son of Napoleon Bonaparte Johnson, cashier of the Home National Bank of Milford, and his wife Mary Tufts (*née* Stone). He studied the classics at Milford High School and helped his father in the bank during the afternoons and holidays. However, his interests were more scientific and technical than financial, and he went up to Harvard in 1883 to read mathematics, geology and engineering. He also joined the Phi Eta Fraternity. On passing out in the class of 1887 he attended the Lawrence Scientific School and graduated with a civil engineering degree. Two years of further study in Europe were spent at the *Eidg Polytechnikum* of Zurich and the *École des Ponts et Chaussees* of Paris, and culminated in a period of travel for pleasure in Egypt, Palestine and Greece.

Returning to America in 1890 he was appointed instructor in engineering at Harvard University, then assistant professor, and in 1906 professor of civil engineering. He combined his academic role with engineering consultancy in general structural practice, accepting commissions for work in Boston, New York and Chicago. In research, practice and writing, his professional interests were largely directed to reinforced concrete.

In this field his crowning achievement was his contribution to the design of the Harvard Football Stadium, which was erected in 1902. When it was opened it was said to be the largest reinforced concrete building in the world. Perhaps more contentiously, it has been claimed that 'it is historically significant that this stadium represents the first vertical concrete structure to employ reinforced structural concrete'. It is now listed as a National Historic Landmark.

The *Harvard Magazine* of September/ October 2003 acknowledged Johnson's input: 'The architect was Charles Follen McKim of the New York firm of McKim, Mead & White who took off from a design sketched several years earlier by assistant professor of civil engineering Lewis J. Johnson.' Certainly Johnson's contribution was widely recognised, and there is a plaque dedicating the stadium to his honour on the eastern wall.

Johnson wrote *Statics by algebraic and graphic methods* (published by Wiley) and many articles on engineering subjects, but it is for his *Reinforced concrete* (Moffat, 1909) that he is included here.

He wrote widely on many other matters too: economics and civics, especially taxation and voting systems. In politics he was a Democrat, and with the active participation of his wife – Grace Allen Fitch: a writer whom he married in 1893 – was a member of the Anti-Imperialist League, the American Free Trade League, the Men's League for Woman Suffrage, the Massachusetts Direct Legislation League and the Massachusetts Single Tax League (of which he was president in 1913). He and Grace had two sons: Jerome Allen and Chandler Willard.

A pen-portrait of Lewis Johnson at the time his *Reinforced concrete* was published appears in the *Biographical history of Massachusetts*, edited by Samuel Atkins Eliot (Boston: Massachusetts Biographical Society, 1909).

Johnson, L. J. *Reinforced concrete*. New York: Moffat, 1909 [Peddie]

1909/UK

Middleton, George Alexander Thomas (1861–1935)

ARIBA

The effects of reinforced concrete building was a rare foray into the subject by a prodigious author of construction books. Middleton's subjects ranged across architecture, building materials, drainage, structural engineering, surveying and town planning. Perhaps best known was his six-volume work *Modern buildings: their planning, construction and equipment,* published by Caxton in 1906 and 1921, which also touched on, among many other subjects, reinforced concrete.

George Alexander Thomas Middleton was born in Bridgend, Glamorgan, early in 1861. He was the son of Thomas A. Middleton (a solicitor) and his wife Elinor, both of Lambeth. Despite the location of his birth and time spent studying in Colchester, South London was to be home for most of his life: Balham (1882); Streatham (1901); Wandsworth (1911) and Camberwell (1935).

Middleton was an architect's pupil in 1881, resident at 53 Crouch Street in Colchester, and by March the following year had qualified to become a candidate for election as an associate of the RIBA. He was duly recommended for admission on 15 May 1882. Once established as an architect, Middleton married Anna (or Annie) in early 1887.

His professional interests were diverse and he wrote widely. Beside work on concrete his titles included:

- Building materials: their nature, properties and manufacture
- Drainage of town and country houses
- House Drainage: a handbook for architects an building inspectors
- The principles of architectural perspective
- Surveying and surveying equipment
- Stresses and thrusts

He enjoyed drawing and on at least one occasion joined the Dundee architect John Donald Mills on a sketching tour to France. A collection of his drawings is held at the British Architecture Library and his portrait at the National Portrait Gallery. Later in life he became vice-president of the Society of Architects. He died aged 74 and his obituary was published in the *Builder* of 26 July 1935 (v.149, p.143).

Middleton, G. A. T. *The effects of reinforced concrete building*. London: Francis Griffiths, 1909 [ICE] and New York: Spon, 1909 [Peddie]

1909/USA

Morsch, Emil (1872–1950)

DR ING

Emil Morsch Courtesy of the Archives of the University of Stuttgart

Like Considere – whose *Experimental researches* heads this survey – Morsch's significance as a German engineer and academic lies outside the literature in English, but is included here in translation under the year in which his work was first published in America.

Morsch was born in the newly unified German Empire on 30 April 1872, at Reutlingen, Baden-Wurttemberg. Apart from his university post in Zurich, he was to remain in the southwest of Germany throughout his life, although his writing in German, French and English gave him an international importance.

Between 1890 and 1894 he studied at the Technical School, Stuttgart, and entered the employment of the state government in the roads and waterways department. Having passed further examinations in 1899 he transferred to the bridges office of the Wurttemberg State Railways. In 1901 he joined the firm that was to make his name: Wayss & Freytag.

Wayss & Freytag AG had been a reinforced concrete specialist for over a decade and was perhaps the longest established of such firms on the Continent. Morsch was appointed chief engineer and head of the technical bureau. Impelled to provide generic and authoritative guidance to the scientific principles of reinforced concrete by the emerging market presence of untested (and possibly unreliable) proprietary systems, Wayss & Freytag published its landmark *Der Eisenbetonbau* in 1902. Morsch was the author of its theoretical component. *Der Eisenbetonbau* had an immediate impact and formed the basis of recommendations published – within months – by the *Verbands Deutscher Architekten und Ingenieur-Verine* and the *Deutscher Beton Verein*, and the regulations issued by the Prussian Government.

In 1904 – perhaps arising from his newly acquired reputation – Morsch was offered a position at the Zurich Polytechnic as professor of bridge engineering. Whilst there, Wayss & Freytag issued a second edition of *Der Eisenbetonbau* in 1905, this time jointly with Morsch who was, of course, no longer an employee. In 1905 Morsch became an advisory member of the German Concrete Association – a role he was to play for the rest of his life. He also enjoyed practical experience in a number of construction projects beyond his academic duties, acting as engineer for the Grunwald bridge in 1904 and Chippis railway bridge in 1906, and as designer of the Gmunder Tobel bridge in 1908.

1908 saw the publication of *Der Eisenbetonbau*'s third edition – in Morsch's name alone, revised and expanded during the previous year. The new edition included the results of experiments relating to shear in T-beams and continuous beams (conducted by Wayss & Freytag during the intervening years), along with recent tests undertaken for the Reinforced Concrete Commission by the laboratory at Stuttgart under Morsch's direction.

It was this edition that was translated in 1909 by E. P. Goodrich and published in New York by the Engineering News Publishing Co., and in London by Archibald Constable. *Engineering News* described the original as 'probably the clearest exposition of European methods of reinforced concrete construction that has yet been published. It has for some years been a recognized standard in Europe and has also had a considerable demand in this country, but the comparatively limited usefulness of the German edition to American engineers prompted us to make arrangements with Professor Morsch for the rights of translation and publication of the book in the English language'. The book was well received: 'from the point of view of the engineer-designer of reinforced concrete, Prof. Morsch's treatise is the best that has been published on such construction' (*C&CE* v.5 p.453).

With this third edition Morsch returned to Wayss & Freytag, becoming a director and member of the board for the years 1908 to 1916. Projects included the Hohenzollerischn railway bridge over the Danube (1909), the Tubingen railway bridge (1910) and the Lower Neckarbrucke at Rottweil (1915).

Having been awarded his doctorate in 1912 by the Stuttgart technical college, Morsch was appointed professor in 1916 and remained there until the Second

World War. This was a period of further writing and public recognition. *Der Eisenbetonbau* ran to a fifth edition in 1925 and his work appeared in French:

- Morsch, E., *Le nouveau pont sur le Neckar pres de Heilbronn*, in *Construction et Travaux Public,* September 1933
- Morsch, E., *Lesprescritions officielles et ls relements pour les constructionsen beton arme en France et a l'Etranger: Allemagne*, in *Travaux*, December 1935

Matching his award from Stuttgart, he received his doctorate from Zurich in 1929. In 1938 he was the first recipient of the Emil Morsch medal issued by the German Concrete Association; and in 1942, for his 70th birthday he was awarded the Geothe Medal of Arts and Science. A year later, in the difficult circumstances of the war, his *Spannbetontrager* was published in Stuttgart by Verlag von Konrad Wittwer.

Morsch maintained his interest in concrete to the last, contributing his expertise to testing the Rosenberg Bridge in Heilbronn in 1950. He died on 29 December that year, at Weilimdorf, Stuttgart, but a sixth edition of his *Brucken aus Stahlbeton und Spannbeton* was issued posthumously in 1958. Surely this was a lifetime's commitment to concrete?

Morsch, E., trans Goodrich, E. P. *Concrete-steel construction* (from *Der Eisenbetonbau*, third ed, 1908). New York: Engineering News Publishing Co., 1909 [BL, ICE, PCA] and London: Constable, 1909. Second ed., 1910 [C&CA 1959, BL, TCS. Copy: TCS]

1909/USA

Goodrich, E. P.

MASCE

E. P. Goodrich is included here as the translator *Der Eisenbetonbau* by Emil Morsch, though he was a civil engineer in his own right and author of a number of technical articles in the professional press:

- *Economies in the use of reinforced concrete. Cement Age*, v.2, 1905/06, p.847 and v.3, 1906, p.40
- *Vibrations of concrete floors. Cement Age*, v.3, 1906, p.336
- [*Steam cleaning of concrete*]. *Engineering Record*, 7 March 1907
- *The necessity of continuity in the steel reinforcement of concrete structures. Proceedings* ACI, v.4, January 1908, pp.74–82
- *Safety and economy of concrete as a fireproof material. Cement Age*, v.6, 1908
- *Costs of reinforced concrete bridges especially with regard to maintenance. Proceedings* ACI, v.5, January 1909, pp.219–240 (based on discussions conducted at the ASCE in1906)
- *The bonding of new to old concrete. Cement Age*, v.8, 1909
- *Making forms service as conventional designs. Cement Era*, 1911

He would seem to be in his early career at this time, moving from a junior to a member of the ASCE in 1906, but was already general manager of his firm the Underwriters Engineering and Construction Company. By 1908 he was a consulting engineer, based in New York.

We have described the process of *Der Eisenbetonbau's* evolution and the decision to translate it into English. Goodrich, whose contemporary interest in concrete reinforcement and bridges would have been a useful background to his linguistic skills, was chosen for the task. To help the American (and British) reader, the continuous German text was divided into two parts, and thence into chapters and an appendix. Imperial equivalents were provided to explain the metric measurements, and where possible English captions replaced German in the imported illustrations. The publisher hoped such efforts would make the work available to English-speaking engineers and that 'this valuable work will merit their approval and appreciation'. The reviewer in *C&CE* in welcoming the work considered 'the translation is on the whole well done, though at times little ambiguous'.

1909/USA

Porter, Harry Franklin (1882–1965)

CE

Born on 31 August 1882 in Bridgeport, Connecticut, Porter studied at Cornell. He had graduated by 1905 when he took up competitive athletics and went on to become US and Olympic champion in the high jump, competing at the 1908 Games in London.

Professionally he was associated with *Concrete Engineering* and involved with the ACI at this time – in the days when it was known as the National Association of Cement Users – and so his association with the concrete industry was early in his career. He wrote two papers for the NACU's *Proceedings*: *Evolution of reinforcement for concrete* (v.5, 1909) and *Preparation of concrete – from selection of materials to final deposition* (v.6, 1910).

The second paper is clearly related to his book of 1909 – *Concrete: its composition and use*, which approached the subject from the point of view of a concrete producer. It reproduced the complete text of the Institute of Applied Concrete Engineering's documentation, as well as a 'carefully selected compilation of the best proven tables and dates available'. The brief notice it received in *Concrete & Constructional Engineering* – quoted here in its entirety – sums it up concisely:

> This book is intended to present, in a direct way, the underlying facts and best present-day practice in concrete making. And it certainly does that very clearly. It is concerned with the analysis and manufacture of the materials, and the making of concrete, and it gives a very extensive amount of information that is not to be found collected in this way in other works on the subjects of concrete. Much is new information, and interesting attempts, explaining scientifically the various phenomena connected with the materials and manufacture of concrete.

In 1912 he added to his output with *Consistency and time in mixing concrete*: an article for *Cement Age* (v.14 p.294). He was still in Bridgeport at this date. Indeed he remained in Connecticut for the rest of his long life; he died in Hartford on 27 June 1965.

Porter, H. F. *Concrete its composition and use: a clear, detailed, complete statement of the fundamental principles of the basic process of the concrete industry, including essential up-to-date, proven tables and data for the users of concrete.* Cleveland: Concrete Engineering, 1909. [IStructE; Peddie. Rev: *CA*, v.10; *C&CE*, 1910, v.5, p.163]

1909/USA

Radford, William Addison (1865–1943)

Both a supplier to the building trade and a publisher of architectural plans, trade magazines and technical books, Radford was hugely influential in spreading the 'Prairie Style' of house design throughout the American Midwest in the years before the First World War. His importance may be gauged by his inclusion in Daniel D. Reiff's *Houses from books: treatises, pattern books and catalogs in American architecture, 1783–1950*.

William Addison Radford was born on 14 September 1865 in Oshkosh, Winnebago County, Wisconsin. He was one of nine children born to William (an Englishman) and Elizabeth (*née* Robinson) from Vermont. Influenced by the family lumber business Radford Bros & Co, the young William founded on his own building supplies company, setting up the Radford Sash & Door Co. with himself as secretary and treasurer. Establishing a subsidiary in Kansas – the Wichita Sash & Door Co. in 1886 – he met Helen Mary Manual (1868–1952) and they were married there four years later on 17 June 1890. He sold the Wichita company and consolidated his operations in Wisconsin, but in 1892 the couple moved to Riverside (a 'garden suburb' of Chicago) in order for William to establish a branch of the Radford Bros Sash & Door Co.

In 1902 he founded the eponymous Radford Architectural Company, taking for himself the roles of president and treasurer. This firm was to become a major provider of house plan catalogues and books over the next quarter of a century.

During its first year of business Radford forged a working relationship with his future rivals, Fred T. Hodgson and the Frederick J. Drake publishing firm, and by the end of 1903 the Radford Architectural Co. had published five technical books, written variously by Radford and Hodgson. These included: *Modern carpentry* (Hodgson, 1902); *The Radford ideal home* (Radford, 1902); *Common sense stair building and handrailing* (Hodgson, 1903); *The Radford American home* (Radford 1903); and the two-volume *Practical uses of the steel square* (Hodgson, 1903). The Frederick J. Drake Co. printed these with green leather bindings, trimmed in bright red.

Radford went on to establish his own publishing business and the story of the relationship with his former associates is told in Floyd Mansgerber's book *A tale of a great Chicago rivalry: the Radford Architectural Company versus Fred Hodgson and the Frederick J. Drake Co.* (UP, 2000).

Over the years of Radford's active involvement until 1926, his companies produced over 1000 house plans, 40 technical books and three monthly trade magazines, drawing on the expertise of architects and draughtsmen such as G. W. Ashby, W. H. Schroeder and Alfred Sidney Johnson who lived and worked among the Arts & Crafts community in Chicago. Radford himself was editor of

American Builder – a magazine that ran from 1905–1929 – and editor-in-chief of *Beautiful Homes* and *Farm Mechanics*.

Besides launching *American Carpenter and Builder* in 1905, his major undertaking in the early years was compilation of the multi-volume works *Radford's Cyclopedia of Construction* (12 volumes) and *Radford's Cyclopedia of Cement Construction* (five volumes), the latter completed in 1911. Then, from 1908 until 1921 his publishing activities tended towards books on specific topics. Among these were two that related to concrete and cement:

- *Cement houses and how to build them: perspective view and floor plans of concrete block and cement plaster houses* (1909)
- *Cement and how to use it: a working manual of up-to-date practice in the manufacture and testing of cement: the proportioning, mixing and depositing of concrete...* (1910)

The first of these gives a strong indication of the marketing approach Radford took in promoting his subjects. It suggests an eager demand that the author was keen to meet, claiming to present essential information in plain language to those who were contemplating construction of their own homes, but presented with an eye to beautiful design. The prose style is breathless; full of hyperbole and certainly worth quoting to give a flavour of the work:

> The history of the human race represents no parallel to that of the marvellous development during the present generation of home architecture by the use of concrete hollow blocks and cement plaster ... Home builders the world over are hungry for information about cement houses. The demand for such information is unprecedented in the annals of building. "Tell us how to build a home of cement" is a popular cry heard throughout the land.
>
> It is to meet this great demand that this book is presented to the building public. As it has been prepared by the most expert and skilled architects and cement experts of the country it is certain that the full requirement of those who want a home built of cement in any of the forms – blocks, monolithic, reinforced or plaster – have been met in his pages. The purpose has been to only to show beautiful designs and plans ... but also to tell in plain language divested of high sounding and technical words exactly how each part of the work should be done. To that extent the volume will serve as a text book on the subject.

Radford's acknowledgements included 'all possible sources of accurate information' – comprising cement companies, the AAPCM, machinery manufacturers and other architects – allowing the book to be considered an 'epitome of all the latest investigations'.

A full bibliography of Radford's numerous other books lies outside the scope of this survey – many of them cover carpentry – but would constitute a lengthy list. As an indicative sample we might take the following arbitrary selection:

- *Radford's combined house and barn plan book* (1908)
- *Architectural details for every type of building* (1921)

Radford retired in 1926, although he maintained an involvement with the publishing business that continued under his sons Roland and William. In 1931 Helen and he moved to Seven Springs Ranch at Capatino in Santa Clara, California, and by the end of the decade were resident in the nearby Alameda Hotel, San Jose. There William died on 20 May 1943 and Helen in April 1952.

Radford still commands an interest – to judge by websites such as www.antiquehomestyle.com. Some of his work has also been reissued in facsimile as *The old house measured and scaled – detail drawings for builders and carpenters: an early twentieth century pictorial sourcebook*, complete with an introduction by John J. Mojonnier (NY: Dover, 1983).

Radford, W. A. *Cement houses and how to build them: perspective view and floor plans of concrete block and cement plaster houses*. Chicago and New York: Radford Architectural Co., 1909 [Copy: Internet Archive]

Radford, W. A. *Cement and how to use it: a working manual of up-to-date practice in the manufacture and testing of cement: the proportioning, mixing and depositing of concrete...* Chicago: Radford Architectural Co., 1910 [BL]

Radford, W. A. *Radford's cyclopaedia of cement construction: a general reference work on up-to-date practice in the manufacture and testing of cements, the selection of concreting materials, tolls and machinery; the proportioning, mixing and depositing of concrete, plain, ornamental and reinforced*. 5 vols. Chicago: Radford Architectural Co., 1911 [Peddie]

1909/USA

Trautwine, John Cresson (1850–1924)

John Cresson Trautwine was not one man, but three! The first of that name (1810–1883) was a distinguished railway engineer working on America's eastern seaboard in Panama (1850) and Honduras (1857). Remembered for stating that a canal through Panama would be impossible, he was the author of two books on railway construction and of the *Civil engineer's pocket book* of 1871. This became known as the 'engineer's bible' and under the editorship of John Cresson Trautwine Jr. (1850–1924) and his son J. C. Trautwine III, was reissued in many editions. At the end of our period of review it had reached its 21st edition – under the title *Civil engineer's reference book* – and had sold 180,000 copies.

In 1909 the Trautwines of Philadelphia – son and grandson – extracted newly written information from the 19th edition of the *Civil engineer's pocket book* and prepared a spin-off, entitled simply *Concrete*. Although presented as a slender volume for the pocket, it consisted of nearly 200 densely packed pages summarising contemporary modern practice and even found space for the results of recent experiments. Specifications and recommendations from the ACI, ASCE, ASTM and the Board of US Engineer Officers were included, and even the Engineering Standards Committee (precursor to the British Standards Institution). Ignoring theory, the emphasis was on practical information, costs and quantities, suitably 'condensed and collated for convenience of reference' (*CA*, v.9, p.445). The first edition (published by Wiley) ran to a thousand copies and was distributed in the UK by Chapman & Hall.

Trautwine, J. C. Jr. and J. C. III. *Concrete*. New York: John Wiley & Sons, 1909 [Copy: TCS], and London: Chapman & Hall, 1909 [Peddie]

1909/USA

Turner, Claude Allen Porter (1869–1955)

MASCE, MCanSCE

Claude Allen Porter Turner Courtesy of the American Concrete Institute

Inventor of the 'mushroom' shear head, Turner was a pioneer of flat slab construction and therefore one of the more significant technical advances brought about by reinforced concrete. However, he had to contend with other claimants in America and with parallel developments by Robert Maillart in Switzerland, and managed to miss some of the credit and fame due to him. A long overdue assessment of his achievements was made by D. A. Gasparini in the ASCE's *Journal of Structural Engineering* (October 2002, pp.1243–52).

Turner was born in Lincoln, Rhode Island on 4 July 1869. Twenty-one years later, he graduated from the School of Engineering at Lehigh University in Bethlehem, Pennsylvania, and embarked on career of structural engineering. Throughout the following decade he worked for various bridge companies in the industrial belt stretching from the east coast to Ohio. These included the Edgmore Bridge Co. (Wilmington, Delaware), Columbus Bridge Co. (Ohio), Pittsburgh Bridge Co. (Pennsylvania), Berlin Iron Bridge Co. (East Berlin, Connecticut), and the Pottsville Iron & Steel Co. (Pennsylvania). In 1897 he came to Minneapolis to work for the Gillette Herzog Co., which in 1900 was incorporated into the American Bridge Co. By coincidence Milo Ketchum [1908] was also working in Minneapolis at this time, and entered the employ of American Bridge. In the autumn of 1901 however, Turner left to establish his own practice.

It was during the next few years that Turner's distinctive contribution to concrete engineering was made, and at the time when he turned his attention to buildings rather than bridges. Whether he knew of them or not, George M. Hill and O. W. Norcross had also experimented with the flat slab concept; the former designing several structures in 1899–1901 – as described in the *Railway Gazette* (December 1901 and January 1902) and reported later in *Cement Age* – and the latter filing an unexploited patent in 1902.

In 1904 Turner was responsible for the floor design at Building Four of the Northwest Knitting Mill Co.'s factory in Minneapolis and then, in 1905, for the Minneapolis Paper Co. He publicised his device of the 'mushroom' shear head in *Engineering News* on 8 June and 12 October 1905 and described the results of load tests on both buildings in a discussion published in ASCE *Transactions* (v.56) the following December. A drawing of the mushroom design was reproduced soon after in books by Reid [1907] and Turneaure & Maurer (1908 edition, above).

Turner's first complete building to use the flat slab system was the Johnson-Bovey building in Minneapolis, erected in 1906. This was also described in *Engineering News* (4 October 1906). His next – the Hoffman building in Milwaukee – was erected in 1907 after the architect had witnessed the load tests at Johnson-Bovey. The load tests at Milwaukee, in their turn, were so favourable they formed the basis of an article in the new British journal *Concrete & Constructional Engineering*.

Following such success Turner applied for a patent on 11 June 1907 and throughout the years 1906 to 1909 was responsible for the slab design of 33 buildings. These are listed by Gasparini (after H. T. Eddy) and some would have been described by Turner in a series (*Concrete in the Northwest* [from v.4, 1907]) for *Cement Age* that featured typical buildings with which the author was familiar in cities such as Milwaukee, St. Paul, Minneapolis and Winnipeg. *Engineering News* of 18 February 1909 carried an article by Turner (*Advance in reinforced concrete construction*) that claimed he had designed 400 acres of flat slab floor. Another article of – *The mushroom system as applied to bridges* in *Cement Age* (v.10, 1910) – proposed a total of nearly a thousand acres since its introduction 'a little over five years ago'. His publicity, perhaps even more than his than his invention, had bred success.

By this date Turner had written a book – *Concrete-steel construction* – to consolidate his credentials, and 'written mainly with a view to extolling the author's patented method of construction; in fact, it may be termed a glorified catalogue of the 'mushroom' system'. The reviewer in *C&CE* (v.5, pp.145–146) hints at an opinionated approach by Turner. He also comments that it was 'written from the contracting engineer's point of view, and consequently differs very materially from the usual treatise on reinforced concrete. Practical considerations in construction are given the preference over theoretical formulae'. Included in this respect was a chapter on surface treatment to achieve an 'artistic' finish. Critics in America conveyed similar views. 'This volume', observed *Cement Age*, 'is in the nature of a protest combined with practical information on the subject … and in the former respect is quite unique as an engineering work'. Its price was (according to the author), 'sufficiently high to induce purchasers to read it if they wish to get their money's worth'!

Much of the value of the book lay in the prominence of a large amount of test data, against which Turner's conclusions could be checked, supported unusually by illustrations of building failures caused by faulty design. It also set out to address the costs of construction (an aspect generally ignored by the literature) and to discuss safe and unsafe details from practical experience. That safety was of concern to Turner – which, for a contractor, it would have to be – was suggested by his recent article *Plain facts on fireproof construction* (*CA* v.6, 1908)

The safety of reinforced concrete was also the subject of an address read to the Canadian National Association of Builders' Exchanges, and reported in *Cement Age* (v.2) for March 1911. In it Turner argued that: 'practically all failures have occurred with one-way reinforcement and generally where there was an insufficient lap in the reinforcing metal over the support.' His theme was developed in extracts from the paper published during the year, including *Field notes for concrete constructors* (p.165) and *Concrete honesty* (p.195).

He continued to appear in *Cement Age* and its successor *Concrete-Cement Age*, taking part in discussions between 1911 and 1914 on aspects of concrete construction, including guidance on the mushroom design.

Having provided the appendices for H. T. Eddy's *Theory of the flexure and strength of rectangular flat plates* in 1913, he collaborated with the older man on a second book, confusingly bearing the same title (though with different subtitle and contents) as his first (see Eddy [1913] for a discussion of this work). *Concrete-steel construction* was further revised in 1919.

Turner's early success however, was short-lived. He soon became involved in litigation – *A federal Court of Appeals decision on flat slab patents, Engineering News*, January 1915 – and his subsequent decline is succinctly put by Gasparini: 'Turner's revolutionary successes prior to 1910 were, sadly, followed by prolonged, destructive legal battles, devastating defeats in court, and as a result, a consuming lifelong bitterness.' In 1919 he gave a colourful account of his lawsuits this revealing title: *State of the art of reinforced concrete from the patent standpoint and the menace to progress by unscientific decisions*.

A final, privately printed, book in five volumes – *Elasticity and strength* (1922) – reflects his state of mind, though it should be said that he had enjoyed a flourishing consulting practice throughout the 1920s, registered in Minneapolis, Chicago, New York and Canada. In 1934 he revised his previous book with the expanded title *Elasticity, structure and strength of materials used in engineering construction*.

Retiring in 1936 he moved to Columbus, Ohio, and died there on 10 January 1955. His remaining professional papers are held at the Northwest Architectural Archives, University of Minnesota libraries.

Turner, C. A. P. *Concrete-steel construction. Part I – building: a practical treatise for the constructor and those commercially engaged in the industry*. Minneapolis: Farnham Printing & Stationery Co., 1909 [PCA, Peddie]

Eddy, H. T. & Turner, C. A. P. *Concrete-steel construction. Part I – building: a treatise upon the elementary principles of design and execution of reinforced concrete work in building*. Minneapolis: Heywood Manufacturing Co., 1914. Second edition, 1919

1909/USA

Tyrrell, H. G. (b.1867)

Tyrrell's bibliography points to his interest in bridges – as his work on *Concrete bridges and culverts* in 1909 might also suggest. This was followed by *History of bridge engineering* (Tyrrell, 1911), *Artistic bridge design* (Myron C. Clark, 1912) and *The evolution of vertical lift bridges* (1912).

H. Grattan Tyrrell was a graduate of Toronto University and his wife, Maude, a graduate of the Chicago Art Institute who practised architectural design. With such twin influences perhaps it is not surprising that he should be described as 'bridge engineer and aesthetic critic'.

An early involvement with ornamental bridges prompted an article in the *American Architect* entitled 'American park bridges' (March 1901), but by 1909 his interest had widened. A 128-page *Hand book on concrete bridges and structure* was announced in *Cement & Engineering News* (1909 v.21, p.233), but a much more substantial work appeared from the press of Myron Clarke in November. This was the 242-page *Concrete bridges and culverts: for both railroads and highways*. Tyrrell was emphatic in his support of concrete, declaring in his preface that 'bridges of solid concrete are superior to those of any other material'. Against the trend of the times he went to some lengths to 'eliminate mathematical formulae' and was careful to present his subject in 'the simplest possible manner'.

In parallel was his earlier work on mill buildings: *Mill building construction* (Engineering News Publishing Co., 1901); and *A treatise on the design and construction of mill buildings* (Myron C. Clark, 1911). He was finally published by McGraw-Hill in 1912, with his *Engineering of shops and factories*. Of all his writing, only *Bridges and culverts* specifically addressed concrete.

After his death the H.G. Tyrrell Library was bequeathed to the Toronto University Library by Joseph Burr Tyrrell, the Canadian geologist and explorer.

Tyrrell, H. G. *Concrete bridges and culverts: for both railroads and highways*. Chicago: Myron C. Clarke, 1909 [ICE]

1910/UK

Becher, Heinrich

Becher – an engineer from Berlin – contributed the volume on reinforced concrete to the Deinhardt-Schlomann series of technical dictionaries, published in six languages by R. Oldenbourg of Munich. It was a substantial work of 415 pages, with 900 illustrations and numerous formulae. The translation into English, published in London by Constable, 'was chiefly done by Dr. A. B. Searle of Sheffield' [1913].

In his practice as a consulting engineer (in 1926), Becher was associated with the *Ullsteinhaus* in Templehof, Berlin – significant as the city's first reinforced concrete frame high-rise building.

Becher, Heinrich. *Reinforced concrete in sub- and superstructure*. (The Deinhardt-Schlomann series of technical dictionaries in six languages, 8). London: Constable, 1910 and New York: McGraw, 1910. Translated chiefly by Dr. A. B. Searle from *Der Eistenbeton im Hoch-und Tiefbau* (Schlomann-Oldenbourg *illustrierte technishe* Worterbucher, 8. Munich: R. Oldenbourg, 1910. [BL, Peddie. Rev: *C&CE* v.5, 1910, p.454]

1910/USA

Boynton, Walter Channing (b.1876)

A graduate of the University of Michigan (featured in the *Michiganensian Yearbook* of 1899), Boynton pursued a career in the press, being 'identified with the editorial and business management of several successful publications'. He was described in 1912 as 'an old newspaper man' (*C-CA*, v.1, p.30), having served on the *Detroit Free Press* in a number of capacities from 1896. With business partner E. R. Kranich, he bought the recently established journal *Concrete* in 1906 and took on the post of editor. In 1910 he and associate editor **Roy Marshall (b.1885)** jointly compiled

How to use concrete from articles that had appeared in the journal, including some by A. A. Houghton [1910]. According to its authors, 'a large portion of this book is given up to descriptions of easily made moulds and instructions for their use in the manufacture of concrete products'. The text was written in a 'plain homely way for the lay user', according to *Concrete*'s British counterpart *C&CE*, which considered that along with the 'smaller manufacturer of artificial stone', the 'farmer and the builder in country districts will find it specially valuable. It is almost of the nature of a collection of recipes for concrete work'.

With the merger of *Concrete* with *Cement Age* in 1912, Boynton became general manager of the new enterprise (while Marshall became business manager). He might easily be confused with his namesake and contemporary C. W. Boynton of the Universal Portland Cement Co., who contributed to the pages of *Concrete-Cement Age*, and wrote on the subject of concrete sidewalks for the ACI's *Proceedings* in 1908 and in *The decorative possibilities of concrete* for *C&CE* in 1914 (v.9, pp.266–276).

Boynton, W. C. & Marshall, R. (comp). *How to use concrete*. Detroit: Concrete Publishing, 1910. [PCA. Rev: *CA*, v.10; *C&CE* v.5, pp.612–13]

1910/USA

Davison, Ralph C.

Davison was assistant secretary of the Concrete Association of America and presumably well placed to notice market trends. His book of 1910 – *Concrete pottery and garden furniture* – presaged an increased interest in domestic concrete in the years immediately before and after the Great War, and its reviewer in *Cement Age* noted that it provided 'information for which there is great demand'. He continued: 'The book should be especially welcome to the large numbers just beginning work of this character, as it places in their hands information not to be obtained from concerns engaged in ornamental concrete work on a large scale.' It suggested designs and gave practical instructions together with 150 illustrations. The book was reissued in 1917 at the time of J. T. Fallon's book on the same subject.

Davison, R.C. *Concrete pottery and garden furniture*. New York: Munn & Co, 1910. Second edition, 1917 [C&CA 1959. Rev: *CA*, v.9]

1910/USA

Dodge, Gordon Floyd

Diagrams for designing reinforced concrete structures came early in Dodge's career. By the 1930s he was employed as an engineer in the construction machinery division of Jeffrey Manufacturing Co., Columbus, Ohio. He secured a patent (312194) on 9 June 1931 on a drive mechanism for conveyor belts. And at the 32nd Annual Convention of the ACI in Chicago (25–27 February 1936), he presented a paper entitled *Aggregate production for Grand Coulee Dam*. It was published in the ACI's *Proceedings*, v.32.

Dodge, Gordon Floyd. *Diagrams for designing reinforced concrete structures, including diagrams for reactions and strengths of steel beams*. New York: Clark Publishing Co., 1910 and London: Spon, 1910 [Peddie]

1910/UK

Fleming, Lt Col. John Gibson, CBE (1880–1936)

RE

Though not published by a trade publisher, *Reinforced concrete* follows that of Colonel Winn [1903] and was issued in book form by W. & J. Mackay & Co. for sale beyond a military audience. It also drew in part on publications by Marsh & Dunn [1904]. Fleming – a captain at the time – was Colonel Winn's junior officer, in both age and rank, whom the older man had in earlier days 'initiated into the mysteries of construction'. And after publication of his book, Fleming took the now retired Winn's place on the influential RIBA Committee on Reinforced Concrete when it reassembled for its second report in 1911.

John was born on 9 January 1880 in Glasgow, the second son of Dr W. J. Fleming and named after his grandfather, a prominent surgeon. He was educated at Haileybury College and the Royal Military Academy, Woolwich. He was commissioned as a second lieutenant in 1898, serving with the Royal Engineers, where he was taught by Lt Col. John Winn, RE (then a Major) at the School of Military Engineering. He went on to serve in South Africa during the Boer War between 1901 and 1902, and was awarded the Queen's Medal with five clasps. His wife Blanche Deglon, whom he married in 1903, was from the Transvaal. They raised a family of four (a son and three daughters), and Fleming was promoted to captain in 1907.

In 1910 he wrote the book for which he is remembered – *Reinforced concrete* – and in a preface contributed by his former teacher, Winn consciously drew attention to how its publication was 'an indication of the great advance made in the use of reinforced concrete in recent years since 1903' – the precise period of the foregoing survey. Winn also characterised the book as a practical one: 'the author has hit upon the happy mean between a mere sketch and a ponderous work which wearies rather than enlightens. Designed primarily for the use of practical men it contains enough detail to satisfy the needs of such without being overloaded with needless dissertations on abstract points, which may never arise, or cumbrous calculations difficult to apply.'

Fleming's own preface acknowledged 'great assistance' from William Dunn, who had helped in 'every possible way'. This included reproduction of a section on domes from Marsh and Dunn's collaboration. He also reproduced illustrations from the catalogues of Mouchel, Truscon, Considere, Coignet and the new Expanded Metal Company, thanking them for allowing him to inspect construction work in progress.

C&CE's review was not kind however, regarding the graphs as 'not quite so ready a means of designing as we could wish' and the formulae for beams still too cumbersome. The information about arches was described as 'meagre'. If not an original contribution to knowledge, the book was however, a 'handy summary' and 'commendable as an introduction to the subject'.

In 1911 Fleming acted as military representative on the RIBA's committee when it was reconvened to prepare its second report on the use of reinforced concrete.

During the First World War he was promoted to Major in 1915, mentioned in dispatches and awarded the DSO in 1916. At the end of hostilities in 1919 he was made a CBE. He retired from the Army as a lieutenant colonel in 1924. The final years of his career was spent at 4th Crown Agent for the Colonies and from 1932, engineer-in-chief. In private life he enjoyed motoring, yachting and shooting. He died on 14 September 1936.

As a footnote, one might add that the Royal Engineers at Chatham were to make one further contribution to the literature, with the School of Military Engineering's publication of two volumes of *Notes on cement and concrete* in 1919 and 1922. The author **A.F. Day**, when a lieutenant shortly before the Great War, had designed the Devil's Peak Redoubt – a concrete emplacement in colonial Hong Kong, and presumably honed his craft during the hostilities of the intervening years.

Fleming. J. G. *Reinforced concrete*. Chatham: Royal Engineers Institute, 1910 [BL, C&CA 1959. Copy: TCS. Rev: *C&CE* 1910]

1910/USA

Houghton, Albert Allison (b.1879)

Albert Houghton was born on 25 February 1879 in Orleans, Ionia County, Missouri and married Carrie Elizabeth Barkett in 1902. His career was notable for the profusion of low-priced practical guides to the use of concrete, particularly ornamental concrete.

The first title to make it into print was *Ornamental concrete without molds*, which was published by Norman W. Henley in 1910. Henley was not the usual choice of publisher for books on concrete, so perhaps reflected the market these particular titles were aimed at. Certainly this first book must have been a success for it was followed by *Concrete from sand molds* – a technique for casting ornamental concrete in rammed sand moulds – and then, the following year, by a series of concrete workers' reference books. These were aimed at the 'ordinary worker', and *Cement Age* (v.11, p.413) imagined that they would be 'especially appreciated by the contractor who has not had extended experience with concrete'. There were four other titles, but these are likely to have been a continuation of the concrete workers' books. Of the books available to the *C&CE*'s reviewer in 1911 (v.6, p.634), the first two were considered to be of 'most interest and value, especially as no other books on the subject were known. Interestingly, nearly a hundred years later, they have both been reprinted to serve a niche market.

Houghton, A. A. *Ornamental concrete without molds*. New York: Norman W. Henley Publishing Co., 1910 [C&CA 1959. Rev: *C&CE*, v.6, p.634]

Houghton, A. A. *Concrete from sand molds*. New York: Norman W. Henley Publishing Co., 1910 [Rev: *C&CE* v.6, p.634]

Houghton, A. A. (concrete workers' reference books). New York: Norman W. Henley Publishing Co., 1910–11 [Rev: *C&CE* v.6, p.634], and London: Spon

1. *Concrete wall forms*. [ICE, Peddie]
2. *Concrete floors and sidewalks*
3. *Practical silo construction* [*CA*, v.13, p.44]
4. *Molding concrete chimneys, slate and roof tiles* [*CA*, v.12, p.333]

5. *Molding and curing ornamental concrete* [*CA*, v.12, p.333]
6. *Concrete monuments mausoleums and burial vaults*. [BL, *CA*, v.12, p.333]

Houghton, A. A. *Concrete bridges, culverts and sewers*. New York: Norman W. Henley Publishing Co., 1912 [BL; *CA*, v.14, p.266] and London: Spon

Houghton, A. A. *Molding concrete bathtubs, aquariums and natatoriums*. New York, ND and London: Spon

Houghton, A. A. *Constructing concrete porches*. New York: Norman W. Henley Publishing Co., 1912 [*CA*, v.14, p.266] and London: Spon

Houghton, A. A. *Molding concrete flower-pots, boxes, jardiniers* New York, 1912 and London: Spon

Houghton, A. A. *Molding concrete fountains and lawn ornaments*. New York, 1912 and London: Spon

1910/USA

Ostrup, John Christian (b.1864)

MASCE, MICE

An American of presumably Danish extraction, Ostrup's physical legacy lies in the Boston Elevated Railway, for which he was designing engineer in the years 1898–1902. He was responsible for (at least) the stations at Dudley and Northampton Street, and South, City and Sullivan Squares, and the Lincoln Wharf and Charlestown power plants.

Later in the year, at the time his book *Standard specifications for structural steel, timber, concrete and reinforced concrete* was published by McGraw-Hill, he was professor of constructional engineering at the Stevens Institute of Technology. The book comprised a collection of standard documents 'in the nature of a set of general conditions for competition in design and tendering' from the likes of the ASTM and American Railway Engineering and Maintenance of Way Association. However, in disapproving of this underlying American tendency to 'provide standard specifications for all kinds of work', *C&CE*'s reviewer expressed disappointment in the author's approach:

> On the whole, we cannot say that we are much impressed with the book. We do not see that it serves any really valuable purpose. What it contains is obtainable from other sources within the reach of most of those to whom such a book would appeal. Where the specification has been drafted by person of experience and some authority it is deserving of consideration, but where the specifications have been drafted by the author entirely the dicta of one man cannot carry so much weight.

This view did not reflect the book's popularity in America, where demand for the first edition of September 1910 rapidly exhausted stocks and prompted a revision only three months later for a second edition early in 1911.

We catch a further glimpse of him at the end of 1910, when he participated in debates at the ASCE, including a discussion with Edward Godfrey [1908] over his paper *Some mooted questions in reinforced concrete design* (v.LXX, Dec 1910). Beside the ASCE, Ostrup was a member of the American Society for the Promotion of Engineering Education and of the American Society for the Advancement of Science.

Ostrup, J. C. *Standard specifications for structural steel, timber, concrete and reinforced concrete*. New York: McGraw-Hill, 1911. [Rev: *C&CE*, 1911, v.6, p.153]

1910/UK

Rings, Frederick

MSA, MCI, CE (German Inst)

The London-based architect and engineer Frederick Rings, derives his place in the pantheon of writers on concrete from a short burst of activity between 1910 and 1913 in which he wrote three popular books. He appears not to have been closely associated with the 'establishment' that comprised officers of the Concrete Institute, contributors to *C&CE*, and members of the RIBA committee, though he claimed to have closely followed the development of reinforced concrete construction for 'many years' and did draw on RIBA's 1907 report.

Rings was born in 1874, probably in London where he was to base his later practice. His professional training in architecture and engineering was undertaken (partly) in Germany. An early partnership with John Myers (1881–1915) – trading as Rings & Myers from offices at 21 Railway Approach, London Bridge – was dissolved in March 1908. With a growing interest in the application of reinforced concrete, Rings turned to writing on the subject and committing his professional experience to paper.

In 1910 his *Reinforced concrete; theory and practice* was published by Batsford. Like Fleming's book of the same date, it was intended as an introduction to the subject, outlining and explaining the principles of reinforced concrete design and listing the proprietary systems of the various specialist companies. Its audience differed from Fleming's however, as although concrete had yet to find architectural expression in Britain, the book was aimed at the architect. Rings aimed to help those wishing to become more familiar with the material and lessen the dependence on specialists with vested interests, whilst acknowledging that the book itself was no substitute for specialist expertise. 'The book does not fail in its aim' was the verdict of *C&CE* (v.5, p.455); 'it forms a useful introduction to the subject'. The review continued to state: 'on the whole the book, if not original in its treatment, is of service as an elementary summary of the first principles of the subject', and that 'it contains a good deal of valuable information which the author has collected from various sources'. The review also observed that 'the author is the first to adopt the Standard Notation of the Concrete Institute', whereas other authors' failure to do so was cause of adverse comment in reviews of the period.

One of the features of this first essay into print was a 'ready reckoner' at the end of the main text. This appears to have been the inspiration for a second volume devoted to an expanded series of four such aids. *Ready reckoners for reinforced concrete designs* were published later in 1910 at the author's own expense. This time *C&CE* was not so kind, taking issue with the validity of some of the calculations and claiming to be 'disappointed to find the calculators for beams of such very doubtful utility'. Worse followed as – with sarcasm uncharacteristic of *C&CE* – the review ended with: 'Surely there is some mistake in the title. It should have been *Rough and ready reckoners*' (v.6, p.462).

Whatever Rings felt at such a jibe, he was not dissuaded from writing a further book. Indeed he was encouraged by the otherwise 'favourable reception' of *Theory and practice*, and so in 1912 he tackled the subject of bridges, attempting a more detailed exposition of a key application for reinforced concrete than was possible in his earlier work. The approach was work-a-day, setting aside the 'most remarkable structures of the day' and concentrating on 'the types occurring in the usual practice of the civil engineer'. Among his key sources was *The Elastic arch* by Leffler [1907]. In 1913 Rings' *Reinforced concrete bridges* was published by Constable in the UK, and in the USA by Van Nostrand.

In an historical survey like this it is easy to encounter hostages to fortune and it is with a smile we read the following: 'it seems difficult to believe that the design of the many successful structures of the present day could possibly be improved upon.' Yet Rings was aware enough to realise that 'the last word has not been said yet respecting reinforced concrete bridge construction' and anticipated that in an age of progress concrete would 'supersede the use of iron and masonry altogether'.

Transactions & Notes noted the book without comment, but *C&CE* nit-picked somewhat and criticised Rings' coverage of strains and stresses, only to end with a backhanded compliment: 'The author would be well advised to suppress this page [23] in future editions, lest the demerits of a page should mar the sale of what is otherwise a useful and commendable book.' Across the Atlantic, *Concrete-Cement Age* approved of Rings' concise treatment, adherence to his purpose and to his use the new notation. Even if it did differ from American guidance, 'such notation marks a step in advance, as it its not so long ago that the writer of a book did not consider that his work was well done unless he had devised for it an original scheme of notation!' The commentator thought that: 'Mr Rings' book should be a most useful one to those who have concrete bridges to design … and of service to those whose work is limited to buildings' (*C-CA*, v.5, p.37).

No new edition followed, though *Theory and practice* was reissued in 1918. His practice as 'engineer and architect' continued, with offices at 56–58 Victoria St., Westminster, until it was hit by the effects of post war recession. The business failed and Rings filed for bankruptcy in December 1922.

At the end of our period (in 1935), the elderly Rings wrote a four-page article in *C&CE* entitled *Notes on building regulations* (v.30, pp.120–123): presumably a reflection of increased codification following the London Building Act of 1930. Given the themes of this survey, that seems quite an appropriate subject on which to end his writing career.

Rings, F. *Reinforced concrete; theory and practice*. London: B. T. Batsford Ltd, 1910 [C&CA 1959, ICE, IStructE. Copy: TCS. Rev: *C&CE* v.5, p.455], and New York: Van Nostrand [Peddie]. Second edition, 1918 [BL, ICE, IStructE]

Rings, F. *Ready reckoners for reinforced concrete design*, London: the author, 1910 [ICE, Peddie. Rev: *C&CE*, v.6, p.462]

Rings, F. *Reinforced concrete bridges*. London: Constable, 1913 [BL, C&CA 1959, ICE, IStructE, Yrbk 1934. Copy: TCS. Rev: *C&CE*, v.8, p.441, *Trans CI*, v.5, p.xiv], and New York: Van Nostrand [Rev: *C-CA*, v.5, p.37]

1911/UK

Adams, Professor Henry (1846–1935)

MICE, MIMechE, FSL, FRSanI, FRIBA, Hon. MIStructE

Professor Henry Adams
Courtesy of the IStructE

Professor Adams had a distinguished career in engineering, acquiring a reputation as a technical author and attaining several professional and academic honours. As a teacher he become professor of engineering at the City of London College, while in business he gained valuable practical experience during his 12 years in charge of outdoor contracts for the firm of Sir W. G. Armstrong. He drew on this experience when writing his *Engineers' handbook*, published by Cassell in 1908. As an engineer he acquired the highest professional standing, elected variously as president of the Society of Engineers, president of the Civil and Mechanical Engineers' Society and vice-president of the Association of Engineers-in-Charge.

Born in London, Adams was the scion of an engineering family; his father was John Henry Adams, a resident engineer during the railway construction boom; and his grandfather a civil engineer, too. As a youth Adams studied chemistry and in 1865 joined the firm of Sir W. G. Armstrong & Co., where he was soon promoted to assistant outdoor manager. For a while he had responsibility for 14 foremen and 400 workmen. In 1866 he started evening classes, firstly at the City of London College and then at King's College. After three years, nine prizes and 29 certificates, he was awarded the Queen's Medal. Having achieved first place honours in Applied Mechanics he was invited to join the staff as professor. He fulfilled his teaching duties in the evenings, after a day's work, for 35 years.

Adams senior died in 1877 and Henry took over the practice and his father's position with Messrs Wm Cory & Son as superintending engineer. He also voted in favour of the Institution of Mechanical Engineers' move from Birmingham to London that year. The committee of the Coal Meters Office asked him to report on seaborne coal weighing machines in the Port of London in 1880 and about that time he and his wife moved to Bexley Heath for a period of seven years. In 1891 he visited Belgium with the Society of Architects; a body he was to remain with until its eventual merger with the RIBA. Indeed he spent 30 years as a member of its council and 20 years as superintending examiner.

At the age of 58 (in 1904), he gave up his college position, and though he kept on his practice at Queen Victoria Street he devoted more time to reading and horticulture. Soon afterwards he undertook a brief foray into politics and in 1906 was elected to his local borough council. He became chairman of the Works & Highways Committee in 1907.

Looking to the future he took his eldest son into partnership in 1909, adding water supply, sewerage and central heating to the firm's list of expertise. He was engaged in expert witness work and was retained as an expert by the Home

Office at the time the Regulations of Docks and Warehouses were drawn up.

An interest in reinforced concrete, on which he lectured at the Engineering Exhibition, Olympia (in November 1907) led him into membership of the newly formed Concrete Institute, on whose Science Standing Committee he sat during the opening session of 1908/09 and subsequent years. By 1913 he was also on the Finance & General Purpose Committee. He was active in the life of the institute, attending the evening general meetings and participating in the discussions following the papers presented. He served on the council and from 1914 served as president for two years.

Adams was described at the time as a 'well-known author on technical subjects' and during these years he added to his output. In 1911 he published *Examination work in building construction*, addressing questions set by the Society of Architects and drawing on his role as an examiner in the subject. The book contained examples of both plain and reinforced concrete. *The mechanics of building construction* followed, published by Longmans, Green & Co. in 1912. This was an advanced and extensive treatment of structural design, based on a series of lectures given to science teachers, reworked and illustrated with nearly 600 diagrams. While commended by *Concrete & Constructional Engineering* (1912, p.552), it was deemed rather specialised and likely to appeal far more to the teacher than to the student. In the first year of his presidency (1914), Adams returned to concrete with *Storage of coal with some application of reinforced concrete* for the Institution of Civil Engineers.

But it is for his collaboration with Ernest Matthews – *Reinforced concrete construction in theory and practice: an elementary manual for students and others* – that he earns his place in this review. Despite the title, the authors consciously aimed to be practical – harnessing theory only where necessary to the understanding of calculations required by practical design. The level was introductory, but the coverage extensive: 'the book is certainly very complete, and every aspect of the subject has been dealt with and illustrated, wherever possible, with practical examples.' *Concrete & Constructional Engineering* (1912, p.73) complained that many of these examples were 'merely illustrated without being adequately described', and whilst they requested more on the construction of centring and forms, they complimented the clarity of the diagrams, commended the chapters on stress and welcomed the authors' conclusions on fire-resistance based on test data. As 'a complete treatise on the subject of reinforced concrete construction', the reviewer recommended it to 'students and experienced men alike'.

Having revised the book in 1920, Adams widened his approach with *Structural design in theory and practice* for Constable in 1923.

In 1923 Adams was awarded the first Gold Medal of the newly re-styled Institution of Structural Engineers: an accolade for tireless work on behalf of the institution's forerunner and of a lifetime of achievement. Also made an honorary member that year, Adams' memoirs were considered of sufficient interest to merit a place in the institution's *Structural Engineer*. In them he calculated he had written 40 papers and books – including *Joints in woodwork*, *Timber piling*, *Designing ironwork*, *Strains in ironwork*, and the *Engineers' handbook* – and given 10,000 lectures. He died in 1935.

Adams, H. and Matthews, E. R. *Reinforced concrete construction in theory and practice: an elementary manual for students and others*. London and New York: Longmans, Green & Co., 1911 [C&CA 1959, ICE, Peddie]. Second ed., 1920. [C&CA 1959, BL, ICE, Yrbk 1934. Copy: TCS. Rev: *C&CE* 1912]

1911/UK

Matthews, Professor Ernest Romney (1873 –1930)

AMICE, FRS(Ed), FRSanI, FGS

Professor Ernest Romney Matthews Courtesy of the IStructE

A civil engineer specialising in coastal protection works and sanitation, Matthews was borough engineer and surveyor of Bridlington – where he was to make his name – by the age of 25.

He was born on 16 January 1873 in St Leonard's on Sea, the son of William Henry Matthews, Chief Officer of Coast Guards in the Hastings division. He was educated at St Michael's School, Hastings. His career commenced locally as assistant to the borough engineer in 1890, and in 1898 he became borough engineer and surveyor of Bridlington. In 1900 he married Bessie Barker of Toppesfield, Essex, with whom he had two sons and three daughters.

In the early years of the new century he oversaw construction of Bridlington's sea defences, remodelled the main drainage system, lengthened one of the local bridges and erected an electric light station. His article in *Public Works – Infectious diseases hospital in the smaller towns* (v.2, 1904, p.275) – was based on his work in the borough. By 1912 he had submitted a pioneering town plan; a paper which was reviewed across the Atlantic in the *Engineering Record*. His reputation grew and enabled him to take on the role of consulting engineer for several other local authorities, as well as special lecturer on harbours and coastal protection at Manchester University in 1914. By 1915 he was also Chadwick professor of municipal engineering at the University of London.

During these years he was active in the engineering institutions (like his collaborator Henry Adams), becoming an associate member of the ICE in 1906 and a member of the Concrete Institute, for which he sat on the Science Standing Committee from its inception in 1908. He joined the Society of Engineers, became an associate of the Institution of Electrical Engineers, and in 1913 was elected a member of council for the Institution of Municipal & County Engineers, winning the institution's highest award that year.

Having written papers on concrete reservoirs and chimneys between 1907 and 1910, Matthews (jointly with Henry Adams) undertook a more comprehensive treatment of concrete in 1911, with a book entitled *Reinforced concrete construction in theory and practice*. His work with Adams is described above, but it is worth noting that Matthews' contemporary standing as an author is reflected by this comment from *Concrete*

& Constructional Engineering: 'The authors of this volume are so well known to our readers that they will need no introduction, and we should naturally expect an excellent production from such a combination of authors.' (*C&CE*, 1912, p.73).

Other titles followed (though not specifically on concrete), including *Coast erosion and protection* (1913) and *Refuse disposal* (1915) – the former the subject of a lecture given to the British Association in 1913.

During the First World War he served as specialist sanitary officer with the rank of captain commanding the 34th and 57th sanitary sections of the British Expeditionary Force. He was mentioned in despatches twice and awarded the Imperial Star.

His career resumed in peacetime, with sanitation a continuing theme. Matthews was a fellow of the Royal Sanitary Institute, and in 1922, a member of its council. Two further books appeared: *Logarithms and trigonometry* (jointly with T. G. Howes Thomas); and *Studies in the construction of dams: earthen and masonry* (1919). A paper in 1920 – *Bridge construction in reinforced concrete* (*C&CE* v.15, pp.336–342) – saw a return to his earlier interest in concrete. Recognition followed too, as he was awarded the Silver Medal of the Royal Society of Arts, and was a Bessemer and Nursey prizeman at the Society of Engineers. He died on 6 November 1930.

Adams, H. and Matthews, E. R. *Reinforced concrete construction in theory and practice: an elementary manual for students and others*. London: Longmans, Green & Co., 1911 [C&CA 1959, ICE, Peddie]. Second ed., 1920. [C&CA 1959, BL, ICE, Yrbk 1934. Copy: TCS. Rev: *C&CE* 1912]

1911/USA

Brooks, John Pascal

Described as an 'expert' by *C&CE* (v.7, p.232), Brooks was director of the Clarkson School of Technology in 1911, and formerly associate professor of civil engineering at the University of Illinois. His book was considered by *C&CE* to be 'really interesting' and 'well written and illustrated'. In the reviewer's opinion, Brooks' treatment of retaining walls and arches was better than that of current British works and so, even though it was not wished to displace British text-books, *Reinforced concrete: mechanics and elementary design* was 'a useful addition to the library'. Perhaps in a reflection of the greater range of advanced textbooks available in America, *Cement Age* was more dismissive, describing the book as 'rather brief and somewhat elementary … a good supplementary for engineering students, though one not required when the more complete works are used'. This criticism does seem to overlook the author's aim of avoiding the duplication of overlapping subjects on the college syllabus, and the latest coverage of reinforced concrete design found in the engineering periodicals. Nonetheless, the complete design programmes set out for various structures were commended and evidence of the writer's thorough familiarity with the University of Illinois's 'excellent laboratory work' was noted.

Brooks, J. P. *Reinforced concrete: mechanics and elementary design*. New York: McGraw,1911 [ICE; Peddie. Rev: *C&CE*, v.7, 1912] and London: Hill, 1911 [Peddie]

1911/UK

Cantell, Mark Taylor (b.1869)

With seven distinct volumes to his credit, Cantell was one of the more ubiquitous British authors of his day, though when we say 'British' we should note that his writing reflected American and Canadian experience too. This ubiquity stemmed in part from the books' educational purpose, a role suggested by the title of his first work: *Reinforced concrete construction: elementary course*.

Mark Taylor Cantell in Los Angeles, 1930 Courtesy of Geoffrey Ellis, private collection

Published by Spon in 1911, this budget textbook was derived from Cantell's 14 years of teaching experience at the municipal technical college, Brighton, where, as a professionally qualified architect, he was head of the architecture and building department. The book was welcomed by *C&CE*, which described it as a 'very cheap volume', but considered it would be 'appreciated by students, containing as it does a great deal of information and a very clear explanation of the elementary principles'. One of the book's distinctive features was spelled out in the subtitle – *with examples worked out in detail for all types of beams, floors and columns* – taking the book beyond the descriptive or theoretical. *C&CE's* review, whilst deploring the inadequate size and clarity of the accompanying illustrations, concluded: 'this should prove a very useful book, and is well written, and we are pleased to find that the construction of the various formulae is explained at the end of the volume.'

Its author was born on 1 November 1869, at Newhaven, Sussex, the eldest son of Mark P. Cantell – a boat-builder originally from Jersey. In his youth he was reckoned to be a fine oarsman. At the age of 23 (in 1892), he married a lady from the nearby town of Lewes. Moving to Brighton he appears to have combined his teaching at the technical college with some practical building work in the locality. Records of Steyning East RDC for new building and alteration work in Patcham and Preston, include the following:

- 66 Surrenden Road, 1906, G. W. Higgins / M. T. Cantell
- 52 Surrenden Road, 1907, Mrs E. B. Percy / M. T. Cantell

After publication of his *Elementary course*, Cantell left Brighton to take up a post in Canada as head of technical department at the Kelvin Technical High School, Winnipeg. It was here that he was employed when his second book was published in 1912. Naturally enough the new work was entitled *Reinforced concrete construction: advanced course*, and it too contained 'numerous fully worked examples'. The reception was mixed however, and although the review in *C&CE* (v.7, p.942) applauded the way the derivation and construction of formulae were shown – giving this book an advantage over others – it regretted the uneven treatment of subject matter. Some chapters were simply too perfunctory and the illustrations from and references to specialist systems were conversely so profuse that they 'render the work almost an advertisement pamphlet'. That said, the work was thought to be well written and clear.

Cantell's whereabouts during the following years are obscure, but he became an associate of the Surveyors' Institute in 1917 and a member of the American Society of Mechanical Engineers in 1918. He was evidently still in (or back in) Canada by 1920 for in that year he became an associate of the Engineering Institute of Canada and a member of the Manitoba Association of Professional Engineers. A year later he added membership of the Canadian Institute of Mining and Metallurgy. 1920 and 1921 were also the years of his next publications – *Reinforced concrete construction*, parts I and II.

It is not made explicit whether or not these were intended as revised editions of the *Elementary* and *Advanced* books of a decade before, but they appear to have fulfilled the same function as them and to have had a similar successive relationship to each other. As previously, they was aimed at students, formulae were explained, examples fully worked out, and systems illustrated. As the last lines of Cantell's preface were to thank the editor of *C&CE* for the use of illustrations, it must have been disappointing for him to read the subsequent review in that journal, written by no less an authority than Oscar Faber. In responding to Cantell's statement of purpose, Faber wrote:

> We can only conclude that he has a very imperfect idea as to what constitutes a proper design and a working drawing; since, in our opinion, there is nothing in the book which approximates to our idea of the requirements of either, and without having the advantage of knowing how much experience Mr. Cantell has had with the practical design and construction of reinforced concrete, we get the impression that his energies have mostly been diverted to other subjects. Fig. 90 and Fig. 38 show arrangement of reinforcement, who the very worst possible arrangement which could be devised...

Once again the quality of illustrations was criticised.

Soon after (in 1922), Cantell became a fellow at the Royal Society of Arts and later (in 1930), a fellow of the Royal Institute of British Architects (having been a licentiate since 1910). But much of this decade was spent in Canada and the USA, where he was responsible for the design and construction of a number of civil engineering and architectural structures. He became a certificated architect and registered civil engineer in California, in addition to his Canadian affiliations. Consequently when he came to write his final work, he drew on that combined experience, with reference to the recommendations of the RIBA, Engineering Institute of Canada and the American Joint Committee.

Practical designing in reinforced concrete – written in three parts between 1928–1935 – continued in his characteristic vein, preoccupied by the student's need for clarity and simplicity, and providing fully worked examples and illustrations. Kempton Dyson noted its publication kindly in the *Structural Engineer* (v.12, p.114). Volume III – coming 38 years after he started to teach at Brighton and when Cantell was aged 65 – probably represented the culmination of Cantell's career: a retrospective of past developments with a desire to pass on his experience to new seekers after expertise in concrete. He remained in education a little longer, however – he was described as head of the Winnipeg Technical College in 1944 – and is thought to have died in Los Angeles.

Cantell, M. T. *Reinforced concrete construction: elementary course*. London and New York: Spon, 1911 [Peddie. Rev: *C&CE* v.7]

Cantell, M. T. *Reinforced concrete construction: advanced course*. London: Spon, 1912 [BL, C&CA 1959. Copy: TCS. Rev: *C&CE* v.7, p.942]

Cantell, M. T. *Reinforced concrete construction: part I*. London: Spon, 1918? second ed., 1920 [BL, Yrbk 1934]

Cantell, M. T. *Reinforced concrete construction: part II*. second ed. London: Spon, 1921 [BL, C&CA 1959, ICE. Copy: TCS. Rev: *C&CE*, v.16, p.745]

Cantell, M. T. *Practical designing in reinforced concrete: part I*. London: Spon, 1928 [BL, C&CA 1959, ICE, Yrbk 1934. Copy: TCS]

Cantell, M. T. *Practical designing in reinforced concrete: part II*. London: Spon, 1933 [BL, C&CA 1959, ICE. Rev: *SE*, v.12, p.114]

Cantell, M. T. *Practical designing in reinforced concrete: part III*. London: Spon, 1935 [BL, C&CA 1959, ICE]

1911/UK

Davenport, John Alfred (1878–1959)

MSc, AMICE

John Davenport was born on 3 April 1878 at 48 Castle Street, in Liverpool. He was educated locally at Parkhill Higher Grade School and the Liverpool School of Science and Art, before undertaking practical training with John D. Davies – a mechanical engineer in the city. In 1899 he attended the Victoria University to read for a BSc. in engineering and added a Master of Science from the newly independent University of Liverpool to his degree in 1904. Leaving Liverpool he was appointed lecturer in mechanical engineering at the East London College in Mile End, London, as a post recognised by London University. He joined the Institution of Mechanical Engineers in 1904 and a year later he married. In 1907 he returned to Liverpool, taking a position as assistant lecturer and demonstrator in engineering at the Walker Engineering Laboratories of the University. After a year in post he left to set up as a consulting engineer at 20 North John Street, Liverpool, and then as partner to J. Parr Emett. During the next few years he specialised in reinforced concrete construction, designing and supervising the erection of numerous concrete structures, including factories and coal-bunkers. He became a member of the Concrete Institute in 1909.

Davenport's first venture into print was with a volume of labour-saving design diagrams – *Graphical reinforced concrete design* (1911) – just at a time when the market for these was becoming crowded, or a little 'overdone' as *C&CE* put it in its review of July 1912. The diagrams were appreciated for their ingenuity, but regarded as too involved for general use. The reviewer regretted too that the book had come out just before the second RIBA report and appeared a little out of date in consequence.

Nonetheless, Davenport was apparently just getting into his stride as a writer, submitting numerous articles to the professional journals (particularly *C&CE* and the Concrete Institute's *Transactions & Notes*) in the years immediately following. In 1912 – the year he was elected to the ICE – he concentrated on T-beams in *T-beam design* (v.7, July, pp.520–524) and *Shear force in T-beams* (v.7, October, pp.749–759). The year after he focused on economical reinforcement:

- *Economical slab reinforcement*, v.8, March 1913, pp.166–170 (corresp. H. Kruse, p.732)
- *Economy in reinforced design*, v.8, May 1913, pp.350–360; *Trans & Notes* 5(1) 1913, pp.162–172

These were not without controversy – as the correspondence with H. Kruse testifies. Other articles followed until after the start of the First World War:

- *Sand and coarse material and proportioning concrete*, v. 9(8) August 1914, pp.553–558
- *The testing of reinforced concrete beams*, v.7(9) July 1914, pp.451–456
- *Moment of inertia of reinforced concrete sections*, v.10(4) April 1915, pp.179–186

After the war he wrote *Consolidation of granular fill materials* (published in *C&CE* in April 1919), and a final piece for the journal: *Reinforced concrete silos in America* in 1925. In the post war period however, he transferred to the baths department of his local authority, and was based in the municipal George's Dock Buildings from 1921–1938. His career in the design and management of swimming pools came to a peak in the late 1940s when he published a series of three titles on the subject between 1946 and 1948. He died in early 1959.

Davenport, J. A. *Graphical reinforced concrete design*. London and New York: Spon, 1911 [ICE, Peddie, Yrbk 1934. Rev: *C&CE* v.7, p.311)

1911/UK

Gammon, John Charles (c.1889–1973)

BSc Eng., ACGI, MCI, Assist. Engr. Public Works Dept., India

Gammon published his contribution to concrete literature as a young man and a recent graduate with first class honours in civil engineering from the University of London. In July 1910 he submitted *Design of rectangular beams in reinforced concrete* to *C&CE* (v.5, pp.502–506) and was appointed as assistant engineer at the public works department in Bombay that year. Developing a role as an expert in reinforced concrete he contributed to the structures of the Science College, Prince of Wales Museum and the Customs House in Bombay. Based in part on this early experience, he wrote his *Reinforced concrete design simplified*, which was published by Crosby Lockwood in 1911, with an introduction by H. Kempton Dyson (a prominent member of the Concrete Institute).

Although others were publishing labour-saving diagrams at this date (Twelvetrees and Davenport, for instance) the reviewer in *C&CE* singled out Gammon's work for praise: 'The diagrams are very good in the volume, and we go so far as saying they are the best we have seen, while the inclusion of the examples will give the user a good opportunity of realising the correct method of applying each particular diagram.'

After war service Gammon returned to Bombay in 1919 and three years later established his own firm; J. C. Gammon Ltd. He also returned to writing on the subject of rectangular beams with his *New formulae for economical design of rectangular beams particularly as regards shear resistance*. This was published in *C&CE* in April 1920 (v.15, pp.243–246).

His firm undertook routine work in the construction of factories, whilst bridges such as those at Bonum and Patalganga bear testimony to Gammon's ingenious design. He won recognition for his work on the prestigious Gateway of India and went on to achieve several 'firsts' in India:

- RCC pile foundations for the Gateway of India
- Thin shell structures of the Meerut Garages
- The colloidal grouting process at Mundali Weir
- Hyperbolic cooling towers at Sabarmati
- Pre-stressed concrete spans of up to 200 ft long

The firm expanded and in due course had branches or associated companies in Malaysia, Ghana, Nigeria, the Persian Gulf and Hong Kong, and after partition the Indian business operated in Pakistan and Bangladesh too.

The family settled in Cornwall, near the golf club at Trevose in which Gammon had a share. He became sole owner in 1955 and his son Peter took over management in 1961. The family eventually divested shares in J. C. Gammon and the firm was floated as a public company in 1970. John Gammon died in 1973, but his name lives on in a group of companies jointly owned by Matheson Jardine and a succession of construction companies (of which Balfour Beatty is the most recent [2008]).

Gammon, J. C. *Reinforced concrete design simplified*. London: Crosby Lockwood, 1911 [ICE, Peddie. Rev: *C&CE* v.6, p.881]. Second edition, 1913 [BL]. Third edition. Kingston Hill: Technical Press, 1921 [BL, C&CA 1959, Yrbk 1934. Copy: TCS]

1911/USA

Lewis, Myron H. (b.1877)

Writing for the same publisher as his contemporary A. A. Houghton [1910], Lewis was author of two titles in 1911, both off the mainstream of concrete literature. *Modern methods* built on earlier work entitled simply *Waterproofing* (1909), and was reissued in 1914. His *Popular handbook for cement and concrete users* however, attracted rather more attention, certainly in the British press.

Initially the notice was unkind. Comparisons – not favourable – were made with the British equivalent, *Everyday uses of Portland cement*, which had been published by the Associated Portland Cement Manufacturers a couple of years earlier. *C&CE* predicted that the American book was 'fated to be forestalled by that work'. While it contained a great deal of information little of it was original, 'being reprinted from American catalogues, papers and reports issued from time to time – some of which are out-of-date'. Rather harshly, the review continued: 'it is neither happily-enough presented, nor does it contain sufficient original information to lead us to view its production with any enthusiasm. Its practical part is often skimpy and effete, while the theoretical information, on both cement and plain concrete, is often in error; while in the matter of reinforced concrete, it seems to us too amateurish, and of little use to the student or designer.'

Despite such disparagement the book ran to a second edition ten years later, issued in Great Britain by Henry Frowde and Hodder & Stoughton. This volume was greeted with more warmth: 'It has been felt that there was still a place for a semi-popular book of general type that would prove interesting and, without being too technical, useful, that would bring home to the mind the great economic and artistic qualities of concrete as a building material … We welcome this work, therefore, which meets a definite want … a book for the non-technical reader.' Parallels with the work of the Concrete Utilities Bureau were noted, but American practice was seen to be 'far in advance of our own' and the artistic treatment of concrete – moulding and surface finishes – covered in the book addressed 'an aspect of concrete work [that] has been but little regarded in this country'.

Lewis, M. H. *Modern methods of waterproofing concrete and other structures*. Norman W. Henley Publishing Co., 1911 and 1914

Lewis, M. H. & Chandler A. H. *Popular handbook for cement and concrete users*. Norman W. Henley Publishing Co., 1911 [*CA*, v.12, p.220. Rev: *C&CE* v.6, pp.633–634]. Second edition. Henry Frowde and Hodder & Stoughton, 1921 [BL. Rev: *C&CE* pp.53–56]

1911/UK

Piggott, J. T.

Taking the 1909 London County Council Regulations as a framework, Piggott edited a slim volume of tables, formulae, diagrams and data to provide a 'series of simple practical rules for obtaining the dimensions and main reinforcement of beams and slabs for all classes of reinforced concrete structures'.

Piggott, J. T (ed). *Reinforced concrete calculations in a nutshell*. London: Spon, 1911 [Yrbk 1934]

1911/UK

Scott, Augustine Alban Hamilton (1878–1944)

FRIBA, MCI, MIM, MRSI

Augustine Alban Hamilton Scott Courtesy of the IStructE

Alban Scott was an architect whose interest in concrete was to lead him into active participation in the early life of the Concrete Institute. He was present at the third meeting of the Concrete Institute on 18 Feb 1909 (contributing to discussion of the paper), and at the first annual dinner on 7 June 1911, which was held at the Trocadero Restaurant, Piccadilly Circus. By 1913 he was a member of the Reinforced Concrete Practice Standing Committee and honorary secretary (and later chairman) of the Tests Standing Committee. He was a member of the council for the 1916–1917 session.

Scott was educated at the City of London College & Polytechnic Institute and embarked on a career as an architect. He was articled to Thomas Arnold from 1894 to 1899 and then Assistant at Southend-on-Sea in 1900. In 1905 his first book was published, entitled *The planning and construction of factory buildings*. On Arnold's retirement that year, Scott went into practice on his own account before entering into a partnership with P. M. Fraser in 1908.

Scott's early experience of working with concrete came from his time as architect to the British Aluminium Company's new works at Kinlochleven, and associated hydroelectric power station. This granite-clad reinforced concrete structure, built in the Coignet system during the years 1905 to 1909, was described in his paper for *Cement & Concrete Engineering* in August 1910 – *The British Aluminium Company's works at Kinlochleven* (*C&CE* v.5, pp.585–589) – and in *Transations & Notes* (v.2, pp.219–239). Other contemporary concrete projects included a road bridge at Mauld, Inverness, and, in the Southeast of England, a factory for Messrs J.C. & J. Field, Ltd., and a footbridge for the Erith Oil Works.

At the end of this period Scott and his partner, Percival Fraser, felt sufficiently experienced in the subject to propose *A specification for reinforced concrete work*; a document published by Witherby in 1911, not long after the RIBA report of that year. *C&CE* (v.6, pp.461–462) was cautious: 'it is a useful enough document'; but also – a result of being drafted 'from the architect's point of view' – rather 'inequitable' in places and 'surreptitious' in introducing clauses to the architect's favour. Scott was taken to task at this time in a Concrete Institute debate, after appearing to claim for the architect a monopoly on good design to the exclusion of the engineer. Fraser came to his defence and explained that the impression was not what had been intended.

Scott's closer involvement with the Concrete Institute lay in the Tests Standing Committee. His *Testing of materials used in reinforced concrete* was published in *C&CE* for May 1912 (v.7, pp.362–371) and *Reports of the Testing Standing Committee on the testing of reinforced concrete structures on completion* appeared in *Transactions & Notes* (v.4, pp.240–241). Other articles for wartime issues of *C&CE* followed:

- *The construction and protection of buildings in relation to fire*, *C&CE* v.10, March 1915, pp.153–158
- *Some recent factory constructions*, *C&CE* v.11, March 1916, pp.117–128

His major work during the war years however, was *Reinforced concrete in practice*, which Scott, Greenwood & Son published in May 1915. Scott's preface betrays his interest in testing: 'there is no form of building so lasting and requiring less maintenance than Reinforced Concrete, he considers that it should only be employed when the work is under constant, careful, and experienced supervision and when every material used in the work is tested.' *C&CE*'s reviewer concurred, welcoming the volume as 'based on the right principles and consequently it deserves to be popular'. In contrast with its views on the partiality of Scott's 1911 work, *C&CE* praised this as 'presented in such a manner that the interests of all parties are considered', including helpful hints to the contractor. The book was a success and was reissued in July 1925 when Scott had become a fellow of the RIBA and was based at 13 Old Square, Lincoln's Inn, London. His partnership with Fraser had ended in 1913 and Scott was back in practice on his own.

Like Searle [1913], he wrote about shipbuilding in concrete during the latter stages of the war. His *Reinforced concrete ships* was published in *C&CE* in 1918 (v.13, pp.423–430), and reproduced in *Transactions & Notes* in 1921 (v.9).

In the post-war years Scott was appointed as architect for a series of additions to the Thornycroft motor works, Hampshire. This was an experience which, to judge from the acknowledgements, was incorporated into the revision of *Reinforced concrete in practice*. Commissions over the years included huts, common room, WCs and showers (1919); sandblast house and stock house (1926); repair shop and store shed (1928); and lavatories and drawing office (1931). Also in Hampshire, Scott undertook commissions for Venture Ltd with a petrol station (1931); restaurant (1932); swimming pool (1933) and restaurant extension (1934/35).

During the Second World War he formed a partnership with W. Leslie Twigg, but died on 2 January 1944. Beside his various professional memberships noted at the head of this entry, Scott was a vice-president and medallist of the Society of Architects.

Scott, A. A. H. and Fraser, P. M. *A specification for reinforced concrete work*. London: Witherby & Co., 1911 [Peddie. Rev: *C&CE* v.6, pp.461–462]

Scott, A. A. H. *Reinforced concrete in practice*. London: Scott, Greenwood & Son, 15 [ICE]. Second edition, 1925 [BL, C&CA 1959, ICE, Yrbk 1934. Copy: TCS]

1911/UK

Fraser, Percival Maurice (fl. 1898–1926)

ARIBA, MCI

Partner of Scott (above) in his architectural practice between 1908 and 1913, Fraser was also his collaborator on *A specification for reinforced concrete work* in 1911. Like Scott, he was an active member of the Concrete Institute, and also the Institute of Sanitary Engineers – to which he presented a paper in 1912 entitled *The modern house*. A similar *Suggestions for a well-planned house* appeared in the RIBA *Journal* on 23 March that year. After the partnership with Scott ended, Fraser continued his interest in 'factory construction' and contributed an article under this title to *C&CE* (v.9, pp.193–202) in 1914. This was followed in 1915 with *Factory construction in reinforced concrete*; a paper to the Concrete Institute that was reprinted in *Transactions* (v.5).

1912–1918: Understanding and Acceptance

1912/USA

Hering, Oswald Constantin (1874–1941)

Compared with the British authors of 1912 that follow, Hering and his book on concrete houses were relatively unknown in the UK. He did not feature in the institutional or periodical press, his book was not reviewed in this country, and although distributed here by Batsford, *Concrete and stucco houses* clearly addressed a niche interest that was topical in America but had not yet arisen in Great Britain. Its time would come once the impending First World War was over.

Hering was born in Philadelphia on 12 January 1874. His father Rudolph, was a civil engineer, and his mother Fanny, a librarian and teacher who had written a book on the French artist Jean Leon Gerome. The young Oswald studied at the Massachusetts Institute of Technology from 1893 to 1897, then read art and architecture at the *Atelier Ginain* and *École des Beaux Arts* in Paris for two further years. In 1901 he opened his own architectural practice in New York, operating from 1910 in partnership with Douglass Fitch.

The new practice specialized in country and suburban homes and – as his book of 1912 suggests – pioneered the residential use of reinforced concrete. 'Who can say that from the interrelated masses of aggregate, cement and steel, a new and true architecture shall not be born', he asked rhetorically, suggesting that 'a native style is to be realized through reinforced concrete' (*Town & Country*, May 1908). Clients included the Democratic Party's presidential nominee for 1920, James M. Cox; and besides East Coast residences he designed Brentano's bookshop in New York and Lakewood Theater in Skowhegan, Maine. Some of his commissions, we might suppose, were those that figured in the articles he wrote for *Cement Age* in 1909 and 1910:

- *A concrete residence at Ardsley-on-Hudson, N.Y.* (v.9, 1909)
- *Description of Butler Sheldon Residence, Columbus, Ohio* (*CA* v.9, 1909)
- *Two colonial houses* (v.10, 1910)

Besides reinforcing the essential role of the architect, these case studies emphasise the fire-resistant qualities of concrete and its scope for decoration. In 1912 he presented his views on architectural concrete in a book: *Concrete and stucco houses*. It was 'handsomely bound, printed and illustrated', declared *Cement Age*: 'a strong plea for more durable and carefully thought-out methods of dwelling construction.' Indeed *Cement Age* reproduced the chapter on concrete blocks, subtitled *an architect's view*, inviting architects to comment on masonry's merits and defects. Several responded and Hering welcomed the debate, claiming that it was 'partly the architect's fault that the concrete block has not acquired the dignity of brick or stone as a building unit. He has not given it the study and attention that it deserves. An architect is also an educator and it is his duty to investigate every form of building material ... that will add to the comfort and happiness of life within the structure he designs'.

As well as his *Concrete and stucco houses* (an edition of which was published as late as 1929), Hering wrote other books on architecture including *Economy in home building* (1924) and *Designing and building the chapter house* (1931) – the outcome of the Interfraternity Conference's 1931 committee on architecture which he chaired. *Down the world* (1932) was an altogether different book, describing his world tour of 1927.

He was a member of many social and professional organisations, including the American Institute of Architects, the Beaux-Arts Institute of Design, the Technology Club and the Sons of the Revolution (whose quarterly magazine he edited). From 1919 he became involved in the *Delta Kappa Epsilon Quarterly* (the publication of his university fraternity) and in 1925 became its editor. As architectural commissions declined during the Depression, this work became his main activity.

In 1933 he married for the third time. His wife was Adelaide Heriot Arms, owner of the Sleepy Hollow Bookshop in New York's Irving Place. The couple ran the shop together until they sold up in 1937. They retired to Falls Village, Connecticut, where Hering died on 6 March 1941.

Hering's professional papers are preserved in the New York Public Library.

Hering, O. C. *Concrete and stucco houses: the use of plastic materials in the building of country and suburban houses in a manner to insure the qualities of fitness durability and beauty*. New York: McBride Nast & Co, 1912 [BL] and London: Batsford, 1912 [Yrbk 1934]. Second edition. New York: R.M.McBride, 1922 [BL]

1912/UK

Martin, Nathaniel

AGTC, BSc, AMICE

Martin wasn't an author at all, but a translator. Unlike Moisseiff however, with his translation of Considere in 1903, Martin acquired a prominence beyond this role. Perhaps it is because he acted more as an editor, selecting from and abridging the report of the French Government Commission to serve before a British audience in English translation. For whatever reason, it was as the 'Author's Preface' that his introductory words of December

1911 were published. These had the honour of preceding those of Considere himself (who had worked for the French Government Commission), which were not written until June 1912.

A civil engineer by profession and lecturer in reinforced concrete at the Royal Technical College, Glasgow, Martin was well qualified to recognise the importance of the commission appointed by the French Ministry of Public Works. He acknowledged American practical experience, written up and published in English, yet dismissed all experimental work from Austria, Germany and Italy. He recognised how little experimental research had been conducted in this country and so he was firmly of the opinion that "it is undoubtedly to France and French literature that a student must turn for the most concise and authoritative information on the subject'.

Derived at the outset from a programme of testing to destruction many of the buildings no longer needed after the Paris Exhibition of 1900, the results were combined with the theoretical explorations of leading experts. As Martin put it:

> The report of the Commission ... is unique in the literature of Reinforced Concrete, containing as it does all the necessary scientific data, based on first-hand observations, for the design of reinforced concrete structures, with the observations thereon of a group of engineers of the widest and most mature experience obtainable. The instructions are characteristically French in their clearness and boldness – a boldness derived from intimate knowledge, and entirely justified by results.

In presenting his synopsis, Martin's achievement was to make the full sweep of the commission's work available to an English-speaking audience in a concise and readable manner. To quote *Concrete-Cement Age*, he had 'seized upon the pith and ... excluded the pulp in the voluminous reports' (v.5, p.131). Its publication concluded a phase of dependence on French understanding that Oscar Faber (below) was to transform with his own experimental work that same year.

Martin, N. (trans). *The properties and design of reinforced concrete: instructions, authorised methods of calculation, experimental results and reports by the French Government Commissions on reinforced concrete.* London: Constable & Co., Ltd, 1912 [C&CA 1959, ICE, Yrbk 1934. Copy: TCS. Rev: *C&CE*, v.8, p.142], and New York: Van Nostrand [Rev; *C-CA*, v.5, p.131]

1912/UK

Faber, Oscar (1886–1956)

OBE, DSc, MICE, ACGI

Oscar Faber
Courtesy of IStructE

The young Oscar Faber – London-born, though of Danish extraction – was the first really original contributor to the British tradition of writing on concrete. Like his contemporary E. S. Andrews, he was to dominate the concrete establishment throughout the 1920s and into the 1930s, taking the presidency of the Institution of Structural Engineers in 1935 (the final year of this survey).

Faber was born at 56 Dalberg Road on 5 July 1886, the eldest son of Harald Nicolai Faber (1856–1944) – the Danish commissioner for agriculture – and Sofie Cecilie (*née* Bentzien). He was educated at St Dunstan's College, Catford, where he excelled at sport as well as winning academic prizes. A scholarship took him to the Central Technical College (later the City and Guilds Engineering College) where he studied electrical engineering. He qualified in 1906, then graduated as a civil and mechanical engineer in 1907, emerging from his studies with an interest in the material he was to champion: reinforced concrete. He worked for a while as assistant engineer for the recently amalgamated cement company Associated Portland Cement Manufacturers (better known later by its trademark Blue Circle), and having joined the Concrete Institute in 1911, contributed to debates on tall chimney construction. Such structures would have been one of the responsibilities of his employment. He then went to work for the reinforced concrete specialist Indented Bar & Concrete Engineering Co. in 1909, which would have provided a solid grounding in the subject. Finally in 1912 he rose to chief engineer with the long-established contracting firm Trollope & Colls. But it was his work with the company's assistant engineer P. G. Bowie, exploring the theoretical basis for design that hitherto had been the preserve of (largely foreign) patentees, which propelled him to prominence. His first book – written with Bowie and published in 1912 as *Reinforced concrete design* – announced his arrival.

Here was a British book that did not rely on literary study (though it did quote Taylor & Thompson), translations of foreign theorists, descriptions of patented systems or reports of building projects. Faber and Bowie brought fresh thinking combined with experiment and tempered by practical experience. 'A good deal of the matter is new, and several important considerations are taken into account which have thitherto been ignored, as far as the Authors are aware, in published literature on the subject'. The mathematics of bending moments was seen as more complex than previously treated and the performance of beams in shear was a key consideration of the new work. The latest recommendations from the RIBA committee were also included.

The novelty of Faber and Bowie's approach was highlighted by the reviewer in *C&CE* (v.7, p.390): 'The features of the book which deserve most praise are the original ones, and these are especially the simple and fairly accurate, though approximate, means of calculating the stresses in members under combined bending and direct stress, the detailed discussion of continuous beams and the effect of unequal depression of supports, and also eccentric loading and flexure of columns due to beams built monolithic with them.' He went on to judge that 'the authors have displayed excellent judgement and good commonsense', though he did – interestingly in view of Faber's later thesis and reputation on the subject – make adverse comment on the treatment of shear. Taken in all, the reviewer thought the book 'strong meat for elementary students', but that it was 'exceedingly well produced, with many carefully drawn illustrations, and should be in the hands of all advanced students of the subject'.

In America the book's originality was recognised too:

> This book is the first exposition of a number of very important considerations in the design of reinforced concrete structures that has appeared in print. Previous authors have apparently neglected the fact that the stresses in a monolithic structure of reinforced concrete are not governed by the same laws that apply to a built-up structure (*CA* v.14, p.314).

Cement Age considered it clear in reasoning and language, and considerate in confining much of the advanced mathematics to an appendix.

Like the RIBA recommendations, Faber's formulae for bending moments were accepted for inclusion in the subsequent LCC Regulations of 1915.

By this time Faber had married. His bride was Helen Joan Mainwaring (1881–1967), with whom he had two daughters (Eileen and Barbara) and a son (John, who followed him into practice).

Faber's early career can be charted in *Concrete & Constructional Engineering*, which serialised his doctoral thesis of 1915 under the title of *Reinforced concrete beams: new formulae for resistance to shear*. This work was based on five years of laboratory testing at the Northern Polytechnic Institute, which had also given rise to the earlier *Reinforced concrete design* and presumably responded to the criticism of that book's treatment of shear. In Faber's biography his son John describes the serialisation as 'consistent with his determination for publicity', which with his expertise and an OBE for wartime work on concrete barges, coastal defences and anti-submarine devices of non-ferrous reinforced concrete, made him an ideal candidate for the role of technical advisor to *C&CE* on the death of its founding editor Edwin O. Sachs, in 1919.

During Faber's period of influence over the next couple of years, *C&CE* took a more instructional role. Practical notes and hints 'of especial interest to contractors, supervisors, clerks of works and others who are in charge of the actual execution of work' formed a new feature of the journal. In another direction, students were treated to a series of 'textbook'-type articles entitled *Concrete in Theory and Practice*: 'a practical section especially written for the assistance of students and engineers, and others who are taking up the study of reinforced concrete, or who are interested in the subject on its educative side.' Faber himself provided the text with 'explanations so simple as to be intelligible to anyone desiring to understand the underlying principles of reinforced concrete without wading through a lot of mathematics' (1920, p.701). According to his son, he was a logical thinker and clear speaker, with a natural gift for teaching.

Faber was active in teaching at this time. He had been teaching 4th year students as a lecturer at the City & Guilds Engineering College since 1916, and his lectures at the Architectural Association and University College were given some coverage in *Concrete & Constructional Engineering*. Indeed the series of articles in *C&CE* was borne out of material for the architectural lectures and was later republished as *Reinforced concrete simply explained* (1922). Through his involvement with the Concrete Institute, Faber presented technical papers, and as in January 1921, gave detailed responses to other authors' work.

Besides teaching and consultancy, Faber was busy with revising and extending his and Bowie's *Reinforced concrete design*. It was reissued as Volume I of an enlarged two-volume set, with minor corrections but largely unchanged. Volume II – which was by Faber alone – gave a fuller treatment to bending moments, unequal spans of continuous beams and shearing resistance based on subsequent theoretical and experimental work associated with his doctoral research.

Volume II evidently made an impact as it was reviewed twice in *C&CE*. The reviewer on page 66 of Volume 16 expressed Faber's standing: 'Any publication upon Reinforced Concrete by Dr. Faber deserves careful consideration by structural engineers because he is one of the comparatively few engineers who have had the advantage of advanced theoretical training and scientific research combined with considerable practical experience under commercial conditions.' Faber was seen to hold 'strong views' on bending moments and shear stresses, to the extent of slighting the practices some 'so-called specialist firms'. Despite the reviewer's evident distaste for this passing innuendo, he was clear that the book was one 'which every reinforced concrete engineer should study'. Albert Lakeman [1913] was more fulsome: 'The book can be strongly recommended as an excellent addition to the literature on reinforced concrete design, and it deals with a difficult subject in a new, helpful, and simple manner while making for that prime necessity – accuracy.'

Partly through Faber's participation in both, relations between *C&CE* and the Concrete Institute became even closer. The publication of the institute's *Transactions* had been postponed during the war, so in July 1920 *Cement and Constructional Engineering* became the institute's official journal, subtitled *The officially appointed journal of the Concrete Institute*. The arrangement was to continue until the institute transformed itself into the Institution of Structural Engineers in 1922 and its new journal, the *Structural Engineer*, was issued monthly from January 1923.

As 1921 drew to an end, and despite the economic uncertainty of the times, Dr. Faber turned his attention to expanding his own consulting business as a full time practice. He launched it with a capital of £2000. The new business was to include mechanical and electrical, as well as structural, engineering. This meant there would be less time for work on the journal. His change of circumstances, and acknowledgement of his role in advising the journal, was announced in *C&CE*:

> We have much pleasure in notifying our readers that Dr. Oscar Faber, O.B.E., D.Sc., A.M.Inst.C.E., etc., has

> recently set up in practice for himself as consulting engineer, at 5 South Street, Finsbury Pavement, E.C.2.
>
> Dr. Faber is, of course, well known to our readers by his articles and books on reinforced concrete, as well as the research work he has carried out on this subject. He is a member of council of the Concrete Institute, and in addition to his ordinary consulting work he holds the appointment of consulting engineer to H.M. Office of Works, the Calico Printers' Association, Manchester, etc. He also is a regular lecturer at King's College, University College, the Architectural Association, etc.
>
> For the last two years this journal has had the advantage of Dr. Faber's co-operation as its Technical Adviser (1921, p.612).

He soon found work and by 1924 he was consulting engineer to the Bank of England, H.M. Office of Works and the Hudson's Bay Co. Of these it was considered that 'one of his most important works is the structural, mechanical and electrical engineering of the new Bank of England, which is ranked among the greatest building enterprises of this generation' (*SE*, v.13). He went on to undertake similar work for the London offices of Barclays, Lloyds, Martins and Glyn Mills banks, along with public buildings, hospitals, theatres, cinemas, offices and flats. His grain silos at Cardiff and Avonmouth were highly regarded for their economic design and speed of construction.

Having moved into consultancy in 1921, Faber was to continue his involvement with *C&CE*'s publisher, as an author and co-editor of the *Concrete Yearbook*. In 1922 H. L. Childe was appointed as both editor of *C&CE* and manager of Concrete Publications Ltd, and quickly initiated an expansion of its publishing activities in which Faber was to be an important contributor.

Before Childe's appointment was effective however, Faber arranged to have his instructional articles from the journal published in book form. His *Reinforced concrete simply explained* was published in 1922 and was a critical success. Albert Lakeman (writing in *C&CE* [p.371]) paid tribute: 'One of the most important problems for an author is that of writing a treatise dealing with a complicated subject in a manner which will enable the uninitiated to follow and fully understand the principles … and in this volume the problem has been handled in an excellent manner.' A second edition was issued by the Oxford University Press in 1926.

His next book was derived in a similar, but more convoluted way. *Reinforced concrete in bending and shear: theory and tests in support* was compiled from articles in *C&CE* (back in 1916), but these articles were themselves drawn from his PhD thesis of 1915. Significantly, when the book was published in 1924 it was as the sixth in Childe's new Concrete Series of textbooks; a series that was to dominate specialist concrete publishing in the UK for the next 40 years.

Another title in this series was the *Concrete Yearbook*, jointly edited by Faber and Childe from 1924. Comprising three parts: a handbook summarising current concrete practice; a directory of firms; and a catalogue of products, it was praised as 'a very excellent production'. The review concluded: 'It is hardly necessary to say that no-one interested in concrete should be without this book, as we feel sure that it will soon be found that they cannot afford to be.' The annual series went on to enjoy huge popularity and continued in much the same format until it was taken over by the Cement & Concrete Association in 1969.

Following the doctoral research on bending moments and shear, Faber carried out experiments on foundations in the 1920s, and from 1927 was engaged on the underpinning of Durham Castle. In recognition of this he was awarded the honorary degree of Doctor of Civil Law in 1935. By that time he had also undertaken research into steelwork, specifically into the rigidity of different forms of connection between steel beams and stanchions. His interest in steel frame construction was manifested in his books on the subject, entitled *Constructional steelwork simply explained* (1927) and *Examples of steel design under the New Code of Practice* (OUP, 1934). The latter was presaged by an article in the 1933 volume of *Building*, entitled *The new steel code* (v.8, pp.57–60). His *Recent developments in building* (*RIBA Journal* v.40, pp.389–401) of the same year, looked at new methods of economic construction. Research into long-term plastic yield of concrete under load (mooted in the ICE's *Proceedings* of 1927/28) was written up in 1936; the year when his *Heating and air conditioning of buildings* (written with J. R. Kell) was published too. His interests in both steel and concrete were represented by membership of the appropriate research committees of the Department of Scientific and Industrial Research, and in heating and air conditioning by his membership and eventual presidency of the Institution of Heating and Ventilating Engineers.

Besides membership of the Institutions of Electrical and Mechanical Engineers, Faber took an active part in the life of the Institution of Structural Engineers, acting as chairman of the science committee and presenting many papers at the institution's general meetings. As vice-president he wrote *Modern structures* as part of *The history and progress of structural engineering in the present century* for the Royal Charter issue (1934) of the *Structural Engineer*. In the summer of 1935 he finally succeeded E. S. Andrews to become president and an introduction was published in the June issue of the *Structural Engineer*. The editorial had this to say of the man:

> Perhaps one of the most outstanding characteristics of our new president is his refusal to accept anything which is vague or loosely reasoned. Whether it be a question of long-accepted beliefs or of new ideas and developments, he has always rejected whatever fails to bear searching investigation, but has whole-heartedly accepted and applied whatever passes this test. This characteristic is well exemplified in the part which Dr. Faber has played in the development and use of reinforced concrete in this country. At the outset of his career the use of this material was limited and many engineers looked upon it with distrust. Dr. Faber, however, fully realised its possibilities and set himself the task of mastering the particular problems associated with its use, carrying out theoretical and experimental researches. The results of this work have contributed largely to

> our present understanding of design in reinforced concrete and to the confidence which designers are now able to place in modern accepted methods.

Faber went on to play a continuing role in the developments of concrete design for many more years. He advised Churchill on the design of the Mulbury Harbours and just after the Second World War was appointed consulting engineer for the restoration of the House of Commons. A remark attributed to him is typical of Faber: 'we shall never please 600 members, so we will do it properly and please ourselves'. For this work he was awarded the CBE in 1951. In 1948 he invited his five senior assistants into partnership but continued to work. Work did not, however, prevent him from exercising his talents in watercolour painting and playing clarinet, organ and piano. He was still closely involved with his work in 1956, developing a design for the Wales Empire Pool, when he was taken ill. He died shortly after on 7 May 1956. His passing was recorded with an obituary in the August issue of the *Structural Engineer* (v.34(8), pp.302–304).

His firm outlived him and continued under the Faber name until merging with Maunsell at the end of the century. His name was further commemorated in the Concrete Institution's Oscar Faber Bronze Medal. His son John wrote a full biography, while J. R. Harrison paid tribute to him in the *Dictionary of National Biography*, in which he characterised the man as follows:

> His technical mastery of his subjects was impeccable and his approach simple and direct, his clear thinking never obscured by technicalities. To his fellow engineers he was sometimes an enigma and always a challenge. To his staff he was a stimulating if exacting master, impatient of inexact thinking or tardy action; often critical of a proposal, but always willing to spend much time putting it right ... and to everything he brought an apparently inexhaustible energy.

Faber, O. and Bowie, P G. *Reinforced concrete design* (later Vol. I – theory). London: Edward Arnold, 1912 [BL, ICE. Rev: C&CE v.7, p.390], and New York: Longman, Green & Co., 1909 [PCA. Rev: CA, v.14, p.314]. Second edition, 1919 [BL, C&CA 1959, ICE, Yrbk 1934. Copy: TCS] and 1924 [BL]

Faber, O. *Reinforced concrete design*: Vol. II – practice. London: Edward Arnold, 1920 [BL, ICE, Yrbk 1934. Rev: C&CE v.16, p.66 and p.602] and 1924 [BL, C&CA 1959]; and Oxford University Press, 1929 [BL]

Faber, O. *Reinforced concrete simply explained*. London: Henry Frowde and Hodder & Stoughton, 1922 [Rev: C&CE, p.371]. Second edition. Oxford University Press, 1926 and 1929 [Yrbk 1934]

Faber, O. *Reinforced concrete in bending and shear: theory and tests in support*. London: Concrete Publications Ltd, 1924 [C&CA 1959, ICE, TCS, Yrbk 1934. Copy: TCS]

Faber, O. and Childe, H. L. *Concrete Yearbook*. London: Concrete Publications Ltd, 1924 et seq [Yrbk 1934. Copies TCS]

1912/UK

Bowie, Percy George (b.1889)

ACGI, AMICE, MIStructE

Bowie was notable principally for his collaboration with the young Oscar Faber (above) at the Northern Polytechnic in writing *Reinforced concrete design*. He appears not have been as involved in its reissue as Volume I of a two-part set in 1919.

Percy George – always known in print as P. G. – was born in London to William and Eliza Bowie. His father was a merchant and having retired lived with his family in Hemel Hempstead. Growing up, Bowie found a position with the reinforcement specialist Indented Bar and moved to Lewisham, and then Croydon. There he met Faber and the two men worked on a series of experiments at the Northern Polytechnic and the consequent book that made their reputations. In the year the book was published both he and Faber were engaged by the London contractor, Trollope & Colls. Bowie's role was assistant to Faber's chief engineer.

During the First World War Bowie volunteered for the army. He was on probation as a second lieutenant in May 1915 and served throughout the war as an officer in the Royal Scots.

After a long absence from the record in the 1920s, he reappears in 1935 as a member of staff at the newly constituted Cement & Concrete Association in Victoria, London. His early duties were in the structural and architectural section, but in the early years of the Second World War he was 'appointed technical manager of the Association, with effect from 1st February 1941'. The C&CA's work was directed by the Government during the war, but by 1944 he had started to write professionally again. He appears in the *Journal* of the ICE in April 1944 and with A. R. Collins wrote *Pre-cast concrete in Britain* for the American Concrete Institute (*Proc* 46, 1 March 1950, pp.541–556). He also tried his hand at broadcasting: his *Elastic concrete – the possibilities of prestressed concrete* was broadcast as part of *Science Survey* on Thursday 27 May 1949.

We last hear of him when the C&CA's 19th *Annual Report* recorded: 'Mr. Bowie has retired from full-time employment by the Association after 20 years devoted service.'

Faber, O. and Bowie, P. G. *Reinforced concrete design* (later Vol. I – theory). London: Edward Arnold, 1912 [BL, ICE. Rev: *C&CE* v.7, p.390], and New York: Longman, Green & Co [Rev: *CA*, v.14, p.314]. Second edition, 1919 [BL, C&CA 1959, ICE. Copy: TCS] and 1924 [BL]

1912/UK

Andrews, Ewart Sigmund (1883–1956)

BSc, MIStructE, AMICE

Ewart Sigmund Andrews, President of the IStructE, 1934–35 Courtesy of the IStructE

Anyone who has waded through his bibliography can be in no doubt as to the importance of Andrews to the contemporary literature of concrete. To judge from his presence in the periodicals – with 58 main articles in *C&CE*, *Transactions & Notes*, and *Structural Engineer* – he was one of the most prolific British writers of the 1910s and 1920s. He was also a dominant figure in the Institution of Structural Engineers who, by the time he took the presidency in 1934, was regarded as 'one of the foremost authorities on structural engineering in the country'.

He was born at Stoke Newington in June 1883 and attended Owen's School, Islington. He read engineering at University College London, and in 1903 was awarded a BSc (Eng) with first class honours in the University of London's first degree examinations in the subject. On graduating he remained at University College as assistant (demonstrator and lecturer) to Professor J. D. Cormack, and also in 1903 wrote the Drapers' Company Research Memoir on *The strength of curved beams*. A year later he took a position as assistant examiner at H.M. Patent Office and in 1905 was appointed lecturer in theory and design of structures at Goldsmith's College, New Cross. The twin thrusts of his career were now in place.

1907 saw his advent as a writer of engineering books with the *Theory and design of structures* – coincidentally the title of his lecturer's position at Goldsmith. Given that coincidence, it is unsurprising to discover that the book was based largely on Andrews' lectures at the college. Chapman & Hall reissued this first work with a companion volume – *Further problems in the theory and design of structures* – in 1913. Both titles were reviewed in the specialist journals. *C&CE* (v.8, p.733) noticed that though Andrews acknowledged his indebtedness to Professor Karl Pearson, the book betrayed a gradual transition from the latter's methods to mathematical analyses, 'which the author uses with a native skill'. *Transactions & Notes* (v.5) explains that the second book was intended as an extension of the first, addressing omissions in the subject matter. The reviewers accepted that the mathematical calculations that abound in both volumes – though long – were necessary to ease of understanding.

By this date (1913) Andrews had been a member of the Concrete Institute for a year and had written the first of his books on concrete. *Elementary principles of reinforced concrete construction: a text-book for the use of students, engineers, architects and builders* was published in 1912 by Scott, Greenwood & Son (and in America by Van Nostrand). In this work Andrews endeavoured to supply an 'inexpensive book on Reinforced Concrete, written from an elementary standpoint in an explanatory manner, without the use of advanced mathematics, and is, at the same time, in accordance with the best-accepted theory'. 'He has succeeded', declared *Concrete-Cement Age* (v.2, p.248): 'Mr Andrews has given a brief, concise and easily understood explanation.' He addressed the theory behind a series of problems and then illustrated it with worked examples, sometimes providing alternative methods to individual problems. *C&CE* (v.7, p.391) was of the opinion that 'for a lecture text-book this method is admirable', and though less straightforward for private study, it noted that Andrews had helpfully marked more advanced sections for omission at first reading. 'Considering its small size', the reviewer continued, 'the book is well written' though with a certain 'compression of style'. The information and explanations were given in 'clear and comparatively simple language, though the author seldom, if ever, descends from his academic rostrum to 'sit by the side' of his pupil'.

To those seeking a good grounding in the subject, *C&CE* could 'recommend it heartily'.

In the years leading up to the First World War Andrews took to writing for *C&CE*, mainly on concrete, but also more widely on constructional engineering:

- *Materials under compound stress*, February 1911 (v.6, pp.95–97)
- *A problem in deflections of beams*, November 1911 (v.6, pp.821–823)
- *Secondary stresses in riveted joints*, November 1912 (v.7)
- *Alignment charts for constructional formulae*, January 1914 (v.9, pp.32–39)

In particular he wrote an ongoing series of articles under the heading *Problems in the theory of construction*, which was to continue until April 1922. Before the War these included:

- *Concrete beams with double reinforcement*, February 1912 (v.7, pp.106–109)
- *Wind pressure on roofs*, September 1912 (v.7, pp.668–673)
- *The oblique loading of beams and columns*, April 1913 (v.8)
- *Slab formulae for reinforced concrete design*, June 1914 (v.9, pp.396–404)

1915 was a landmark year in the control of concrete construction as the London County Council regulations were finally introduced. Andrews was quick to comment, preparing for publication by B. T. Batsford Ltd the following guidance: *Regulations of the London County Council relating to reinforced concrete (London Building Act, 1909, Amendment) and steel framed buildings (L.C.C. General Powers Act, 1909): a handy guide, containing the full text, with explanatory notes, diagrams and worked examples*. The reviewer in *C&CE* (v.11, p.267) was emphatic in his approval of the accompanying diagrams: 'The diagrams are a very valuable adjunct to the regulations, and save the price of the book in the time saved during the first day of its use.'

Andrews also prepared his *The Strength of materials* for publication in 1915. Chapman & Hall was the publisher of this and the related *Elementary strength of materials* that was issued a year later. At 604 pages, the first title was an extensive treatment of its subject and although 'specially intended for students', Andrews endeavoured to present it in 'sufficiently practical form to be of greater assistance in practical design than is the case with an ordinary text-book'. 'In this respect it is successful', declared *C&CE* (v.10, p.590); 'we can thoroughly recommend the volume as one possessing considerable merit.' The second book (of 216 pages) was really an abridged edition of the first and may be described (according to *C&CE* [v.11, p.502]) 'as one which contains a very good combination of theoretical and practical matter, well arranged and expressed, and it should prove extremely useful to all students in engineering and building'.

Whilst these three books were being prepared for the press, Andrews continued to write for *C&CE*. His *Problems in the theory of construction* series became quite a prominent feature in the journal:

- *Column bending moments in frame construction*, January 1915 (v.10, pp.10–16)
- *Column bending moments in frame construction*, February 1915 (v.10, pp.68–76)
- *A semicircular arch carrying its own weight only*, May 1915 (v.10, pp.230–232)
- *Influence lines for beam deflections*, August 1915 (v.10, pp.393–397)
- *Wind stresses in building frames*, November 1915 (v.10, pp.555–559)
- *Bending and twisting moments in beams curved in plan*, February 1916 (v.11, pp.66–72)
- *Wind stresses in building frames II*, January 1917 (v.12, pp.13–17)
- *The effect of varying moment of inertia upon continuous beams*, May 1917 (v.12, pp.253–258)

Other articles were self-contained:

- *Some modern methods of arch calculation*, March 1915 (v.10, pp.151–153; also in *Transactions & Notes*, 1918, pp.255–284)
- *Percentage reinforcement diagram for reinforced concrete beams*, August 1916 (v.11, pp.433–434)
- *A bending moment problem*, March 1917 (v.12, pp.130–131)

In 1917 Andrews was seconded to the Air Board for war service and appointed subsection director in charge of aeroplanes, instruments and accessories. Possibly his interest in wind stresses would have been of relevance to his new post. Certainly the war did not prevent him writing and throughout the two years of his service he supplied *C&CE* with seven successive parts of his *Detail design in reinforced concrete* (February, April, June, September 1917 and January, August, December 1918). His review of Taylor & Thompson's 1917 edition has already been commented on above. He also submitted two parts of *The strength of pillars* to his *Problems in the theory of construction* series (March 1918 and June 1919).

This period saw three new books: one on concrete and two of them issued by a publisher previously not encountered in our subject area, James Selwyn & Co. Both of these were commentaries on provisions of the LCC Act of 1909, for steel and concrete respectively:

- Andrews, E. S. and Cocking, W. C. *Tables of safe loads on steel pillars*. London: James Selwyn & Co., 1918
- Andrews, E. S. *Detail design in reinforced concrete: with special reference to the requirements of the reinforced concrete regulations of the L.C.C.* London: James Selwyn & Co., 1919
- Andrews, E. S. *Elements of graphic dynamics*. London: Chapman & Hall, 1919

On the conclusion of peace, Andrews left government service and embarked on a new career in private practice as a chartered patent agent and consulting engineer, specialising (as one might assume) in steel frame and reinforced concrete construction. The new practice operated from 201–206 Bank Chambers in High Holborn, London. Not long after commencing business, Andrews wrote his final solo book: *The structural engineer's pocket book*. This 'indispensable volume', as the *Structural Engineer* (v.12[6]) put it, subtitled *a handbook of formulae, figures and facts* was published by B. T. Batsford Ltd in 1921. He later wrote – jointly with A. E. Wynn – a simple guide entitled *Modern methods of concrete making*, as part of the newly inaugurated Concrete Series published by Concrete Publications Ltd. It explained the variables inherent in mixing concrete and publicised the recently accepted water/cement ratio method of controlling strength. The book sold out in six months and was reissued in 1928.

If the production of books had dried up, the 1920s witnessed a prodigious output of articles for *C&CE*, and after 1923, the *Structural Engineer*. *Problems in the theory of construction* continued until 1922:

- *Rigid reinforced concrete frames*, February 1920
- *Rigid reinforced concrete frames continued*, March 1920
- *Calculations for continuous beams with third-point loading*, November 1920
- *Calculations for continuous beams with third-point loading*, October 1921
- *Some problems in deflection of beams*, April 1922

Other articles during the 1920s included:

- *Some methods of securing impermeability in concrete*, June 1921 (v.16, pp.369–377)
- *Structural steelwork reinforced with concrete*, January 1923 (v.19, pp.5–8)
- *The design of steel-frame buildings*, February 1923 (v.19, pp.140–145)
- *Structural steelwork reinforced with concrete*, August 1923 (v.19, pp.505–508)
- *The strength of concrete*, April 1924 (v.19, pp239–245)
- *A study of loading tests on reinforced concrete structures*, June 1924 (v.19, pp.369–375)
- *A study of loading tests on reinforced concrete structures pt 2*, July 1924
- *Engineering failures*, March 1926 (v.21, pp.227–229)
- *Problems in the theory of construction*, May 1926 (v.21, pp.353–358)
- *Reinforced concrete construction in Italy*, October 1926 (v.21, pp.661–663)
- *High tension reinforcement*, November 1926 (v.21, pp.723–726)
- *Concentrated loads on reinforced slabs*, May 1927 (v.22, pp.297–302)
- *Concrete retaining walls*, October 1930 (v.25, pp.572–573)

A similar pattern appeared in the *Structural Engineer*. A 12-part series entitled *The essentials of reinforced concrete design* ran from March 1924 to April 1928. Other articles, with a greater emphasis on steelwork, included:

- *Structural steelwork reinforced with concrete*, July 1924 (v.2, pp.290–291)
- *The strength of steel frame buildings*, December 1924 (v.2, p.383)
- *The strength of filler joint floors*, February 1925 (v.3, pp.38–41)
- *Structural engineering and the colleges*, October 1925 (v.3, pp.347–348)
- *Production of structural steel sections*, May 1926 (v.4, pp.146–153)
- *Characteristic points and continuous beams*, November 1928 (v.6, pp.339–343)
- *The strength of compression members*, March 1931 (v.9, pp.95–99)
- *Steelwork for buildings*, June 1932 (v.10, pp.265–269)
- *The actual strength of steel I-beams*, March 1934 (v.12, pp.118–129)

The 1920s were a time of increasing involvement with the Institution of Structural Engineers. His papers were presented at general meetings – attracting on many occasion, 'a well-filled lecture room and a keen discussion' (*SE*, v.12) – and he was as regular a reviewer in the *Structural Engineer* as he was for *C&CE*. He was a member of council from 1919 to 1948 and of all the various standing committees at different times. He was also a representative of the Institution on the Steel Structures and Reinforced Concrete Structures committees of the DSIR. He was awarded the institution's Silver Medal, and in 1931, being a long-standing member of the board of examiners, he instituted the Andrews Prize – awarded twice yearly to the most successful candidate in the examinations for associate membership. He was elected vice-president in 1926 and became president of the institution in 1934. He appears to have been a popular choice: 'The new president's genial personality and kindly courtesy, which have already endeared him to many members of the Institutions, will do much to increase the number of his friends and admirers.' Likewise he was vice-president of the International Association of Bridge and Structural Engineers, and president of the British section from 1930.

Thereafter his writing tended to matters of policy rather than technology. *The trend of structural engineering* (*SE* v.12, pp.410–414) reviewed recent developments in steel and concrete and argued for official encouragement of new methods and materials. His presidential address was published in the *Structural Engineer*: *Looking forward* in January 1935 (v.13, pp.35–39) and *Inexpensive research work* in June 1939 (v.17, pp.311–324).

In private practice Andrews was responsible for the engineering input to the Reuters Building in Fleet Street and the reconstruction of Portobello Power Station. He was also consulting engineer to the Prison Commissioners. During the Second World War he advised several government departments on aspects of war work and in 1946 founded the firm of Andrews, Kent & Stone. He retired in

October 1954, just two years before his death. Andrews (like Oscar Faber) died in 1956. His obituary, which was published in the *Structural Engineer* (1956, p.428), praised him for his achievements and pointed to a quite different personality from that of his contemporary: 'He did much to advance the science and art of structural engineering, and the Institution is indebted to him in a multitude of ways. His genial kindly personality will be greatly missed by all those who knew him.'

Andrews, E. S. *Elementary principles of reinforced concrete construction: a text-book for the use of students, engineers, architects and builders*. London: Scott, Greenwood & Son, 1912 [C&CA 1959, ICE. Copy: TCS. Rev: *C&CE* v.7, p.391; *Trans C.I.* v.5, p.xii]. Second edition, 1918 [BL], and New York: Van Nostrand [Rev: *C-CA*, v.2, p.248]. Third edition, 1924 [BL, ICE, Yrbk 1934]

Andrews, E. S. *Regulations of the London County Council relating to reinforced concrete (London Building Act, 1909, Amendment) and steel framed buildings (L.C.C. General Powers Act, 1909): a handy guide, containing the full text, with explanatory notes, diagrams and worked examples*. London: B. T. Batsford Ltd, 1915 [ICE]. Second edition, 1924 [BL, Yrbk 1934. Copy: TCS. Rev: *C&CE* v.11, p.267]

Andrews, E. S. *Detail design in reinforced concrete: with special reference to the requirements of the reinforced concrete regulations of the L.C.C.* London: Pitman, 1921 [BL, ICE, Yrbk 1934]

Andrews, E. S. and Wynn, A. E. *Modern methods of concrete making*. Second edition. London: Concrete Publications Ltd, 1928 [Yrbk 1934]

1912/USA

Hool, Professor George Albert

SB

George A. Hool was one of the major textbook authorities of the period, in terms both of the number of titles written or edited, and the number of copies printed and distributed. He follows in the tradition of Turneaure & Maurer and Taylor & Thompson, whom he acknowledged as a constant source of reference. As an educator as well as practitioner across the spectrum of structural engineering, he provides a parallel to his contemporaries E. S. Andrews and Oscar Faber in Great Britain. Indeed both British authors reviewed Hool's work favourably in the professional journals.

Hool achieved his ubiquity largely from his position at the University of Wisconsin, where (unlike Turneaure and Maurer who worked at the school of mechanics and engineering), he was associated with the University Extension Division. His role at the time his first book on concrete was published in 1912, was associate professor of structural engineering, and as such was also a contributor to the 'Consultation' column in the journal *Cement Age*. Under the direction of the dean at this time (Louis E. Reber), the division was actively developing a series of industrial and engineering textbooks. In 1912 Hool contributed his *Elements of structures* (McGraw-Hill) and the first of his three-volume set, *Reinforced concrete construction*.

Volume I was sub-titled (appropriately for the first) *Fundamental principles* and was published for the university by McGraw-Hill. Indeed, Hool's entire output was to be issued by this publisher – the principal publishing house for books on concrete in America. Two companion volumes followed: *Retaining walls and buildings* in 1913; and *Bridges and culverts* in 1916. A second edition of Volume I came out in 1917 and ran to at least ten impressions, with a total issue of 30,500 copies.

These books were designed to provide content for courses offered by the Extension Division, and were written with the needs of correspondence students in mind – doubtless a consideration in their commercial success. The books were certainly well received. *Concrete-Cement Age* pronounced that: 'Mr. Hool has succeeded in producing a work which is clear, logical and concise. The subject is taken up from an initial assumption and carried forward step by step, directly to a conclusion.' He had 'evidently availed himself of many sources of information … analysed each item … and concentrated them into a thoroughly worthwhile book' (v.2 p.147). Reviews in *C&CE* were favourable too, while accepting that American notation was inapplicable to the British market. The content was considered comprehensive, the treatment thorough and sufficiently simple for an introductory work. Volumes II and III were much larger and more advanced in their specialist fields and the reviewers observed that they had found a further audience among practising engineers. The quality of diagrams and use of photographs was praised in each case.

Although Hool had given his address as Madison, Wisconsin, in 1917 he no longer claimed an affiliation with the university when he brought out his next book – *Concrete engineers' handbook* – in 1918. This was a collaborative work with Nathan C. Johnson (a consulting engineer from New York), assisted by Samuel C. Hollister [1918] and a further five contributors: Harvey Whipple [1915]; Adelbert P. Mills; Water S. Edge; A. G. Hillberg; and Leslie H. Allen (who had contributed to Volume II above). Like Johnson, Hool described himself as 'consulting engineer'. This again was a great success commercially, with no less than 21 impressions of the first edition and a total issue of 44,500 copies by 1945.

The book was intended as a concise working manual for the engineer, supplying tables, diagrams and other data for practical application, and 'practically no subject matter' that would be suitable for students. Consequently, little was drawn from Hool's earlier books. It was reviewed in Britain by no less an authority than Oscar Faber (*C&CE* v.16, p.746), who praised the excellent diagrams by Clifford Ives and considered it 'excellently got up', achieving 'what it set out to do as well as is possible'. He 'strongly' recommended the book as long as it was understood it was simply a labour-saving device that for British readers would possibly be of limited assistance given that it addressed American regulations.

Back at the university as professor of structural engineering, his collaboration with Johnson continued with the publication of a handsome, expensive, two-volume set entitled *Handbook of building construction*. Although concrete represented only one aspect, it might be interesting to hear a reviewer's opinion of the work as a whole:

> These two volumes bound in flexible leather, clearly printed and well illustrated, predispose the reader to a satisfactory judgement, and after a closer

> acquaintance, one's only regret is that they are based on United States practice and are not universally applicable to the requirements of the British reader. As a handbook of building construction, the work is most comprehensive, and in some sections exhaustive, so that one envies his American confrere who can have such a mine of information by his side.

1921 saw a partnership with another collaborator – C. S. Whitney [1921] – with whom he wrote *Concrete designers' manual*. This warranted a second edition five years later. In the intervening years Hool joined William S. Kinne as joint editor-in-chief of the Structural Engineers' Handbook Library. Of six volumes in this series, only one specifically addressed concrete: *Reinforced concrete and masonry structures* (1924). The other titles (starting with *Foundations*) comprised:

- *Foundations, abutments and footings*. (1923)
- *Stresses in framed structures*. (1923. second ed, 1942)
- *Structural members and connections*. (1923. second ed, 1943)
- *Movable and long-span bridges*. (1923. second ed, 1943)
- *Steel and timber structures*. (1924, second ed, 1942)

Individual titles were reviewed in the *Structural Engineer* and firmly recommended, not least because 'the publishers' name is a sufficient guarantee that the work has been well produced in every way'. E. S. Andrews went much further in commenting on *Stresses in framed structures*: 'the style is concise and condensed, the illustrations are excellent, and the whole book is so good that we think that it is sure to find a regular place upon the shelf which engineers reserve in their bookcases for the books to which they will often turn' (*SE* v.2, p.62).

At the date *Concrete designers' manual* was reissued (in 1926), Hool entered his final partnership in print. This time it was with Harry E. Pulver; a colleague who had also contributed to the University of Wisconsin Extension Division's series (*Materials of construction*). Hool was back with the division, now as professor of structural engineering, while Pulver had taken Hool's former role as associate professor. Together they wrote *Concrete practice: a textbook for vocational and trade schools* – the title of which is probably sufficiently explicit to indicate its intention. The book provided a resource from which instructors could select assignments for students on vocational course to undertake, varied according to suitability or the availability of equipment.

Having gone full circle from writing student textbooks to editing compilation works and back again, Hool's oeuvre was complete. His work lived on through frequent reprints supporting continuing educational endeavours into the 1940s.

Hool, G. A. *Reinforced concrete construction*. Vol. I: *fundamental principles*. (Engineering education series). New York, McGraw-Hill, 1912 [Rev: *C-CA*, v.2, p.147; *C&CE* v.7, p.788; *Trans*, v.5, pt.1, p.xv]. Second edition, 1917 [BL, C&CA 1959, ICE. Copy: TCS]. Third edition, 1927 [BL, Yrbk 1934]

Hool, G. A. *Reinforced concrete construction*. Vol. II: *retaining walls and buildings*. New York, McGraw-Hill, 1913 [BL, C&CA 1959, ICE. Rev: *C-CA* v.4, p.139; *C&CE* v.9, p.210]. Second edition, 1927 [BL, Yrbk 1934]

Hool, G. A. *Reinforced concrete construction*. Vol. III: *bridges and culverts*. New York, McGraw-Hill, 1916 [C&CA 1959, ICE. Rev: *C&CE* v.11, p.267]. Second edition, 1928 [BL, IStructE, Yrbk 1934]

Hool, G. A. and Johnson, N. C. *Concrete engineers' handbook: data for the design and construction of plain and reinforced concrete structures*. New York: McGraw-Hill, 1918 [C&CA 1959, IStructE, Yrbk 1934. Copy: TCS]

Hool, G. A. and Whitney, C. S. *Concrete designers' manual*. New York: McGraw-Hill, 1921 [Rev: *C&CE* v.16, p.746]. Second edition, 1926 [BL, C&CA 1959, Yrbk 1934]

Hool, G. A. and Kinne, W. S. (eds). *Reinforced concrete and masonry structures*. New York: McGraw-Hill, 1924 [BL, ICE, Yrbk 1934]

Hool, G. A. and Pulver, H. E. *Concrete practice: a textbook for vocational and trade schools*. New York: McGraw-Hill, 1926 [C&CA 1959, ICE, TCS, Yrbk 1934. Copy: TCS]

1912/USA

Ransome, Ernest Leslie (1844–1917)

AASCE, MWSE, MAIA, MRSA

Ernest Leslie Ransome Courtesy of the American Concrete Institute

Like Thaddeus Hyatt of an earlier generation, Ransome was one of the great 19th Century pioneers of reinforced concrete construction, writing at the end of his career. He was Hyatt's corollary – an Englishman settled in America, whereas the American Hyatt wrote his work in London.

Ernest was born in 1844 into a family long established in Ipswich, that ran the famous iron foundry and machinery business founded by his great grandfather. His father, Frederick, was interested in concrete and in diversifying the business in that direction. He developed the first rotary kiln for cement manufacture and took out a patent for Ransome's Stone – a form of concrete masonry. Ernest came to the USA in 1869 with a view to exploiting his father's patent. On becoming superintendent of the Pacific Stone Company in 1870, he settled his family in San Francisco, California. He soon set up a factory to manufacture concrete blocks and after a period of experimentation and discovery, pioneered a series of technological innovations in reinforcement, mixing equipment and construction systems. He established several firms to implement these developments. In turn these were: Ransome Concrete Co.; Ransome Smith & Co.; Ransome Concrete Machinery Co.; and Ransome Construction Co.

His first patent (US 694,580) was awarded in 1884 and became the cornerstone of the eventual 'Ransome System'. He developed a special machine to twist square-section iron bars (up to 2" in diameter), which he found conferred a greater tensile resistance than conventional round bars. His patent soon found practical expression. In 1884 he built the Arctic Oil Works in San Fransisco, then experimented in 1886/7 with two small reinforced concrete bridges in the Golden Gate Park. Of these the Alvord Lake Bridge still bears a plaque as a National Historic Civil Engineering Landmark. Other structures followed:

- 1886/9 – Greystone Cellars, St Helena, CA
- 1893 – Art Museum, Stanford University
- 1897 – Pacific Coast Borax factory, Bayonne, NJ

In 1902 his final patent allowed full-scale factory frame construction using the combination of exposed frame and curtain wall, soon to be 'the workhorse of the new industrial architecture'.

This new approach to industrial architecture was accepted by the United Shoe Machinery Corporation in preference to brick for its new factory at Beverly, near Boston. Between 1903 and 1906 a huge industrial complex was built, described by Reyner Banham in his *A Concrete Atlantis* (1986) as 'that ultimate masterpiece of Ransome's declining years'.

Along with Albert Kahn's Packard factory in Detroit (which followed immediately after), this was the first fully developed factory built in reinforced concrete, and until the 1930s it was the largest reinforced concrete structure anywhere in the world. It was extended in 1911 in a development that was significant because of its extensive use of pre-cast concrete sections. It was the largest use of pre-cast elements anywhere – a point highlighted at the time by the trade magazine *Popular Mechanics*. The construction work was described in *Cement Age* in an illustrated article entitled *New developments in unit work using a structural concrete frame and poured slab* (v.12, p.129). Many years later (in 1997), the architectural critic Ada Huxtable wrote in the *Wall Street Journal*:

> The Shoe was, and is, the single most important, and generally unrecognised, concrete landmark in this country, pre-dating the Detroit auto factories by the engineer Albert Kahn that have been widely credited with the structural and engineering innovations that actually appeared here first ... Of undeniable interest for their technical innovation, these buildings are also remarkable for their impressive and pleasant proportions, direct expression of structure, and comparative freedom from conventional decorative details that were to 'enhance' even the best of the later industrial architecture ... As engineering and design, Ernest Ransome's work deserves a prominent place in the story of American architectural advance.

Ransome's system coincided with the great expansion of American industry and so became widespread throughout the USA. To crown his commercial success he and his manager Alexis Saurbrey (below), wrote a book in 1912. *Reinforced concrete buildings* described his patented system and was suffused with personal reminiscences. By this date Ransome – based in Plainfield, New Jersey, with a consulting practice in New York – was getting on in life and he died not many years later in 1917.

Ransome, E. and Saurbrey, A. *Reinforced concrete buildings: a treatise on the history, patents, design and erection of the principal parts entering into a modern reinforced concrete building*. New York: McGraw-Hill Book Co., 1912 [Copy: TCS]

1912/USA

Saurbrey, Alexis (1886–1967)

Assoc Member ASCE; Member, Dansk Ingenior Forening

Danish-born manager and chief engineer of The Ransome Engineering Co., Alexis Saurbrey was the co-author of *Reinforced concrete buildings* (above) with his employer, E. L. Ransome. In October 1912 he wrote to *Concrete-Cement Age* to publicise a court hearing that had upheld an appeal by the Ransome Concrete Co. against an infringement of Ransome's patent of 1902. The offending party was the German-American Button Company of Rochester, New York. We might surmise that the lawsuit may have had a bearing on the book's publication as a vehicle for corporate justification and promotion.

1913/UK

Ball, James Dudley Ward

AMICE, ACGI

Ball had already demonstrated an interest in railway engineering when he wrote his book in 1913: his article *Distribution of locomotive axle loads in relation to bridges* had been published in January 1909 in the *Proceedings* of the ICE.

With *Reinforced concrete railway structures*, Ball addressed a use for the material that appeared to observers at the time to be accepted with reluctance by engineers. *C&CE* (v.9, pp.69–70) noted that: 'Mr. Ball gives reasons for this cautious attitude, and he wisely advocates its use in the best possible way – not be preaching reinforced concrete in season and out of season, but by showing where it is of the greatest service to the railway engineer, and, on the other hand, where its older rivals still hold their own.' The reviewer went on to comment that: 'after reading his book one is struck with the variety of ways in which reinforced concrete can be used in railway work with the greatest convenience and with true economy.' Indeed, the railways have since proved to be a major market for concrete. The book was well illustrated and quoted comparative prices and excerpts from specifications, though as the reviewer disapprovingly noted, failed to adopt the Concrete Institute's standard notation.

C&CE (v.35, pp.479–483) published Ball's *Simplification of design diagrams* in October 1940, in what may be considered a coda to his career in concrete.

Ball, J. D. W. *Reinforced concrete railway structures*. (Glasgow Text Books of Civil Engineering). London: Constable & Co., 1913 [BL, C&CA 1959, ICE, Yrbk 1934. Rev: *C&CE* v.9, pp.69–70)

1913/USA

Cochran, Jerome

BS, CE, MCE

An engineer from Detroit, Michigan, Jerome Cochran had already written *A treatise on cement specifications* (Van Nostrand, 1912) and *A treatise on the inspection of concrete construction* (1913) by the time he compiled his *General specifications for concrete and reinforced concrete* later in the year. This work, which he hoped would be 'of increasing influence toward good concrete construction', drew on the standardising work of the preceding decade, commencing with the American Railway Engineering Association in 1903. Cochran examined the committee recommendations from ASTM, ASCE and the forerunners of ACI and PCA, most of which collaborated in the Joint Committee on Concrete and Reinforced Concrete. His work in this area had been trailed in an article for the *Cornell Civil Engineer* (November 1911) entitled *A comparison of seven building code requirements for the design of reinforced concrete*. It was reprinted in *Cement Age* (v.14, p.297).

Besides addressing the subject of concrete, Cochran also wrote *Principles of municipal refuse collection and disposal*; *Inspection of sewer construction* and *Principles of camp sanitation*.

Cochran, J. *A treatise on the inspection of concrete construction*. Chicago: Myron C. Clark Publishing Co., 1913 [PCA]

Cochran, J. *General specifications for concrete and reinforced concrete: including finishing and waterproofing*. New York: D. Van Nostrand Co., 1913 [Copy: Internet Archive]

1913/UK

Coleman, T. E.

Based on a series of articles from *Building News*, Coleman aimed his *Estimating for reinforced concrete work* of 1913 at the engineer and architect, as a guide for assisting the preparation of estimates and bills of quantities. *C&CE* was pleased to describe it as 'a very useful little book' and 'well worth the price', though did note that the coverage of the various proprietary systems of reinforced concrete was incomplete. The *Transactions* of the Concrete Institute drew attention to a distinguishing new feature:

> The items and prices are based on the average cost of materials and labour in the London district. A new feature that has been introduced into this price-book is the recording in brackets the prime costs basis adopted for the respective materials used in each class of concrete immediately after the general description of the concrete, so that any adjustments which may be required to suit varying local conditions can be easily made.

Knowing the niche for price-books occupied for so long by the publisher E&FN Spon, it causes little surprise to note that Coleman's later volume of 1924 – *The civil engineer's cost book* – was published by that house.

Coleman, T. E. *Estimating for reinforced concrete work*. London: Batsford, 1912 [IStructE. Rev: *C&CE*, v.8, 1913; *Trans CI*, v.5, 1913]

Coleman, T. E. *The civil engineer's cost book*. London: Spon, 1924 [IStructE]

1913/USA

Eddy, Professor Henry Turner (1844–1921)

CE, PHD, DSC.

Professor Henry Turner Eddy Courtesy of the University of Cincinnati

Born in Massachusetts on 9 June 1944, Henry Turner was raised on the East Coast. He attended university at Yale, and graduated with a BA in 1867. For the following year (1867– 1868), he became an instructor in engineering fieldwork at Sheffield Scientific School, where he gained an MA and his first doctorate. He moved to the University of Tennessee where he taught mathematics and Latin in the 1868–69 session and then transferred to Cornell until 1873, where he added civil engineering to his teaching of maths. During these years he was awarded his CE and a PhD in mathematics, and he married Sebella Taylor. In 1873–1874 he taught maths at Princeton and then headed west for Cincinnati.

Arriving in Ohio he became one of the first three professors employed by the new University of Cincinnati, where he taught maths, astronomy and civil engineering. Very shortly after arriving he was appointed dean of faculty and served two terms for the years 1874–1877 and 1884–1889. Between these terms he spent a sabbatical year (the 1879–1880 session) in Berlin and Paris. In 1890 'Eddy's Method' – for investigating the maximum shear at a section due to the distribution of a travelling load – was published in the *Transactions* (v.22) of the ASCE. Acting-president that year, he retired from the university to assume the presidency of Rose Polytechnic Institute for the period 1891–1894.

After 20 years of stability since the restless days of his youth, Eddy accepted another long-term post at the University of Minnesota, where he was to serve as professor from 1896 to retirement in 1912. He was professor of mathematics and mechanics in the College of Engineering, and dean Emeritus of the Graduate School when he finally left the university.

Based in Minnesota and with a professional interest in engineering, Eddy encountered the local innovator C. A. P. Turner [1909] and spent much of his retirement collaborating with him on developing the mushroom head technique for flat slab design. In 1913 he wrote *The theory of the flexure and strength of rectangular flat plates applied to reinforced concrete floor slabs*, for which Turner supplied a standard specification as the appendix. The book approached flat plate theory on 'broad general principles, while having careful regard for the experimental results revealed by numerous tests that have been made on the floors of buildings already constructed'. However it may have been received in America, it was favourably reviewed in Great Britain: 'The book is well written and very interesting and instructive, and every endeavour has been made by the author to deal with the subject in a straight-forward and simple manner.' (*C&CE*, v.8, p.589).

The collaboration deepened and Eddy and Turner jointly wrote their *Concrete-steel construction* a year later. Confusingly published under the same title as Turner's book of 1909, it was differentiated by a subtitle: *Part 1 – building: a treatise upon the elementary principles of design and execution of reinforced concrete work in building*. *C&CE*'s reviewer approved:

> There is no doubt that a large amount of time and energy has been spent by those responsible for this volume in the preparation of particulars and in investigation, in order to put forward evident in support of their contentions, and the matter presented is both theoretical and practical in character. Several tests are described and illustrated,

and many buildings actually executed in reinforced concrete are dealt with (Rev: *C&CE*, v.10, pp.106–7).

Although much of the text concentrated on the mushroom system, it contained plenty of information of more widely applicable use and the review concluded that 'the volume is well written and arranged, and the matter is expressed in a manner which renders it easy to understand and interesting to read'. Doubtless the market thought so too, because in spite of Turner's defeats in court, a second edition was issued in 1919.

After a long life in academia, and participation in the activities of the American Mathematical, Philosophical and Physical Societies, Professor Eddy died in 1921. Besides the unquantifiable impact of his teaching, his legacy lies in a retirement project – helping to bring the new flat slab technology to professional notice.

Eddy, H. T. *The theory of the flexure and strength of rectangular flat plates applied to reinforced concrete floor slabs*. Minneapolis: Rogers & Co., 1913 [PCA. Rev: *C&CE*, v.8, p.589]

Eddy, H. T. & Turner, C. A. P. *Concrete-steel construction. Part 1 – building: a treatise upon the elementary principles of design and execution of reinforced concrete work in building*. Minneapolis: Heywood Manufacturing Co., 1914 [Rev: *C&CE*, v.10, pp.106–7]. Second edition, 1919

1913/USA

Foster, Wolcott C.

The fourth edition of this 1891 work is included for 1913 because after revision, a telling addition to the title – *and their concrete substitutes* – indicates the arrival of concrete into this field of engineering. The book provides an American parallel to the British practice described by Ball.

Foster, W. C. *A treatise on wooden trestle bridges and their concrete substitutes according to the present practice of American railroads*. Fourth edition. New York: Wiley, and London; Chapman & Hall, 1913 [ICE. Rev: *C&CE* v.8, p.883]

1913/UK

Geen, Albert Burnard (1882–1966)

AMICE, MSE, MCI, CONSULTING ENGINEER

Albert Burnard Geen
Courtesy of the IStructE

A civil engineer of nearly 60 years standing, Geen specialised and 'acquired a remarkable reputation' in foundations and structural engineering.

Born on 30 July 1882, Burnard (as he was known), was educated at Thanet College, Margate, and Whitgift Grammar School, Croydon. Specialising for his career, he then studied at the Crystal Palace School of Practical Engineering from 1899 to 1902.

His early work was as assistant civil engineer to Smith's Dock Co., Ltd, in North Shields and in 1904 as chief designer to a large steelwork firm in Yorkshire. There he designed a 105-ft span-curved girder roof for the Edinburgh and Leigh Gas Company. He then took a position on the staff of Sir Benjamin Baker, engaged on the design of Walney Bridge and other bridge work. In 1905 he secured a position with the bridges department of the London County Council, and under W. C. Copperthwaite, he worked on tramway bridges, steamboat piers and the Holborn-Strand subway. Like Oscar Faber a couple of years later, he became chief assistant engineer to the Patent Indented Steel Bar Co. and the three years from 1907 gave him valuable experience designing in reinforced concrete. He undertook a wide range of structures: reservoirs; water-towers; arch and girder bridges; jetties; and retaining walls. Ten years ahead of Faber (in 1910), he entered private practice as consulting engineer, specialising in foundations. One of his earliest commissions was appointment as consulting engineer to the Festival of Empire Exhibition at Crystal Palace in 1911. Amongst other features of the exhibition, two 'gargantuan' reinforced concrete staircases were built to his designs.

Geen staircase at the Festival of Empire Exhibition at Crystal Palace in 1911
Courtesy of The Concrete Society

In 1912 Geen addressed the problem of calculating bending moments and shearing forces in concrete beams, being unaware of any such method in publication. In this he was pursuing in parallel a subject more fundamentally researched by Faber and Bowie, though from a different perspective. His *Continuous beams in reinforced concrete* appeared in 1913 (published by Chapman & Hall), and contained diagrams and tables to give more definite and practical guidance to the engineer than provided by the existing rules of thumb, or the more cautious statements of the RIBA committee that had reported in 1911. *C&CE* (v.8, p.512) welcomed the author's attempt to deal with the contemporary vagueness on this subject in a 'thorough and comprehensive manner', and applauded the clarity of his diagrams. The only adverse comment in its review was on the brevity of explanatory notes, but otherwise described it as 'an excellent book of reference'.

Shortly before war broke out, Geen reported on a scheme for ship repairing and construction of a ship-building yard on the Canadian Great Lakes. After the commencement of hostilities he was appointed to the Metropolitan Munitions Committee. In 1918 he transferred to the Mechanical Warfare Department (responsible for tanks), where he handled the liquidation of contracts amounting in total to £10m.

The war over, he returned to practice and following in the steps of Henry Adams [1911] and Noble Twelvetrees [1905], was elected president of the Society of Engineers for 1920. He was one of the youngest to hold that office. Apart from a visit to the USA in 1924 (when he was aged 42), we see Geen active in the design and specification of a succession of prestigious major projects including Southampton Civic Centre, the extension of the Bodleian Library, Cambridge University's new library and Cairo, Liverpool and Guildford cathedrals. His work with the architect Edward Maufe on Guildford Cathedral was described in an article of that title in *C&CE* (v.47, 1952, pp.165–183). It followed an earlier partnership with Maufe on Saint Thomas the Apostle – a new church at Hanwell – that was completed in 1932. More routine work was in the design of foundations for gasworks, factory buildings and schools. One of the better-known clients was Ampleforth College, for which he undertook the foundations of 'The Lower Building'. He was described at the time as 'Burnard Geen, M.Inst.C.E. of Westminster'.

In the 1940s he served the war effort for a second time. In the Second World War he worked for the Ministry of Aircraft Production, during which one notable episode – saving a factory from conflagration – was described in the ICE's *Journal* (*Fire under a factory*, v.23, p.91) in December 1944.

Following his earlier ecclesiastical work at Hanwell and Guildford, the early 1950s saw Geen as consulting engineer for SS Mary & Joseph Roman Catholic Church in Upper North Street, London, between 1951 and 1954. In 1952 he was appointed consulting engineer to the City Corporation of London for the reconstruction of the Guildhall, including underpinning the north wall, which he recounted in *Guildhall: provision of new foundations to the north wall by underpinning and other means* in *Proceedings of the ICE* (p.201) and in *Concrete Quarterly* (No.21, p.5), March 1954. By now of advancing years, he entered into association with Messrs W. S. Atkins & Partners of Westminster in 1958. He died in 1966 after a 60-year career in which he had completed 600 commissions.

At the time of his book's publication he was an associate member of the ICE (elected in 1908), but became a member in 1921 and a 'Chadwick' lecturer in 1937. He was also a member of the Institution of Structural Engineers and a fellow (as well as president for a year) of the Society of Engineers.

An obituary (published in April 1967) in the *Proceedings* of the ICE (v.36), noted that he was offered the OBE. It also recalled that he 'learned from his wife to paint landscape in watercolour up to exhibition standard' and described him as 'generous by nature and an amusing speaker'.

Geen, B. *Continuous beams in reinforced concrete*. London: Chapman & Hall, 1913 [C&CA 1959, ICE, Yrbk 1934. Copy: TCS. Rev: *C&CE* v.8, p.512, *Trans & Notes* v.5, pt.1, xiv]

1913/UK

Jones, Bernard E.

Bernard Jones – while author of the introduction and historical section of his book *Cassell's Reinforced Concrete* – was first and foremost its editor, supported by a staff of specialist writers. In this function he was like the American G. A.

Hool [1905] and his collaborators [1912] and (closer to home) Noble Twelvetrees in his revision of Rivington's *Notes on Building Construction* (London: Longmans, 1916). Here Jones was equipped for the task by his role as editor of *Building World* – the 'illustrated weekly trade journal' from Cassell & Co. (issued 1895–1920), which was aimed at architects, builders, carpenters, bricklayers, plasterers and the like. Cassell & Co. initially published the book in 1913, while the second (revised) edition of 1920 was issued by the Waverley Book Co. in a handsome gold-blocked binding.

Besides Jones, the various contributors included: **Frank B. Gatehouse**, FCS (Portland cement); Thomas Potter, MCI (concrete); A. B. Searle (concrete, cement, waterproofing); **E. L. Rhead**, MSc, Tech. FIC (steel); Albert Lakeman, MCI (theory); **V. Sussex Hyde**, MCI (practical erection); **F. Charles Barker**, MSA (examples of forces); **T. Elson Hardy** (architectural treatment); **A. Seymour Jennings**, FIBD (painting concrete); and **W. H. Brown**, FSI (estimating, quantities). Of these, we will consider Lakeman and Searle separately as authors of books in their own name.

The purpose of the book was 'to provide a practical, simply-worded guide to construction in reinforced concrete, as distinct from the theoretical and often obscure treatises that have appeared in large numbers during the past few years', though the contribution by Lakeman did address basic theory. Jones expressed the hope that it would to be regarded as both a textbook for teaching and a work of reference. He drew heavily on *Concrete & Constructional Engineering* for his own elucidation as well as the loan of 'a large number of blocks' for reproducing images. In keeping with Cassell's tradition of lavishly illustrated non-fiction, the illustrations in this volume ran to 171 photographs and 500 diagrams.

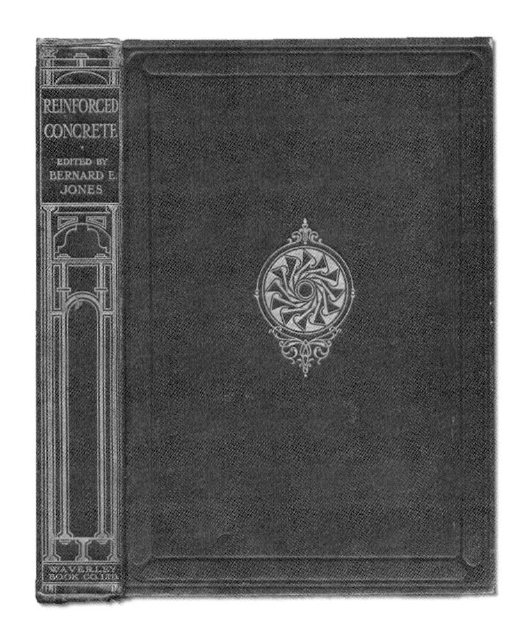

Gold-blocked cover of Cassell's Reinforced Concrete *(Second edition) edited by Jones*
Courtesy of The Concrete Society

C&CE may have had an interest in the outcome, and predictably approved of the new book when it appeared on the market (v.8, pp.212–213):

> This new volume certainly deals with the subject of reinforced concrete in a manner unlike any other book that has been published in the country up to the present, as great attention has been paid to the practical side of the

> subject ... The fact that various portions have been contributed by writers possessing a special knowledge of that portion of the subject with which they had to deal should naturally tend to produce a reliable volume, and we feel that the value of this book will be due in a great measure to this fact ... It should form an excellent book of reference for architects, and prove very useful to students preparing for examinations, and would, we consider, be a very text-book for class purposes.

In a similar way, whilst editor of *Work* (an illustrated weekly on the subject of handicrafts and mechanics), Jones produced another book for Waverley: the three-volume *Amateur mechanic: a practical guide for the handyman*.

Jones, B. E. (ed) and Lakeman, A. *Cassell's Reinforced Concrete: a complete treatise on the practice and theory of modern construction in concrete-steel*. London and New York: Cassell & Co, 1913 [ICE, PCA. Rev: *C&CE* v.8, pp.212–213]. Second edition. London: Waverley, 1920 [BL, C&CA 1959, Yrbk 1934. Copy: TCS]

1913/UK

Lakeman, Albert (1883–1969)

MSA, LRIBA, MISTRUCTE

We have already encountered Lakeman as a reviewer of Faber's book in 1912, but his name first appears on a title page in 1913 as assisting Bernard Jones (above). He is described as 'Honours medallist in building construction' and was asked by Jones to provide the theoretical section of an otherwise practical volume. In *C&CE* (v.8, pp.212–213) the reviewer considered that: 'the theoretical portion of the subject is dealt with in a very simple manner, every definition and symbol being explained as fully as possible, in order that the reader may be quite conversant with the elementary principles. The reader should experience no difficulty in following the various formulae and calculations given, and more especially as the construction of the former are clearly shown step by step, and no knowledge of advanced mathematics is necessary.' Such an approach was repeated years later, as we shall see, when Lakeman wrote his *Elementary guide* in 1928.

Lakeman was born in Chertsey, but his youth was spent in Somerset and he married in Exeter in 1908. As 'honours medallist', Lakeman clearly was successful, but his career developed as a promoter of concrete rather than as a particularly original thinker or as a researcher or educator. He became a regular contributor to *C&CE* throughout the 1910s and 1920s, initially reviewing new concrete buildings (often prefixed *Reinforced concrete at/in...*), then turning to articles of practical guidance after the First World War. Lakeman's reports were a regular feature of the journal before and during the early years of the war:

- 1911 (v.6)
  Free Church, Hampstead (September); Royal Liver Building, Liverpool (October).
- 1912 (v.7)
  Messrs R. Clay & Sons, Ltd, Brunswick Street (May); Messrs William Whitley Ltd (July); King's College hospital (September); silos, Immingham (November);

Messrs Peek, Frean and Co., factory extension (December).

- 1913 (v.8)
  Waterford Bridge (March); South Lambeth goods depot (June); improvements in the Port of London (October); dome of public library, Melbourne (December).
- 1914 (v.9)
  Zoological Gardens, (January); HM new stationery office (June); Cross Hill service reservoir (August); prison buildings (September); *Theatre des Champs Elysees*, Paris (October); Wallace-Scott Tailoring Institute, Glasgow (December).
- 1915 (v.10)
  Racecourse stand, Cheltenham (January); Bristol General Hospital (April); Empire and India Houses, Kingsway (June); Crossley Lads' Club, Manchester (October).
- 1916 (v.11)
  Canterbury Theatre (February); Stockport Grammar School (March).

At the end of the First World War the pattern of reporting significant projects was briefly resumed – with articles on a factory in Birmingham (October 1918); a Midland munitions factory (May 1919); and a canteen (November 1919) – but by then a newer preoccupation was absorbing the construction industry's attention. Along with upgrading the road network for the motor age, there was a significant national interest in expanding the house building programme – providing 'homes fit for heroes'. The prospects of concrete to meet this demand received impetus from America where concrete housing (see Hering [1912], and the architect Irving Gill) had been promoted in the *ante bellum* years. Concrete Publications Ltd (publisher of *C&CE)* took the step of issuing a book on this subject – edited by Lakeman – as the first volume in what would grow to be an extensive booklist on concrete. *Concrete cottages* was intended as 'propaganda' (to use the idiom of the day), initially published by CPL for the Concrete Utilities Bureau. The promotional material declared:

> This book has been prepared and published for the purpose of giving useful and reliable information as to the use of concrete for cottages and other small buildings. The problem of the cheap and well-built cottage is daily becoming more pressing, while the use of concrete is becoming more extensive, and there is no doubt that considerable scope exists for the use of this material in cottage work if its possibilities are thoroughly understood.

The book appears to have been very successful, to judge from the press comments. *C&CE* (v.13, p.217) printed a selection of testimonials from local and national newspapers, trade magazines and professional journals – even the *British Medical Journal*. A second edition, with modified subtitle was published in 1924.

Lakeman returned to writing articles for *C&CE*. In the 1920s however, they became more instructional:

- *Concrete stone manufacture: a review*, August/September 1919 (v.14, pp.529–355)
- *Economical house building: a review*, December 1919 (v.14, pp.711–715)
- *Erecting tower for concrete works*, November 1920 (v.15, pp.738–731)
- *American methods in concrete work*, April 1923 (v.18, pp.209–223)
- *The foreman's guide to concrete*, (a series of articles starting in October 1924)
- *Saw tooth roofs: their design and erection*, January 1929 (v.24, pp.66–79)

In 1928 Lakeman was commissioned to write another book for Concrete Publications Ltd, and in *Elementary guide to reinforced concrete* he reprised his theme of 1913 and doubtless made use of his series of *C&CE* articles aimed at the foreman:

> Every effort has been made to keep the explanations as simple as possible, and highly technical language has been avoided in the desire to appeal to the reader who has received only a moderate education and to the student who is commencing the study of construction ... Simplicity has been the writer's aim throughout ... It is necessary to give the student some simple guidance at the commencement of his study, and to assist the practical man on the site of operations, if they are to be led on to greater interest in scientific construction. There does not appear to be any treatise available which answers this purpose and the present book has been written in the hope that it will successfully fill the important gap which now exists in the literature of reinforced concrete.

That the book filled this gap is clearly demonstrated by the sale of 10,000 copies in ten editions during the next 11 years of publication. A 12th edition was published as late as 1950.

A further volume – *Concrete construction made easy* – written with Leslie Turner (below), followed in 1929. This book containing designs, tables and illustrations – was aimed at the builder carrying out reinforced concrete work for which no designs had been prepared, by providing the guidance necessary in a standard format. At the other end of the scale, Lakeman also speculated on the outlook for structural engineering in the pages of the *Structural Engineer* (January 1932).

The 1930s were a time when Lakeman devoted more attention to practice as an architect. In 1934 for instance, he went to South Africa to design the Dunlop factory at Congella, near Durban, having recently developed a concrete paving block with embedded rubber skeleton (British patent No.390950) for the Dunlop Rubber Co. But it is for his 1939 landmark Imperial Airways headquarters of 1938 in Buckingham Palace Road – with its accompanying sculpture *Speed* – that he is chiefly remembered today.

Jones, B. E. (ed) and Lakeman, A. *Cassell's Reinforced Concrete: a complete treatise on the practice and theory of modern construction in concrete-steel.*

London: Cassell & Co., 1913 [ICE. Rev: *C&CE* v.8, pp.212-213]. Second edition. London: Waverley, 1920 [BL, C&CA 1959, Yrbk 1934. Copy: TCS]

Lakeman, A. *Concrete cottages: small garages and farm buildings* [later subtitled *bungalows and garages*]. London: Concrete Publications Ltd, 1918 [Rev: *C&CE* v.13, p.146]. Second edition, 1924 [BL, C&CA 1959. Copy: TCS]. Third edition, 1932 [C&CA 1959, Yrbk 1934]

Lakeman, A. *Elementary guide to reinforced concrete*. London: Concrete Publications Ltd, 1928 (12 editions to 1950) [BL, C&CA 1959, ICE, Yrbk 1934. Copy: TCS]

Turner, L and Lakeman, A. *Concrete construction made easy*. London: Concrete Publications Ltd, 1929 [Yrbk 1934]

1913/USA

Leonard, John Buck (1864–1945)

MASCE

A pioneering bridge engineer and architect, Leonard was early advocate of reinforced concrete in California after the earthquake of 1906.

He was born in Union City, Michigan and was educated at the University of Michigan. He moved west to California in 1888, visiting San Diego and Los Angeles before settling in San Francisco in 1889. There he was employed on engineering projects, establishing his own consulting office in 1904.

After the 1906 earthquake, and inspired by the performance of Ransome's buildings at Stanford University, Leonard became an articulate advocate for concrete's qualities in earthquake and fire. He wrote extensively on this theme in *Architect and Engineer of California*, and professionally, he appeared in the *Proceedings* of the ACI with *Use of reinforced concrete in San Francisco and vicinity* (v.7 1911).

After forming a partnership in 1913 with William Day (below), which was to last for three years, the two men produced an illustrated booklet – *The concrete bridge* – to argue for concrete construction and to display Leonard's work in that medium. Its propagandist role was stated unashamedly: 'The purpose of this publication is to supply information on the vast strides made in California in the use of reinforced concrete for permanent bridge construction.'

This was picked up in *Concrete-Cement Age* (v.4, January 1914), which reported that: 'the authors open the book with a brief, emphatic argument for the use of concrete, saying why it is the material in bridge construction. This is followed by series of excellent photograph reproductions coving in detail many types of bridge construction.' The editors approved, declaring it to be, 'a most artistic and well prepared booklet'.

Though in later life he was to concentrate more on inspection, Leonard designed 20 buildings in post 1906 San Francisco and 45 bridges throughout the state of California. They include:

- *1906 Clune's Auditorium, Los Angeles*
- *1906 Sheldon Building, San Francisco*
- *1911 Fernbridge, Eel River, Eureka*
- *1911 Gianella Bridge (a steel bridge on State Highway 32)*

- *1922 Van Duzen Bridge, Humbolt*
- *1922 Chili Bar Bridge, El Dorado*

Leonard, J. B. & Day, W. P. *The concrete bridge: a book on why the concrete bridge is replacing other forms of bridge construction.* San Francisco, 1913 [Copy: Internet Archive. Rev: C-CA, v.4, January 1914, p.42]

1913/USA

Day, William Peyton (1883–1966)

Born 8 April 1883, our interest in William Day lies in his architectural partnership with John Leonard (above), in which he acted as co-author of the firm's booklet, *The concrete bridge* (1913). In 1916 he formed a new practice under the name of Weeks & Day, and finally traded as W. P. Day Associates, architects. Projects in which he was involved included:

- *1920 Lowe's State Building and Theatre, Los Angeles*
- *1926 Pensinsula Theatre, Burlingame*
- *1927 California Theatre, San Jose*
- *1929 Fox Theatre, San Diego*
- *1929 Dufwin Theatre, Oakland*
- *1931 Joseph Magnin Dept Store, Oakland*
- *1936 Marina Middle School, San Francisco*

He died on 5 August 1966 in San Francisco. His professional papers are held by the California Historical Society.

1913/UK

Searle, Alfred Broadhead (1877–1967)

Alfred Broadhead Searle
Courtesy of Quarry Management

Whereas Lakeman (above) comes over as the young publicist, Searle appears as a continuation of the tradition of more mature writing on general construction materials (in the vein of Potter, Sutcliffe and Middleton), in which reinforced concrete is accepted as a new material among others. However, it should be said that he enthusiastically embraced the new technology of boat building in concrete. He also was versatile linguistically, undertaking the translation of foreign work such as Emile Bourry's *A treatise on ceramic industries* (Scott Greenwood, 1911) and contributing to Heinrich Becher's *Technical dictionaries in six languages*. *C&CE* in 1910 (v.5, p.454) noted that the English element of Volume VIII – *Reinforced concrete in sub- and superstructures* – 'was chiefly done by Dr. A. B. Searle of Sheffield'.

Searle was born in Sheffield in 1877 and raised there as a child. He attended Ackworth School – a Quaker institution established in 1777. In 1904 he was elected to the Royal Society of Arts.

We have seen Searle as a contributor on 'concrete, cement, waterproofing' in *Cassell's Reinforced Concrete* of 1913 (ed. Jones, above). In the same year he had his own book published: *Cement, concrete and bricks*. This was an exploration of the chemical and physical changes experienced by the three materials during their production, arguing for the role of testing during the process of manufacture. Whilst the book was mainly a theoretical treatise, the practical side of testing was described and accompanied by illustrations of the relevant apparatus. *C&CE* (v.8, p.885) described the volume as 'very complete' and writing by 'an author who is quite conversant with the subject'.

A year later he revised and adapted Van der Kloes's *Manual for masons, bricklayers, concrete workers and plasterers* – again tackling a broader spectrum of materials than simply concrete. As *C&CE* (v.9, pp.210–211) put it: 'this volume has been prepared by Mr. Searle to put before English and American readers the researches of Professor Van der Kloes … modified and adapted to render them more suitable to the readers in this country.'

Despite tackling concrete, Searle's early output was dominated by studies of clay (as the inclusion of 'bricks' in the above two titles hints at), and by the time he had produced *Clays and clay products* the following year (in 1915), he had already written: *The clayworker's hand-book* (Griffin, 1906); *The natural history of clay* (Putnam, 1912); and *An introduction to British clays, shales and sands* (1912).

At the end of the First World War, *C&CE* published a supplement specifically on shipbuilding – an application of concrete that enjoyed a brief vogue during the steel shortages of the time. Searle was its author, and the supplement took the form largely of a monthly serialisation of an extended piece of writing.

Other writing followed, with books on ceramic and refractory materials related to, but not about, cement and concrete: limestone, clay and sand. They included the following titles, many of them revised successively throughout his life:

- *The ceramic industries' pocket book* (Pitman, 1920)
- *Sands and crushed rocks*, 2 volumes (Frowde / Hodder & Stoughton, 1923)
- *Refractory materials* (Griffin, 1924, 1940, 1950)
- *The chemistry and physics of clay and other ceramic materials* (Benn, 1924, 1933, 1959 and revised by R. W. Grimshaw, 1971, and in the USA by Van Nostrand, 1926)
- *Clay and what we get from it* (The Sheldon Press, Macmillan, 1925)
- *An encyclopaedia of the ceramic industries* (Benn, 1929)
- *Limestone and its products* (Benn, 1935)
- *The glazer's book* (Technical Press, 1935)
- *Refractories for furnaces, kilns, retorts, etc* (Lockwood, 1939, 1948)
- *Modern brickmaking* (Benn, 1956)
- *Bricks and artificial stones of non-plastic materials: their manufacture and uses* (N/D)

When (at the end of our period of review) a new edition of his *Chemistry and physics of clay and other ceramic materials* was published in 1934, it was described as 'a well-known standard work' (*SE* v.12, p.59).

Formal recognition came in 1926 when he became the first recipient of the Maybury Medal – the Institute of Quarrying's award for a technical paper (named after the institute's founding president, Sir Henry Maybury).

As president of the Sheffield Rotary Club in 1920, Searle was still associated with his home city, but business was increasingly based in London. By the 1930s he was operating there as a 'consulting advisor to the lime, cement and clay products industries' (*CLG*, 1930, p.111), advising manufacturers and calling on his experience of over 20 years in the business.

We close in 1935 with Searle looking forward to a *Proposed new system of cement testing* (*Cement*, v.8(4), 1935), and arguing for the introduction of mineralogical analysis.

Searle, A. B. *Cement, concrete and bricks*. London: Constable & Co, 1913 [Rev: *C&CE* v.8, p.885]. Second edition, 1926 [BL, ICE]

Kloes, J. A. van der, and Searle, A. B. (rev. and adapt.) *A manual for masons, bricklayers, concrete workers and plasterers*. London: Churchill, 1914 [BL, ICE]

1914/UK

Kloes, Jacobus Alida van der (1845–1935)

Van der Kloes was an architect and professor in 'the Science of Materials of Construction' at the University of Delft. The work that warrants his inclusion here – *A manual for masons, bricklayers, concrete workers and plasterers* (which we have already learned was adapted for the British market by Alfred Searle) – concentrated on the use and performance of mortars for brickwork and masonry. It drew on research conducted by an international committee to which Van der Kloes had been appointed chairman in 1909. 'The notes on reinforced concrete', however – according to *C&CE* (v.9, pp.210–211) – 'are very meagre, and furthermore do not express the nature and function of the component parts in a satisfactory manner'. He addressed the subject of surface treatments, but was of the (widely-held) opinion that architectural treatment required the employment of brick or stone facings.

His previous career had commenced in 1863 as a draftsman for the Dutch railways and then for the engineering firm, Stork. He worked on land reclamation projects in the 1870s and moved to Dordrecht where he designed the county hospital, two schools, a church and the wine museum. His water tower at Wantij was a landmark structure. In 1882 he joined the staff at the polytechnic, later the University of Delft. There he wrote a series of six volumes entitled *Our building materials*, and practical guides aimed at the working man. Later he edited a monthly magazine for nearly 20 years, which he was still involved with up to the time of his death in 1935.

In these latter years he appeared occasionally in other journals, too. In *Zement* (v.16, 1927, pp.789–94), his *Cement mortar, cement concrete and reinforced concrete: notes of Portland cement and cement testing* criticised the prevailing methods of testing materials and describes methods used in his own laboratory. His practical work featured in *Master*

Builder (1930, p.27), where in *A means of improving concrete*, he described a design of formwork in which stretched gauze within a frame replaced solid timber, and the gauze itself was incorporated into the face of the resulting concrete element on removal of the frame. The stipples of paste protruding from the apertures in the gauze made an ideal key for stucco, render or plaster, and thus the method was particularly suitable for casting masonry blocks. The article also appeared in *Cement, Lime & Gravel* (April 1930, p.184), in which he explained that, 'as there is no patent, the method is open to all who care to make use of it'. This magazine incidentally, also featured work by Searle – suggesting continued contact between the two men years after their work together in 1914.

Kloes, J. A. van der, and Searle, A. B. (rev. and adapt.) *A manual for masons, bricklayers, concrete workers and plasterers*. London: Churchill, 1914 [BL, ICE]

1915/UK

Hammersley-Heenan, John (b.1879)

AMICE

John Hammersley-Heenan was born in Cape Town in 1879. His father – an Irishman called Robert Henry – had emigrated to South Africa. He was an engineer, and working on the railway north of East London he met and married Edith Fuller (the daughter of an English '1820' settler), and raised a family. John was the eldest of a family of six, his siblings born during the 1880s. Robert Henry entered the colonial civil service as resident engineer and became involved in the contentious issue of establishing a reliable water supply for Cape Town, called as an expert witness to the Water & Drainage Committee. As general manager of the Harbour Board he wrote *The Harbour of Algoa Bay, Cape Colony* in 1897, and later became a member of the Cape Society of Civil Engineers when it was founded in 1903. On retirement he and his family moved to first to London, then Harpenden, and finally settled in California at Hollywood.

John moved to England – where he was a resident of Dover in 1901 and a civil engineer like his father. He became an associate member of the ICE and in late 1914 presented a paper on winter concreting in Canada. A month later (in January 1915), it was published in the *Proceedings* (v.199, pp.141–146) and was abstracted in *C&CE* (v.10, January 1915 pp.47–48). His more substantial 158-page work of the same title – *Concrete in freezing weather and the effect of frost upon concrete* – was published that same year.

In 1916 he enlisted in the British Army and on 8 August his commission as a second lieutenant was announced in the *London Gazette*. Presumably he served in the Royal Flying Corps, for when he married in 1918 (aged 39), he was a captain in the newly formed RAF.

His bride was Ailsa Linley Robinson (23), youngest daughter of Mr. & Mrs. Sydney Jessop Robinson (a manufacturer and his wife) of Tapton House Road, Sheffield. The wedding took place on 28 September in Trinity Church, Hampstead. Like his parents, they too travelled to America after the war.

Hammersley-Heenan, J. *Concrete in freezing weather and the effect of frost upon concrete*. London, 1915 [C&CA 1952]

1914/UK

'Hollie'

The author styling himself by the pseudonym 'Hollie' is unknown, though a pencil annotation of 'Baumgarten' [see Childe, 1924] on the Concrete Society copy of *Concrete products* may be a clue to his identity (if it is not that of a library benefactor). Either way, Baumgarten is an appropriate name to be associated with a book from the Anglo-German Publishing Company issued poignantly just before the outbreak of war.

The book is certainly a highly distinctive one, partly for its subject matter – which differs from the contemporary treatments of reinforced concrete – and partly for its idiosyncratic style, which is both personal and partisan.

For a start, this was a book about pre-cast concrete products and not about the design or construction of buildings or structures *in situ*: 'I am not writing on the subject of reinforced Concrete construction, but on Concrete Products, which is quite a different subject. I am dealing with Bricks, Blocks, Tiles and other similar materials which simply form preferable substitutes for Clay Bricks Stone, Clay tiles, etc.' The foregoing survey will indicate the paucity of coverage of this aspect of concrete up to this date, and Hollie certainly recognised this for he declared in his foreword, 'there is not a single English book dealing with this branch of industry'. This, he believed, was a contributory factor in Britain lagging so far behind the Continent and United States in the manufacture of concrete products, along with Britain's characteristic conservatism. Hollie hoped that in writing his book, he would stimulate others to follow and assist the development of the concrete industry in the UK.

Then there is the matter of style. Hollie was undoubtedly an enthusiast: 'I am always eager to learn something new with regard to Concrete, and regret passing a day without having gained at least a little.' He also spoke directly to his readers: 'probably you are just the one who could teach me a number of invaluable things about Concrete. In the same way I too, may be able to give you a few hints which may prove of some value to you, and which I hope you will find in this book.'

The book was welcomed by *C&CE* (v.9, p.286), though the reviewer observed that Hollie had failed to consider aesthetics and the influence of artistic taste, as well as pointing out an evident misprint. The review continued:

> This little volume has much to recommend it on account of the very practical manner in which it is written, and the author has imparted a freshness to his work which is seldom met with, and which renders the book very interesting to read … [It] is well illustrated throughout and despite our criticisms we recommend it to our readers as being well written and full of interesting information, which is presented in a style quite different from that found in most technical works.

"Hollie". *Concrete products*. London: Anglo-German Publishing Co., 1914 [Copy: TCS. Rev: *C&CE* v.9, p.286]

1915/USA

Melan, Josef (1854–1941)

With the outbreak of war, the supply of continental volumes for review in the British market dried up, though several Germanic names appeared among the authors published in the USA. Some were translations, others by Americans of European ancestry. Melan and his translator Steinman (following) represent these two strands.

Josef Melan was an Austrian engineer who had gained a reputation for his bridges. He specialised in the arch bridge and was a pioneer in the use of reinforced concrete for such structures. His development of the 'Melan System' – employing rigid truss arches made of iron – was patented in 1892. It was an extension of Monier's ideas applied to larger structures than the former allowed. The system achieved international recognition in 1898 after completion of a shallow arch bridge at Steyr. At 42.4 metres this was the largest reinforced concrete bridge in the world. The famous Dragon Bridge in Ljubljana (1901) is another notable example of his work in the Austro-Hungarian Empire.

Fritz von Emperger introduced the Melan System to the USA where it was patented in 1893. America's first Melan bridge was built in Rock Rapids, Iowa, by William S. Hewett & Co. of Minnesota. Hewett (1864–1951) soon specialised in reinforced concrete bridges and built many in Minnesota into the early years of the 20th Century. His Interlachen Bridge of 1900 remains an historically significant survivor from this pioneering period. Other notable examples include Topeka's 540ft-long Melan arch bridge (built in1896). About 100 were built in America during the patent's first decade of application.

With such a presence in the market it is no surprise that the theory behind the shallow arch bridge should interest American engineers and influence their own thinking. Turneaure for instance, developed his deflection theory from Melan's work, extended in turn by Moisseiff to the theory of elastic distribution (*Transactions* ASCE 1933). But the means of reaching American minds was the translation of two of Melan's works into English by fellow bridge engineer D. B. Steinman.

The first of these was the *Theory of arches and suspension bridges* – a translation of Melan's contribution to the third edition of *Handbuch der ingenieruwessenschaften* (v.2, pt.5, 1906). It was published in Chicago by Myron C. Clark in 1913. Two years later Wiley published *Plain and reinforced concrete arches* – a translation of *Theorie des Gewolbes und es Eisenbetongewolbes im besonderen*, originally published as part of von Emperger's *Handbuch fur Eisenbetonbau* (Berlin, 1908).

Melan – by then professor of bridge design at the German Technical School at Prague – was described as occupying the 'recognised leading position … in this field of structural design', while his system of arch construction enjoyed 'extensive use … in all parts of the world'. Steinman assessed Melan's writing as 'one of the most thorough treatments of reinforced concrete arches in any language', and a source of methods new to American designers. He also expressed his thanks to Melan for 'helpful courtesies' extended during the process of translation.

Just as Steinman identified new thinking in Melan's work, so the review in *C&CE* (v.10, pp.424–425) realised that it 'contains the treatment of many problems that we have not seen dealt with in any other books on the subject'. The treatment depended on advanced mathematics and so while 'the book will undoubtedly be of considerable assistance to engineers who are already familiar with the formulae ... a large number of engineers, however, will in our opinion find the book very difficult to follow on account of the comparative lack of explanation...' Consequently, the reviewer concluded: 'we think there is still great need of a book which develops simpler formulae that will express the results of the standard cases of loading with sufficient accuracy.' Nonetheless a thousand copies were sold in the first year or so and a reprint of another thousand was ordered in 1917.

Melan, J. and Steinman, D. B. (trans). *Plain and reinforced concrete arches*. New York: John Wiley & Sons, 1915 [BL, ICE] and 1917 [BL]. London: Chapman & Hall, 1917 [Yrbk 1934. Copy: TCS. Rev: *C&CE*, v.10, pp.424–425]

1915/USA

Steinman, David Barnard (1887–1960)

Our interest in Steinman as a writer on concrete is captured by his translations of Melan (above), though as a distinguished bridge engineer himself he also wrote 600 professional papers and 20 books throughout his career. Indeed, Steinman is regarded as one of America's greatest bridge builders of the 20th Century, and his output reflects his reputation.

David Barnard Steinman
Courtesy of the J.Burns Library, Boston College

The son of immigrants from Khomsk, Brest Voblast in the Russian Empire, David Steinman was born in New York on 11 June 1887. He was brought up in a poor tenement district on New York's East Side, living in a poverty from which only he of six siblings was destined to escape. Their home was within sight of the recently completed Brooklyn Bridge, and it was this landmark that was to inspire his choice of career.

Education was his route to success. While still at school he started to attend classes at the City College of New York. Undertaking odd jobs earned enough money to continue his studies. Aged 20 he graduated *summa cum laude* with a BSc in 1906. He enrolled at Columbia University and with a combination of scholarships and night-time jobs, supported himself through a course of study leading to three degrees: a first degree in Civil Engineering and masters in Physics and Mathematics (both in 1909), and a PhD awarded in 1911.

It was during these years at Columbia that Steinman received his first publishing contract – translating two books by Melan. The first was *Theory of arches and suspension bridges*, for Myron C. Clark (published in 1913), and the second (see Melan above) was *Plain and reinforced concrete arches* for John Wiley & Sons, in 1915. This experience stimulated a pleasure in writing that was to become almost as much feature of his career as his practical engineering.

He also taught evening classes at City College and Stuyvesant Evening High School and so when his coursework at Columbia was completed in 1910, he accepted a teaching position at the University of Idaho, becoming the youngest professor of Civil Engineering in America. Whilst at Moscow, Idaho, his doctoral thesis – *Suspension bridges and cantilevers: their economic proportions and limiting spans* – was published by Van Nostrand. However, despite learning to ride and becoming one of Idaho's first scoutmasters, he missed city life and longed to return to New York.

His opportunity came in 1914 when he was taken on as a special assistant to Gustav Lindenthal, along with the Swiss bridge designer Othmar Ammann – with whom he would share a professional rivalry for the following 40 years. While working for Lindenthal, Steinman was engaged on the Hell Gate Bridge and the Sciotoville Bridge over the Ohio. He also met his future bride at this time – Irene Hoffman and he were married on 9 June 1915.

Leaving Lindenthal, Steinman found several posts. He obtained work on the Rondout Creek Bridge and as assistant engineer for the New York Central Railroad. He was also taken on by the distinguished bridge builder John Waddell of Kansas and put in charge of the newly established New York office where he designed the Marine Parkway Bridge. Between 1917 and 1920 he was part time professor of civil and mechanical engineering at his *alma mater*, the City College of New York. Then in 1920, he opened his own engineering office.

At first the new business struggled, but Steinman's break came when he was invited by Holton D. Robinson (engineer of the Manhattan and Williamsburg Bridges) to join him in a bid for the design of Hercilio Luz Bridge at Florianopolis in Brazil. Their proposal was accepted and the two men embarked on a partnership – as Robinson & Steinman – that was last until Robinson's death in 1945.

The 1920s and 1930s were busy years for the partnership, partly due to an emphasis on economy of design that took them through the lean years of the early 1920s and the Great Depression. The firm was responsible for such innovations as using eyebar chains as the upper chord of the stiffening truss (as on the Hercilio Luz Bridge) rather than wire cable, and pre-stressed twisted wire rope-strand cables (as on the St Maurice River Bridge in Quebec).

Steinman's designs for construction projects completed between the wars include:

- *1926 Hercilio Luz Bridge (Florianopolis, Brazil)*
- *1927 Carquinez Strait Bridge (San Francisco)*
- *1929 Mount Hope Bridge (Narragansett Bay, Rhode Island)*
- *1929 Grand Mere Suspension Bridge (St Maurice River, Quebec)*
- *1931 St John's Bridge (Willamette River, Portland, Oregon)*
- *1931 Waldo-Hancock Bridge (Penobscot Bay, Bucksport, Maine)*

- *1933 Sky Ride passenger transporter bridge (Chicago Century of Progress Exposition)*
- *1936 Henry Hudson Bridge (New York)*
- *1938 Wellesley and Hill Islands Bridge*
- *1938 Wellesley Island Suspension Bridge*
- *1938 Georgina Island Bridge*
- *1939 Deer Isle Bridge (Maine)*
- *1939 Sullivan-Hutsonville Bridge*

During this period, Steinman became president of the American Association of Engineers, for which he campaigned for higher educational and ethical standards. Then in 1934, he founded the National Society of Professional Engineers to protect legitimate engineers against competition from the unqualified, from unethical practices and from inadequate compensation. He served two terms as president – a tireless and inspiring speaker on behalf of the profession, seeking to raise public recognition and appreciation of the engineer.

By the mid 1930s he was one of the pre-eminent bridge engineers of the USA, with a high professional reputation eclipsed in the public eye only by Ammann's George Washington Bridge (1931) and Joseph Strauss's Golden Gate Bridge (1937).

Besides his bridges and work on behalf the profession, Steinman's reputation was conveyed by his extensive writing. His output comprised more than 600 professional papers and 20 books (of both prose and poetry). Titles include: *A practical treatise on suspension bridges: their design, construction and erection* (Wiley, 1929); *The Wichert truss* (Van Nostrand, 1932); *Bridges and their builders* (with Sarah R. Watson, 1941); *The builders of the bridge* (1945), a bestselling biography of the Roeblings (designers of the Brooklyn Bridge); and *I built a bridge and other poems* (1955). In all he had 150 poems published.

After the Second World War Steinman was responsible for the rehabilitation of the Brooklyn Bridge (1948), and developing a proposal for the Messina Straits. Although the latter was not built, he had two major construction projects left to complete: the 1957 Kingston-Rhinecliff Bridge; and in the same year, his crowning achievement, the Mackinac Bridge in Michigan. This – known affectionately as the 'Mighty Mac' – was the longest suspension bridge of its day.

In these final years Steinman established the David Steinman Foundation to disburse funds for the assistance of engineering students, including the Irene Steinman and Holton Robinson Scholarships. The school of engineering at City College is named in his honour, as are various awards programmes internationally. His own career was capped with 19 honorary degrees and numerous awards from organisations in Canada and France, as well as America.

Steinman died on 21 August 1960, while in office as president of the Society for the History of Technology. His firm was eventually incorporated into the Parsons Transportation Group, but his bridges – and his many books – constitute his lasting memorial. Biographical essays on Steinman abound, but an article in *Structure* (Oct 2005, pp.48–51) is a recent tribute.

Melan, J. and Steinman, D. B. (trans). *Plain and reinforced concrete arches*. New York: John Wiley & Sons, 1915 [BL, ICE] and 1917 [BL]. London: Chapman & Hall, 1917 [Yrbk 1934. Copy: TCS. Rev: *C&CE*, v.10, pp.424–425]

1915/USA

Whipple, Harvey (b.1884)

Originally a Massachusetts family, the Whipples had been resident in Michigan for four generations before the young Harvey took up the occupation of journalist. He started out as a news reporter on the *Port Huron Times* in 1903, then the *Detroit News* in 1904, but in due course became associated with the local magazine – *Concrete* – of which he duly became managing editor. In 1912 *Concrete* merged with *Cement Age* to form the unimaginatively named *Concrete-Cement Age*. This specialist periodical promoted itself as: 'devoted to the manufacture and uses of Portland cement'. It was based in Detroit and published by the Concrete-Cement Age Publishing Co.

Harvey Whipple, Secretary of the American Concrete Institute ('Mr ACI') Courtesy of the American Concrete Institute

Whipple's first book came in February 1915, with the publication of *Concrete stone manufacture*. This, based on personal visits to pre-cast concrete factories and on letters and articles submitted to *Concrete-Cement Age*, was described in Whipple's preface as 'a report of progress' aiming to inspire manufacturers to 'important industrial growth'. *Cement & Constructional Engineering* liked what it saw: 'the author … displays a thorough knowledge of the whole subject. The book is well written, and is a comprehensive treatise on concrete block construction which can be thoroughly recommended' (v.10, p.259).

Two years later he collaborated with his associate editor C. D. Gilbert [1917], in writing *Concrete houses and how they were built*. Concrete houses comprised a compilation of articles from *Concrete* prepared in response to a 'demand greater than could be met with copies of Housebuilding Numbers of that magazine'. The text was accompanied by plans and detailed drawings.

In 1918 Whipple contributed a chapter in G. A. Hool's *Concrete engineers' handbook* and in 1919 wrote the vaguely entitled *How to do it*, published by the Concrete-Cement Age Publishing Co.

After this brief prominence as an author, Whipple was appointed secretary and treasurer to the American Concrete Institute in 1919, in tandem with his continuing duties as managing editor of *Concrete-Cement Age*. He was to remain in this post until 1952, his role becoming full time as the senior employee of the institute. In a history of the Institute, Whipple is described as being considered 'Mr ACI'; his years in charge are remembered as a golden age.

Whipple, H. *Concrete stone manufacture*. Concrete-Cement Age Publishing Co., 1915 [Rev: C&CE v.10, p.259]. Second edition, 1918

Whipple, H. and Gilbert, C. D. *Concrete houses and how they were built*. Detroit: Concrete-Cement Age Publishing Co., 1917 [C&CA 1959. Copy: TCS]

1916/INDIA

Mehta, N. L.

We know nothing of N. L. Mehta, but his book on concrete serves to remind us that several works were written in India under the British Raj, by both Indian and British authors, such as: **Percy Hawkins** (1906); **Antione DeChazal** (1906); **W. H. James** (1907); **Alan Moncrieff** (1926); **Robert Mears** (1931). Much of this writing was undertaken by servants of the presidency governments in Madras, Bombay and Calcutta, whereas Mehta's *Notes on reinforced concrete structures* appear to have been issued by a commercial publisher.

Mehta, N. L. *Notes on reinforced concrete structures*. Madras: Higginbothams, 1916 [BL]

1916/USA

Seaton, Roy Andrew

BS, MS, PHD.

Seaton's book – *Concrete construction for rural communities* – addressed a subject that was highly topical in Great Britain during the war (as agriculture started to industrialise to meet wartime shortages), and was welcomed by *C&CE* as being of 'particular interest'. The book, described as 'well written', covered both plain and reinforced concrete. Its reviewer was particular taken with the fourth section, which discussed casting in moulds and surface finishing, and the fifth, which illustrated typical applications of concrete in agriculture.

Seaton's family had been long-settled in North America, his forebears having made the passage from England in 1635, ending up in Virginia. Many members of the family had been (and still are) associated with newspapers, both as owners, editors and writers. W. W. Seaton – Mayor of Washington DC in 1840–1850 – was owner of the *National Intelligencer* from the time of the War of 1812 until just before Lincoln's assassination in 1864, and counted Charles Dickens among his friends. His cousin's family moved west into upstate New York, and thence to Kansas in the 1870s. Oren Seaton had two sons, the elder of which bought the *Manhattan Mercury*, which – five generations later – is still in the family. The younger was Roy, who was to have a distinguished career as a university educator in engineering.

As a young man he attended the Kansas State Agricultural College, from which he graduated in 1904. His first appointment at the college was as an assistant in mathematics, but in 1906 he transferred to the department of mechanical engineering. It is easy to see how the combination of subject and institution was to influence his choice of writing in the years following. In 1920 he became the dean of the Division of Engineering & Architecture and director of the Engineering Experimental Station. He was to enjoy an 'illustrious' reputation in his capacity as dean – a role in which he served until 1949. He was involved in many aspects of engineering (though not specifically in concrete) and for this work (like his contemporaries Turneaure and

Mauer), was presented with the Benjamin Garver Lamme Award by the American Society of Engineering Education.

A collection of printed papers relating to Seaton are preserved in the Kansas State University library and the campus contains a building in his name – Seaton Hall – which was originally built in 1922. There is also a Seaton Society that aims to celebrate the achievements of distinguished alumni.

Seaton, R. A. *Concrete construction for rural communities*. (Agricultural Engineering Series). New York: McGraw-Hill, 1916 [BL, PCA, Yrbk 1934. Rev: *C&CE* v.12]

1917/USA

Campbell, Henry Colin (b.1868)

CE, EM.

Campbell was director of the editorial and advertising bureau at the Portland Cement Association in 1917, who later claimed a 'long association with the PCA in the preparation of its many informative booklets on concrete' (1920). His career as a journalist included roles as contributing editor to both *Cement World* and *American Carpenter and Builder*, a member of the editorial staff of *Farm Engineering* and contributor to 'numerous farm, trade and technical periodicals'.

He was raised on a farm and continued to operate one throughout his writing career, certainly until 1920, and this background intertwined with his interest in concrete and building. Besides the books listed below, for instance, he wrote about silo construction for *Hoard's Dairyman* (21 Feb 1919).

His first concrete book, *Concrete on the farm and in the shop*, published in 1916 and described as 'very fully illustrated', exemplified this combined interest. Campbell acknowledged the burgeoning supply of books on concrete, but claimed a distinctive approach in addressing the novice:

> 'There are concrete books galore. No apology need be made for this one ... It has, however, been primarily prepared with the beginner first in mind – the man who knows nothing of concrete but wants to learn ... The writer has endeavoured to translate technical expressions and technical terms into plain everyday English'

Practical concrete work for the school and home followed in 1917, written with Beyer, and a 200-page *How to use cement for concrete construction for town and farm*, published by Stanton and Van Vliet in 1920. His final book, *Concrete products: their manufacture and use*, was written with Walter Rutherford Harris [1921].

In an intriguing coincidence of name and dates a Henry Colin Campbell known as 'the Torch Murderer' was executed in 1930 aged 61. He had worked as a civil engineer and advertising executive, besides posing as a physician. His crimes were bigamy during the period 1910 to 1928, and the murder of one (and probably at least one other) of his wives.

Campbell, H. C. *Concrete on the farm and in the shop*. New York: Norman W. Henley, 1916 [Copy: Internet Archive]

Campbell, H. C. and Beyer, W. F. *Practical concrete work for the school and home*. Peoria, Illinois: Manual Arts Press, 1917 and London: Batsford, 1917 [Yrbk 1934]

Campbell, H. C. *How to use cement for concrete*

construction for town and farm, Stanton and Van Vliet, 1920 [C&CA 1952]

Harris, W. R. & Campbell, H. C. *Concrete products: their manufacture and use*. Chicago: International Trade Press, 1921 and 1924 [C&CA 1952]

1917/USA

Beyer, Walter F.

Campbell's co-author had formerly acted as assistant engineer to the Isthmusian Canal Commission, charged with the engineering aspects of the Panama Canal.

1917/USA

Ekblaw, Karl John Theodore (b.1884)

AMASAE.

Karl Ekblaw was an associate member of the American Society of Agricultural Engineers and his published output of two books reflects this professional interest: *Farm structures* came out in 1914; and *Farm concrete* in June 1917. *Farm concrete* was published by Macmillan in New York, with British Empire editions handled by the firm's own offices in Toronto and London. Assisted by his wife (Alma Heuman Ekblaw) in preparing the manuscript for publication, he drew heavily on the bulletins of the Universal and Atlas Portland Cement Companies, and acknowledged the assistance of Messrs Curtis, Brewer and Campbell of the newly formed Portland Cement Association. Extensively illustrated, the book described the use of concrete in a range of applications that reflect the preoccupations of the cement industry's marketing efforts at that time, among them pavements and blocks, along with agricultural topics such as silos, troughs and cisterns, and fence posts.

Ekblaw, K. J. T. *Farm concrete*. New York: Macmillan, 1917 [PCA]

Fallon, John Tiernan (b.1896)

1917/USA

Fallon was writing in the footsteps of R. C. Davison [1910] – the second edition of whose *Concrete pottery and garden furniture* was also published in 1917.

Fallon, J. T (ed). *How to make concrete garden furniture*. New York: Robert M. McBride, 1917 and London: Batsford, 1917 [Yrbk 1934]

1917/USA

Gilbert, Clyde Dee (b.1881)

Harvey Whipple's collaborator and associate editor of *Concrete* followed up their joint book of 1917 with an article on the same subject – *The concrete industrial house: a record of achievement* – in the ACI *Journal Proceedings* of 1 June 1918 (v.14, pp.389–407).

Whipple, H. and Gilbert, C. D. *Concrete houses and how they were built*. Detroit: Concrete-Cement Age Publishing Co., 1917 [C&CA 1959. Copy: TCS]

1917/USA

Post, Chester Leroy (b.1880)

Beyond his book on building superintendence, nothing is known of this author.

Post, C. L. *Building superintendence for reinforced concrete structures*. American Technical Press, 1917.

Second edition. American Technical Press and London: Crosby Lockwood, 1927 [Yrbk 1934]

1917/USA

Talbot, Arthur Newell (1857–1942)

Arthur Newell Talbot
Courtesy of the American Concrete Institute

Talbot was one of the great names in early concrete research – one of only eight included in the ACI's *Selection of Historic American papers on concrete 1876–1926* to mark the US bicentennial. Though his output was mainly through the university press (notably the bulletins of the University of Illinois Engineering Experiment Station), the commercial publishers Chapman & Hall made available to the wider market in Britain one particular title: *Tests of reinforced concrete flat slab structures*. It will serve admirably as an example of the Illinois bulletins and as a window on the illustrious career of Arthur Newell Talbot.

He was born in Cortland, Illinois, and arrived at the state university in 1877 in his 20th year. Except for four years as a young man working on the railways in Colorado after graduating in 1881, he spent his entire career at the University of Illinois. Back from the Rockies in 1885, Talbot was appointed to the position of assistant professor of engineering and mathematics. In 1890 the university set up the theoretical and applied mechanics department with Talbot as its head – a role he was to fulfil for the rest of his working life. He was professor of municipal and sanitary engineering from 1890 to 1926. Early research from 1905 to 1907 was also summarised for a British audience by Charles Marsh [1904] in *Concrete & Constructional Engineering* (v.2 Jan 1908, p.173). There he secured an early reputation in water, sanitary and railway engineering, but perhaps his greatest legacy was the tradition of excellence he established at Illinois and the group of outstanding young engineers he assembled about him. According to Howard Newlon, 'the roster of those who specialized in concrete and reinforced concrete reads like a *Who's Who* of the discipline: Arthur Lord; Duff Abrams; C. A. P. Turner; Harald Westergaard; Hardy Cross and many others', (ACI SP–52). Arthur Lord later characterised his mentor's influence: 'the Talbot tradition did not countenance loose thinking, nor partial data, nor careless composition' (*ACI Journal*, February 1931, p.11).

In 1904 Talbot wrote the first bulletin of the University of Illinois Engineering Experiment Station, entitled *Tests of reinforced concrete beams*. The station was established on 8 December 1903 to carry out investigations and study problems of importance to engineers, and Talbot was one of six professors charged with overseeing the programme of research. The experimental work was expected from the start, to result in contributions of value to engineering science, and the bulletin (published by the university), was the means of dissemination.

The example chosen here (Bulletin No.84, dating from 1917) was described by *C&CE* as forming 'a useful addition to the excellent and educational works which have been published under the auspices of the [university]'. Its co-author – W. A. Slater – could with perfect justification also appear in Newlon's roster of Talbot's acolytes. Distributed in the UK by Chapman & Hall, the bulletin brought the results of a university testing programme to a wider audience, and the action of flat slabs under loading was a topic of interest and differing opinions at the time. The programme was inconclusive but in the opinion of *C&CE*, the value of Bulletin No.84 lay in its summary of test data.

The years spanned by these two bulletins capture the principal period of Talbot's research into concrete, and include the memorable Bulletin No.64 on *Tests of reinforced concrete buildings*, written with W. A. Slater. His surviving correspondence includes an exchange with Britain's Charles F. Marsh. Further correspondence links him with the National Association of Cement Users in 1910 to 1913 and subsequently the American Concrete Institute. He conducted research and tests on cement and concrete throughout the years 1911 to 1916, including corresponding with C. A. P. Turner [1909] over the Carter Building test in 1911 and various other building tests in 1912.

After publication of Bulletin No.84, Talbot's attention was engaged by the war effort and in 1918 he was involved in reviewing army construction projects. 1918 also saw him elected president of the ASCE.

Like several other American professors in this survey, Talbot was awarded the Lamme Medal for engineering education in 1932. In the same year – in the company of three other of our authors – Talbot was made an honorary member of the ACI. Further honours followed when the Materials Testing Lab was renamed the Talbot Laboratory in 1938.

Shortly after his death in 1942, Talbot's name was recalled by one of the wartime concrete ships built by McCloskey of Tampa Bay. SS *Arthur Newell Talbot* was launched on 15 July 1943, and sunk after the war in December 1948. Rather more lasting is the collection of his professional papers maintained by the University of Illinois Archives.

Talbot, A. N. and Slater, W. A. *Tests of reinforced concrete flat slab structures*. (Bulletin No.84 of the University of Illinois Engineering Experiment Station). London: Chapman & Hall, 1917 [Rev: *C&CE* v.12, p.203]

1917/USA

Slater, Willis A. (d.1931)

Co-author with Talbot (above) of *Tests of reinforced concrete flat slab structures*, Slater is another example of a prolific contributor to the literature, but one whose output was not generally in the form of commercially available books. In a vignette that appeared in the *ACI Journal* of February 1931, Arthur Lord describes Slater's 'immense output ... of bulletins and papers and reports relating to the testing and design of concrete; all prepared with reckless expenditure of time, effort

Willis A. Slater
Courtesy of the American Concrete Institute

and thought; all written over and over until he was somehow satisfied to release them – and then finally entirely revised'.

Slater's career in concrete commenced at the University of Illinois under Arthur Talbot, where as a young man he mingled with the likes of Abrams and Hollister (whom we describe below) and other disciples. There, as assistant professor, he was involved in the early tests for actual stresses in concrete buildings – tests that were written up in the ACI's *Proceedings*:

- *The testing of reinforced concrete buildings under load* (1 March 1912, and later as University of Illinois Bulletin No.64)
- *Tests to determine lateral distribution of stresses in wide reinforced concrete beams* (1 December 1913 and 1 January 1914)
- *Test of a flat slab reinforced concrete floor at Shredded Wheat factory, Niagara Falls, N.Y.* (1 November 1914)
- *Test of a flat concrete tile dome reinforced circumferentially* (with C. R. Clark, 1 February 1917)

He moved on to the US Bureau of Standards in early 1917 where he continued to work on testing and analysis of concrete, initially for the war effort. Lord's sketch conveys a flavour of his work-rate in this role too: 'His union brain and body balked at the ceaseless grind of sixteen-hour days, but he dove them on, refused to rest even when headaches filled twenty hours out of the twenty-four. Attending conventions, he registered at hospitals rather than hotels, determined to cure his body without sacrificing precious time.'

War service in 1917/18 was devoted to the design of concrete ships, demand for which soon arose. Slater was placed in charge of design and realising that research was an essential precursor to success he assembled the 'largest force, by far, ever concentrated upon concrete research up to that time'. The requirements for concrete in this application were onerous and in several aspects beyond the state of contemporary practice. Many innovations were introduced by the Concrete Ship Section – including the use of vibration to aid compaction, the development of lightweight clay aggregate and high strength cements, and the achievement of high levels of water-tightness. An account of this work – *Structural laboratory investigations in reinforced concrete made by the Concrete Ship Section* – appeared in the ACI's *Proceedings* (1 June 1919). Slater's role was subsequently recognised when he – like his mentor, Talbot – had a concrete vessel named after him in the Second World War: the 4,690-ton SS *Willis A. Slater* in service 1944–1948.

His employment at the Bureau of Standards after the war was punctuated by secondments to other agencies – notably the Arch Dam Investigation in California for two years – but for much of his time at the bureau he sat on the Joint Committee on Specifications for Concrete and Reinforced Concrete, and in 1930 became its chairman.

By that date had had become director of the Fritz Engineering Laboratory at Lehigh University, where he maintained a close involvement with many of the major research projects – including the programme of concrete column testing at

Lehigh and Illinois, for which he acted as chairman of the steering committee (ACI Committee 105).

Slater had a long involvement with the ACI, writing 18 papers in the Institute's *Proceedings* from 1912 to 1931. His post-war output included:

- *Compressive strength of concrete in flexure* (with R. R. Zipprodt, 1 February 1920)
- *Moments and stresses in slabs* (with H. M. Westergaard, 1 February 1921)
- *Inundation methods for measurements of sand in making concrete* (with G. A. Smith, 1 January 1923)
- *Control of concrete for the University of Illinois Stadium* (with R. L. Brown, 1 February 1924)
- *Field tests of concrete used on construction work* (with Stanton Walker, 1 February 1924)
- *Relation of 7-day to 28-day compressive strength of mortar and concrete* (1 February 1926)
- *Some features of the testing of Stevenson Creek Arch Dam* (1 February 1928)
- *Floor test in the George Mason Hotel, Alexandria, Virginia* (1 January 1930)
- *Compressive strength of concrete in flexure as determined from tests of reinforced beams* (with Inge Lyse, 1 June 1930)
- *First, second and third progress report on column tests at Lehigh University* (with Inge Lyse, 1 February, March and November 1931)

Towards the end of this period, he took on the responsibilities of chairman of the ACI's publications committee and contributed further to the dissemination of knowledge in his field.

Arthur Lord's tribute of 1931 ends with a light touch, balancing the personal against the professional with the revelation that Slater was 'an inveterate gatherer and retailer of jokes and stories'; a weakness that was 'becoming more pronounced with the years'. Sadly this weakness had little time to develop, for within months – in October 1931 – Slater died.

Talbot, A. N. and Slater, W. A. *Tests of reinforced concrete flat slab structures*. (Bulletin No.84 of the University of Illinois Engineering Experiment Station). London: Chapman & Hall, 1917 [Rev: *C&CE* v.12, p.203]

1917/USA

Thomas, M. Edgar

Author of *Reinforced concrete design tables*.

Thomas, M. E. and Nichols, C. E. *Reinforced concrete design tables: a handbook for engineers and architects in designing reinforced concrete structures*. New York: McGraw-Hill, 1917 [PCA, Yrbk 1934]

1917/USA

Nichols, Charles Eliot (b.1884)

Co-author, with Thomas (above) of *Reinforced concrete design tables*.

1918/USA

Abrams, Duff Andrew (1880–1965)

FAAAS, MACI, MASCE, MASTM, MACERS

Duff Andrew Abrahams
Courtesy of the Portland Cement Association

Abram's output was the product of the research institutes and so strictly speaking does not come within the scope of this survey, but his published conclusions were so important, and so widely underpinned the subsequent literature, that his exclusion on a matter of definition seems perverse. Writing of his seminal research on the water/cement ratio, Arthur Lord put Abrams in his academic (and literary) context: 'And then the thunderbolt! Move over Professor Talbot, Dr. Hatt, Dean Turneaure, and make room on the intellectual throne for a newcomer who has made all the old art of concrete proportioning obsolete over night!' As this revelation came in the form of the Portland Cement Association's very first bulletin, Abrams can also be seen here as a figurehead of specialist publishing activity of the cement industry's own representative bodies, of which the British equivalent's birth in 1935 ends this survey.

Duff Abrams was born in Grand Tower, Illinois, on 25 April 1880. Educated locally at the high school in Murphysboro, he entered the University of Illinois in 1900. With a two-year break working on concrete construction projects in 1902–1904, he graduated with BSc in Civil Engineering in 1905.

He secured a post at the university, as 'associate' engineer in the Experiment Station, where he was to be employed until 1914 as part of Professor Talbot's team. Indeed he collaborated with the professor in submitting his first paper for the ACI's *Proceedings* in 1912: *Method of testing drain tile* (v.8(3), pp.713–719). Writing in 1931, Arthur Lord recalled his experiences of the Station in this period:

> In those early days there was an associate – not an associate *professor*, but just a plain associate – in the T. and A.M. Dept., at Illinois, whose job was principally to manufacture all the specimens that the rest of us tested. He bore the strange name of Duff – not an earned nickname but merely the name wished on him by his parents. Everybody in the Laboratory liked him. Even I called him 'Abe' from our first meeting, and it gave me no such thrill as it does today. When not making my specimens, or a host of others, this chap amused himself by sticking short bars in small concrete cylinders and then pulling them out on a small machine over in one corner of the laboratory. How should I know that when he was through, no one would think seriously of making more bond tests for twenty years? That was Duff's way of doing things – not exactly secretively, but quietly … After his famous *Bond Bulletin* appeared – impressive for size even if you have never opened the covers, for Duff insisted on more than a 'shirt tail full of data' even then – people in faraway Chicago began to take notice of Duff and the bright lights of stardom began to call.

This famous bulletin was the 238-page University of Illinois Bulletin No.71 of 1914, entitled *Tests of bond between concrete and steel*. It was an instant success, and in the UK it was reviewed by *C&CE* (v.9, p.704), which described it as 'an excellent volume dealing with a very important question in reinforced concrete construction'.

The call from Chicago came and Abrams was offered a research post at the Lewis Institute (later the Illinois Institute of Technology, Chicago), undertaking research for Universal Cement. In 1916, following the establishment of the Portland Cement Association, Abrams was appointed professor in charge of the Structural Materials Research Laboratory – a joint project between the Lewis Institute and the PCA. He started on 1 September 1916 with a three-year contract and remained in post until 1926. It was during these years that his most famous discoveries were made and the papers that brought them to the attention of the wider world published.

The first (and probably the most fundamental) was the discovery that the water/cement ratio was the prime determinant of concrete strength, and this was published in 1918 as the very first of the PCA's bulletins. Abrams' report was initially printed in the minutes of the PCA annual meeting in December 1918, then published separately as Bulletin 1 (originally under the Lewis imprint) later that month. It has been described by its publisher as 'maybe the most important research paper ever to come from PCA' – three million copies sold by the early 1990s, and it won its author the Wason Medal 'for the most meritorious paper.

Others followed and Abrams' titles over the next few years included:

- *1918 Design of concrete mixtures (Bulletin 1)*
- *1919 Effects of curing condition on the wear and strength of concrete*
- *1919 Effect of fineness of cement on plasticity and strength of concrete*
- *1919 Effect of vibration, jigging and pressure on fresh concrete, ACI Proceedings, v.15(6), pp.63–85*
- *1920 Effect of storage of cement*
- *1920 Effect of tannic acid on the strength of concrete*
- *1920/1 Effect of hydrated lime and other powdered admixtures in concrete*
- *1921 Water tests of concrete*
- *1921 Quantities of materials for concrete (with Stanton Walker)*
- *1922 Flexural strength of plain concrete, ACI Proceedings, v.18(2), pp.20–45*
- *1922 Proportioning concrete mixtures, ACI Proceedings, v.18(2), pp.174–181*
- *1924 Making good concrete, ACI Proceedings, v.20(2), pp.175–181*
- *1924 Tests of impure waters for mixing concrete, ACI Proceedings, v.20(2), pp.442–486*

In 1926 the Structural Materials Research Laboratory was combined with the other activities of the association and in July 1926 Abrams became director of research. He returned to his original theme with *Water-cement ratio as a basis of concrete quality* – a paper for the ACI's *Proceedings* published the following February. His tenure as director was brief however, as he left for a new post on 31 March 1927.

His new employer was the International Cement Corporation of New York and from 1927 to 1931 Abrams was director of research there. This was a time of high professional standing. At the time of his appointment he was a fellow of the American Association for the Advancement of Science, vice-president of the ACI and a member of the ASCE, ASTM, American Ceramic Society, the Western Society of Engineering and the Engineering Institute of Canada. For the previous seven years, he had also been secretary of the Joint Committee on Concrete and Reinforced Concrete and from 1909 been involved with committee work for ASTM.

In 1930 he became President of the ACI – a body he had first joined in 1910. After serving his term, in 1931 his retiring address emphasised the importance of research. Describing it as the handmaiden of design and construction, he argued it was of little value if not translated into use in the office and field. A year later he was awarded the Turner Medal for notable achievement, or specifically as the citation read, 'the discovery and statement of important fundamental principles governing the properties of concrete and reinforced concret'. He was also awarded the Frank B. Brown Medal of the Franklin Institute.

In 1931 Abrams left International Cement to organise his own research laboratory and his prominence (though not his legacy) faded from view. He died on 3 June 1965 in Long Island City, NY, aged 85. An obituary appeared in the ACI's *Journal* in October.

Abrams, D. A. *Design of concrete mixtures.* (Bulletin No.1). Portland Cement Association, 1918

1918/UK

Etchells, Ernest Fiander (1877–1927)

AMICE, AMIMechE, Hon ARIBA, MSCE (France), FphysC., MMathA

Ernest Fiander Etchells, president of the Concrete Institute and IStructE, 1920-23
Courtesy of the IStructE

One of the central figures in the early concrete establishment and first president of the Institution of Structural Engineers, Etchells was the author of a textbook on structural steel rather than concrete, but is included here as his work for the Concrete Institute in developing standard notation for engineering formulae was published for the wider market by E. & F. N. Spon.

E. Fiander Etchells was born in 1877, the eldest son of Edward Ernest Fiander Etchells of Romily, near Marple in the district of Macclesfield. He studied at

Manchester, Glasgow and then King's and University Colleges, London. In practice he was a pupil successively of F. L. Lane of Leeds and John Strain of Glasgow. His early career was spent abroad, where he designed rolling stock for the Pretoria and Pietermaritzbug Railway in South Africa, aerial ropeways in Bolivia and a nitrate plant in Chile. Returning to Britain he undertook structural work in Motherwell, Lanarkshire, and worked on the Clyde Valley Electric Power scheme before heading to Latin America again to work on the City of Mexico Electricity Supply scheme.

The rest of his professional life, from 1902, was spent in the public service, as an official working in the architects' department of the London County Council. Among his roles was arbitrating in disputes between architects and engineers, and the local authority within the provisions of the London Building Acts. His final position was structural engineer for the superintending architect of Metropolitan Buildings at County Hall.

In his early thirties he became an active participant in the proceedings of the newly formed Concrete Institute, a vocal speaker in the discussion of papers read at general meetings, and a member of the science committee. He was a vociferous advocate of standardising the notation of formulae used in reinforced concrete design, and it became his task to undertake formulating the institute's recommendations under the aegis of the standing science committee. *C&CE* commented on this relationship:

> The work under review consists of the Report of the Committee proper, followed by an appendix by Mr. Etchells, who is the prime mover in this matter and deserves every praise for his efforts. We surmise that the true relation of the Committee and Mr. Etchells has been that they have served as the patient upon which Mr. Etchells could experiment. He is responsible for the formulation of the principles and co-ordination and uses the Committee to decide what are truly practical symbols commendable to the rank and file for general adoption ... The Committee has served as a very keen critic. The appendix by Mr. Etchells forms very interesting reading, and we recommend all students of mathematics and engineering to read it.

A draft report was issued on 28 July 1909 and while they were not adopted in continental Europe, its recommendations found favour in America as well as the UK and were used in many textbooks, the second RIBA report of 1911 (for which he was a committee member), and in the LCC regulations for reinforced concrete. Etchells explained the new conventions in *Concrete & Constructional Engineering* in an article entitled *Standard algebraic notation* (Dec 1912, v.7(12), pp.902–908) – and repeated reviews in the journal at this time took authors to task if the notation was not employed in their writing.

The work of Etchells' close friend Noble Twelvetrees [1905] is a case in point. The latter's *The practical design of reinforced concrete beams and columns* (Whittaker, 1911) was reviewed in *C&CE* (v.11, pp.799–800), in which criticism was levelled at Twelvetrees for failing to adopt the Concrete Institute's notation (and this despite Twelvetrees himself deploring the lack of standardisation).

After further revision a new version was published by Spon in 1918, complete with explanatory notes by Etchells. It was this edition that was promoted by Etchells in his support for contemporary authors. Twelvetrees' later *Treatise on reinforced concrete* (Pitman, 1920) featured standard notation and a foreword by Etchells (and claimed to be 'the first book in the English language embodying the improved notation') – as did Harrington Hudson's *Reinforced concrete* of 1922.

These were the years in which Etchells reached the pinnacle of his professional standing. He served as president of the Concrete Institute from 1920 and during the following three years of his extended term, oversaw its transformation into the Institution of Structural Engineers in 1922. More than one presidential address featured in the institution's new journal, the *Structural Engineer*.

Beside these addresses, he was the author of other papers, including *Algebra of magnitudes* (*SE*, 1924, v.2(1), pp.24–25) and *Gauging the constituents of concrete* (*C&CE*, July 1926, v.21(7), pp.473–475). His writing was concluded with a book – *Structural steelwork* (Nash & Alexander, 1927) – that unfortunately, was published posthumously, having been completed by his friend Ewart Andrews [1912]. The book was an anthology of contemporary writing on the application of steelwork to buildings and bridges, and contributors included Andrews and Oscar Faber [1912], both of whom went on to serve as president of the Concrete Institution.

Etchells served as president of other institutions too: the Association of Architects, Surveyors, and Technical Assistants; the Association of Floor Constructors; and the British section of the *Societe des Ingenieurs Civils de France* (for 1926–1927).

Etchells died of heart failure on 5 January 1927, having achieved much in his various fields of interest: physics; maths (he was a member of the Mathematical Association); mechanical engineering; architecture; as well as reinforced concrete. The obituary that appeared in the *Structural Engineer* (v.5, p.12) was fulsome in its warm-hearted admiration for the man and leaves no doubt of the regard in which he was held:

> He, in fact, achieved the rare distinction of being more than a personality – he was what is endearingly known as 'a character'. To his vast circle of friends and acquaintances in the large world of architecture and engineering, Etchells will chiefly be remembered for his tremendously boyish fund of most boundless energy and enthusiasm. He was essentially a Peter Pan of the engineering world; a boy who never grew up, and moreover a boy whose talents in many directions may fairly be said to have attained to the rank of genius. Had he perhaps possessed more power of concentration and been content to confined his enthusiasms within a narrower limit, there is no doubt but that he would have left a real mark for all time upon his profession ... His energy and enthusiasm in the cause of the Institution of Structural Engineers, both before, during, and after his lengthy

period of office as its president, and his essentially lovable, witty, and amusing personality, are too well known to want expounding.

They were expounded however, by his friend E. S. Andrews, who in *C&CE* (v.22, p.149), wrote a heartfelt appreciation of 'a remarkable man'. It is a gracious tribute, perceptive, and warrants quoting at some length:

> Physically he was of massive contour and striking appearance – hewn out of rugged granite, to use an expression that he often adopted himself with a characteristic but rather pleasant rolling of the 'r'. Intellectually he was in many respects also a giant; he had a remarkable memory and an amazing flair for getting hold of obscure matters lying on the borderland of engineering and architecture ... Personality exuded from him, and combined with this quality, which is so difficult to define but so easy to feel, there was in him the power to convince as to has ability and reliability. He possessed the quality, so valuable in a public servant, of being able to 'suffer fools gladly', and was moreover willing to burn much midnight oil in mastering his work.
>
> His most striking quality, which made such a marked impression upon all who came into contact with him, was a strange blend of opposites; he was learned yet jovial – even boyish. Whenever he spoke at a meeting he would be certain to give a sudden descent from the sublime to the ridiculous, which tickled the fancy of his listeners and made them feel that here was a fellow with a quaint sense of humour. Most of us were in the habit of chaffing him, and he always responded in the right spirit – it is a great testimonial to a man's character that his intimate friends chaff him in public.
>
> His disposition was delightfully sunny one – yet his life was not all sunshine; he had to pass through shadows, but they did not darken his outlook and we will remember him for many years as an outstanding and inspiring figure amongst us, always ready to laugh with us and never refusing to help us with our difficulties.

An appeal on behalf of his widow was responded to by several of the authors listed herein: E. S. Andrews; Sir H. Tanner; and R. Travers Morgan among them.

Etchells, E. F. *Mnemonic notation for engineering formulae: report of the Science Committee of the Concrete Institute*. London: E&FN Spon, 1918.

1918/USA

Johnson, Nathan Clarke

MME

N. C. Johnson's first essays into print (of which we have a record) arose from his employment as engineer of tests for the Raymond Concrete Pile Co. of New York. *Further discussion of testing concrete aggregates*' appeared in the November 1914 issue of *Concrete-Cement Age*, in response to a paper at the recent ASTM convention. *The microscope in the study and investigation of concrete* was published in the *Engineering Record*. The latter was reprinted for a British audience in 1915 (*C&CE* v.10, p.288). He appears to have entered practice as a consulting engineer at this time (still in New York), as he was credited as such throughout the rest of 1915 in a further five articles for the journal:

- *Mechanical disintegration of defective concrete*, p.348
- *The microscope as an aid in proportioning concrete for strength*, p.398
- *The microscope as a check on construction*, p.448
- *The microscope shows importance of mixing as a factor in making strong concrete*, p.514
- *High strength concrete produced through lowering of surface tension of mixing water*, p.615.

He expanded his writing activities in a series of collaborations with G. H. Hool [1912]. He was a contributor to Volume II of Hool's *Reinforced concrete construction* (1913), but for the *Concrete engineers' handbook* of 1918 he was co-author. For the 1918 book, the two men were assisted by S. C. Hollister, with specialist contributions by **Leslie H. Allen**, **Walter S. Edge**, **A.G. Hillberg**, and **Adelbert Mills**. Of these nothing further is known except that Allen was actively contributing to the ACI's *Proceedings* at this time – with *Cost accounting for the contractor and is relation to his organisation* (1917) and *Labor turnover and its relation to industrial housing* (1918) – and was to write about *Concrete roofing tile problems* in 1928. Johnson's collaboration with Hool continued with the publication of a handsome, expensive, two-volume set entitled *Handbook of building construction*, which addressed a wider scope than concrete alone.

Later in life Johnson became actively involved in the ACI, having presented *Motion picture studies of the making and placing of concrete* at the 1916 convention (*Proceedings*, v.12, p.394). In 1928 he wrote *Better concrete – do we mean it?* (*Proceedings*, v.24, p.480). By this time he was sitting on Committee 403 and in 1930 reported on *The treatment of monolithic concrete surfaces* in the ACI's *Journal* (May 1930, pp.717–730). This article treated wash and float work, bush hammering, machine grinding, acid treatments, painting, veneered surfaces and chemical surfacings. He also participated in the discussion of C. H. Jumper's *Tests of integral and surface waterproofings for concrete* (*Journal*, December 1931, pp.209–242).

Hool, G. A. and Johnson, N. C. *Concrete engineers' handbook: data for the design and construction of plain and reinforced concrete structures*. New York: McGraw-Hill, 1918 [C&CA 1959, IStructE, Yrbk 1934. Copy: TCS]

1918/USA

Hollister, Solomon Cady (1891–1982)

Solomon Cady Hollister
Courtesy of the American Concrete Institute

Like Slater [1917] and Abrams [1918], Hollister was one of the generation of gifted concrete specialists that assembled around Professor Talbot at the University of Illinois. Individually they achieved distinction over many years, and together they wove a web of excellence around the study of concrete in the period following the end of the First World War.

S. C. Hollister (known familiarly as 'Holly') was born on 4 August 1891 at Crystal Falls, Michigan. Moving as a child he grew up in the Pacific Northwest. In 1909 he enrolled at Washington State University, but took time off to earn money. He completed his final year at the University of Wisconsin where his teachers included Turneaure and Maurer [1907]. He was awarded his BSc in 1916 (and a degree in civil engineering years later in 1932). Newly qualified he taught at the University of Illinois for a year in Talbot's department. In 1917 he took he took up a position at the Corrugated Bar Co. in Buffalo and compiled the handbook *Useful data*, taking over (on his recommendation) from W. A. Slater after Slater had moved to Washington to work on the newly launched concrete ship programme.

Slater, knowing Hollister's capabilities, sought his replacement's assistance and Holly spent his weekends in late 1917 working on ship designs. On 1 January 1918 this was formalised and, at the age of 26, he was appointed chief designer at the Concrete Ship Section, US Shipping Board. With the head of section – R. J. Wig – he wrote up his experiences in his first paper for the ACI's *Proceedings*: *Problems arising in the design and construction of reinforced concrete ships* (June 1918). Also in June he wrote about a related design project: Construction of concrete barges for use of New York State Barge Canal.

At the end of the First World War Hollister turned to private practice, as a consulting engineer in Philadelphia. It is as such that he was acknowledged on the title page for his assistance to Hool and Johnson's *Concrete engineers' handbook* in 1918. This book has already been considered (above) and although Hollister wrote extensively, this represents his only book on concrete.

His papers in the ACI's *Proceedings* (not otherwise cited in this sketch), included:

- *Plasticity and temperature deformations in concrete structures* (June 1919)
- *Construction of concrete barges for use on New York State barge canal* (June 1919)
- *The design and construction of skew arch* (March 1928)
- *Livability of concrete dwellings* (February 1929)
- *Experimental study of stresses at a crack in a compression member* (March 1934)
- *A short method for computing moments in continuous frames* (November 1936)

Hollister specialised in reinforced concrete design, and in 1929 received the first Wason Research Medal for his skew arch bridge at Chester, Pennsylvania. He also contributed much to the development of ready-mixed concrete and the standardisation of specifications, as his papers in the ACI's *Proceedings* indicate:

- *Specifications for the small job* (February 1930)
- *Studies of concrete mixtures* (April 1931)

- *Tests of concrete from a transit mixer* (February 1932)

In 1930 he took up a post at Purdue University, during which time he served as president of the ACI. Four years later moved to Cornell, as professor and director of the School of Civil Engineering. He was appointed dean of the College of Engineering in 1937 and remained in post for 22 years. He became very influential in engineering education, and raised Cornell in to the top flight of American universities in that field. He established schools of chemical and aeronautical engineering and, as VP Development, eventually set up a new campus for engineering. For his role at Cornell, and chairing various committees of the Engineers Joint Council for Professional Development he received the ASEE's Lamme Award in 1952. In 1959 he retired, but remained a trustee at Cornell.

With the advancing years he became the 'grand old man' of concrete in the USA, with honorary doctorates from the Stevens Institute, Purdue University, Lehigh University and Wisconsin, and honorary memberships of six professional societies. As the ACI's longest serving member in 1979 (he joined in 1917), he featured prominently in the institute's 75th anniversary celebrations – he wrote *Sixty-two years of concrete engineering* for *Concrete International* (October 1979) – and was awarded the Turner Medal. Other tributes to him appeared in print – *Pioneering in concrete* in *Engineering: Cornell Quarterly* (Autumn 1980, v.15(2)) – and in person: 'The Solomon Cady Hollister Colloquium: perspectives on the history of reinforced concrete in the United States, 1904–1941' (held at Princeton University, 2 June 1980). Recognition came from surprising quarters; in earlier years President Hoover said of him: 'he is a great engineer, he is a superb teacher, and he knows more about our government than any engineer I know.'

S. C. Hollister died aged 90 on 6 July 1982. He was, in the words of Richard N. White (who wrote a tribute for the National Academy of Engineering): 'a famous man, a distinguished man, a good man, a man of great achievements; to those who know him well, he was a Renaissance man. He was an artist, a [professionally qualified] palaeontologist, a musician, an analyst, an avid reader and collector of rare books, a creative designer, a visionary educator, a most effective promoter, and a great engineer of truly uncommon breadth.'

Hool, G. A. and Johnson, N. C. [assisted by Hollister, S. C.]. *Concrete engineers' handbook: data for the design and construction of plain and reinforced concrete structures*. New York: McGraw-Hill, 1918 [C&CA 1959, IStructE, Yrbk 1934. Copy: TCS]

1919–1929: Exploration and expansion

1919/UK

Ballard, Fred (1860–1941)

JP

Like Lakeman [1918] and Hilton [1919], Fred Ballard was one of several 'prophets' addressing the application of concrete to housing at the end of the First World War, inspired in part by the transatlantic example of architects such as Irving Gill and Oswald Hering [1912]. Perhaps the British distribution of Campbell and Beyer's *Practical concrete work for the school and home* (1917) was a topical influence too.

Ballard's concrete house, 'Maybole', in Colwall, Herefordshire
Courtesy of The Concrete Society

Commemorative plaque to Fred Ballard
Courtesy of Bob Embleton

Although he was a farmer and rural justice of the peace, Ballard did have an engineering background. His father Stephen was a civil engineer and contractor, as well as a farmer of 157 acres, and was a member of the ICE. Stephen was born in Malvern, Worcestershire, in 1804 and having moved to Ledbury in Herefordshire, spent some time in Huntingdonshire where he met his bride, Maria Bird (b.1829, *née* Yaxley). They married in 1854 and returned to Herefordshire, to 'The Winnings' in Colwall, on the outskirts of the Malvern Hills, where Fred was born on 23 December 1860. Known in life as Fred, he was christened Frederick on 20 January 1861.

'The Winnings' was Fred's home until the early years of the century. He was resident there at each census to 1901 – unmarried and recorded as a farmer's son until after his father's death in 1890, when he was described as a farmer and then a fruit and hop grower and cider maker. Appropriately, in view of his future building interests, he also operated a clay pit and brick works in the locality. He learnt the skills of brickmaking at Hamletts of West Bromwich and started the works at Colwall around 1890. His sister Ada designed the decorative bricks that featured alongside the works' hard Ruabon-type bricks, tiles and kerbstones. After producing approximately 4.5 million bricks – enough for 100 to 200 houses in the neighbourhood – the works closed in 1900.

On 21 December 1906 Fred married Beatrice May Macara and during the following year they built their new home – 'Maybole' in Colwall, the subject of Ballard's book. This substantial and rather imposing house was built with walls of hollow concrete blocks and floors of reinforced concrete. Its description in the book accompanied that of cottages and farm buildings, stores and posts, and other such features of farm or estate – written according to a welcoming review by *C&CE*, 'in so simple a manner that it is intelligible to both the practical man and the non-technical reader'. A review of the second edition (in 1921) was equally warm, endorsing Ballard's advocacy of concrete and his view that as timber would be in short supply for the medium term, there should be a ready market for the newer material. It also notes the benefits of the agricultural silo so popular in America, and points to the fact that Ballard's own concrete silo was the first in his native county.

Later life was spent in public service, as a justice of the peace and a member of Hereford County Council from 1901–1937 and chairman from 1926–1937. In apparent contradiction to his earlier interests in brick-making and concrete construction (both of which depend on mineral extraction), he firmly opposed the introduction of quarrying in his beloved Malvern Hills. He was a member and (from 1909) chairman of the Malvern Hills Conservators, securing Parliamentary protection for the hills in 1924. A bench and plaque that was erected to his memory in 1938 describes him as 'Conservator of the Malvern Hills for more than 50 years'.

He was by no means, a typical concrete specialist, but his amateur enthusiasm for a privately funded project was characteristic of its time and helped promote a new application for concrete in a period when it was very much needed.

Ballard, F. *Concrete for house, farm and estate*. London: Crosby Lockwood, 1919 [Copy: TCS. Rev: *C&CE* v.15]. Second edition, 1921 [Rev: *C&CE*]. Third edition, 1925 [BL, Yrbk 1934]

1919/UK

Hilton, Geoffrey William (1880–1922)

A contemporary of Ballard (above), Hilton was a similar contributor to the post-war literature of concrete housing. The subtitle of his book *The concrete house* is sufficient to describe Hilton's circumstances: *an explanatory treatise on how the author, during war time, largely by his own labour, erected and completed a detached, two-storied, mono-bloc, concrete house, designed for his own occupation*. The house in question was Harpfield House. Here was another self-builder of private means providing an example in print to others.

Hilton's concrete house at Harpfield, Stoke-on-Trent Courtesy of The Concrete Society

As an example of domestic concrete work it appears very self-conscious, promoted in the press as well as in the form of a book. An article entitled *A concrete house at Harpfield, Stoke-on-Trent* was published in the May issue of *C&CE*, (v.19, pp.256–260) and the *Sheffield Weekly Telegraph* of 28 June 1919 announced the award of a £20 prize to Hilton for the 'ingenuity and general interest' of his *Concrete house that was built by one man*. This had appeared in *The Telegraph* of 10 May as part of a series of articles on what was widely known at the time as 'the housing problem'.

The book that followed was actually little more than a pamphlet, both in its approach and physical presentation. Written in a terse, spare style – almost in note form – it was polemical in its advocacy of the choice of concrete: 'the experience gained by the occupation of the house convinces the Author that houses of concrete are the homes of the future.' At only 21 pages of text (plus plates and a generous helping of diagrams) it was brief and – presumably as a consequence of wartime shortages – cheaply produced on coarse octavo paper bound within trimmed cloth-covered boards.

Hilton was an architect and surveyor who – before wartime shortages and the lure of a competition in 1917 prompted him to look further – had paid scant attention to concrete. However, after preliminary research he soon discovered the economies and intrinsic benefits of concrete, and painstakingly developed his own construction methods to suit the material. By his own efforts he succeeded in building a substantial home to his family's requirements, creating window frames, pantry shelves and even the staircase banisters from concrete. A plaque attached to an outer wall read: 'such labour in so strange a style, amazed the unlearned and made the learned smile.' Unfortunately for him, Hilton was unable to enjoy his bespoke home for long as he died in November 1922 aged only 42. An obituary appeared in the *Builder* of 1 December, p.822.

His published work is a very personal statement on the value of concrete in overcoming the housing shortage. The book is nowadays considered very scarce. The house was regrettably demolished in 2004, but the architect's name lives on as the designation of a local street.

Hilton, G. W. *The concrete house: an explanatory treatise on how the author, during war time, largely by his own labour, erected and completed a detached, two-storied, mono-bloc, concrete house, designed for his own occupation*. London: E. & F. N. Spon, 1919 [C&CA 1959, Yrbk 1934. Copy: TCS] and New York: Spon & Chamberlain, 1919

1919/UK

Williamson, James

AMICE

Little can be confirmed about the identity of James Williamson, the author of 1919. Sir James Williamson, the former Director of HM Dockyards, is one contender. He appears as a member of the Council of the British Fire Prevention Committee in the company of Edwin O. Sachs, Sir Henry Tanner and other senior figures from the Concrete Institute, according to published reports dated 1915–1918 – only a year before the book *Calculating diagrams for design of reinforced concrete sections*, 'by James Williamson' was published by Constable. However, as an associate of the ICE it is more likely this author was a young man, possibly the eponymous founder of the later firm James Williamson & Partners and the writer of a paper presented in 1936 at the Second Congress on Large Dams, Washington DC.

The book comprised a series of diagrams devised by Williamson as a method of reducing the labour of computation in the office. The diagrams had a more particular aim too: 'to furnish a simpler and more direct method of dealing with the comparatively difficult cases of doubly reinforced members and sections subject to combined stresses, the ordinary formulae for which are too complicated for general office use, and not usually adapted to direct solution.'

Williamson's own experience of computation had led him to combine the alignment chart – a widely known tool in his day – with the logarithmic ordinate diagram, to form a diagram sufficiently flexible to take care of the variables involved in the design of simple elements. To tackle the design of doubly reinforced members, Williamson devised a series of curves to obtain factors that could be resolved by use of 'simple arithmetic'. However, the book was intended 'essentially for the use of those who have a fundamental knowledge of the principles of reinforced concrete design', and though worked examples were provided, the algebraic formulae were not set out in the text.

Williamson, J. *Calculating diagrams for design of reinforced concrete sections*. London: Constable & Co., 1919 [BL, C&CA 1959, ICE, TCS, Yrbk 1934. Copy: TCS]

1921/USA

Baxter, Leon H.

Baxter – in his choice of audience and the identity of his publisher – stands outside the mainstream of specialist writing on concrete, but was a popular author within his own genre. In the early 1920s Leon Baxter was the director of manual training for public schools based at St Johnsbury, Vermont, and between 1920 and 1922 wrote several books on practical handiwork aimed at boys such as those for whose education he was responsible. Titles included *Electro-craft in theory and practice* and *Boy bird-house architecture* (both published in 1920), and *Toy craft* (in 1922). They each ran to a second edition in the mid 1920s. The book that concerns us here – *Elementary concrete construction* – was published by the Bruce Publishing Co. of Milwaukee and distributed in Britain by Batsford.

He was Supervisor of Manual Arts at the Western Reserve Academy in Hudson, Ohio, when he wrote again on concrete. 'How to cast concrete seats' was published in the May 1929 issue of *Popular Science Monthly*.

Baxter, L. H. *Elementary concrete construction*. Milwaukee: Bruce Publishing Co., 1921 and London: Batsford, 1921 [Yrbk 1934]

1921/USA

Harris, Wallace Rutherford (b.1876)

For his one book on concrete – *Concrete products: their manufacture and use* (a substantial 638 pages in length) – Harris collaborated with Henry Campbell [1917] who was director of the editorial bureau at the Portland Cement Association. It ran to a second edition in 1924.

Harris, W. R. & Campbell, H. C. *Concrete products: their manufacture and use*. Chicago: International Trade Press, 1921 and 1924 [C&CA 1952]

1921/USA

Hatt, William Kendrick (b.1868)

'Ken' Hatt was born in Fredericton, New Brunswick, Canada, and graduated from the University of New Brunswick in 1887. Building on his BA, he studied engineering at Cornell University and was awarded a second degree in 1891. Having taught at both his universities he obtained a post at Purdue University in Lafayette, Indiana,

William Kendrick Hatt Courtesy of the American Concrete Institute

He married Josie Belle Appleby and on 17 July 1902 they had a son, Robert Torrens Hatt 1902–1989), who was to become an eminent scientist and whose papers are now held at the Cranbrook Educational Community. The elder Hatt was appointed professor of civil engineering at Purdue and director of the materials testing laboratory. There he took an interest in the development of reinforced concrete reflected in two papers in the recently established ACI's *Proceedings*, dating from the first decade of the new century:

- *Reinforced concrete. Proceedings*, January 1907 (v.3, pp.58–62)
- *Tests of reinforced concrete hollow tile floor spans. Proceedings*, January 1908 (v.4, pp.28–45)

His involvement in the materials testing laboratory is hinted at in the second of these, but made explicit a paper he read to the Indiana Engineering Society in 1911 on *Concrete tables and tanks for a laboratory*. This described the concrete tables, shelves and mixing tanks specially designed for the new cement laboratory at Purdue. The connection with the laboratory was reinforced by the publication of his first book: *Laboratory manual of testing materials*. Originally published by McGraw-Hill in 1913, it claimed to be the 'outcome of the operation, through 18 years, of the Laboratory for Testing Materials'. It was successful and was reissued in 1920, 1922 and 1926. Also in 1913, his work on testing 're-rolled steel rail reinforcing bars' was presented at the 16th meeting of the ASTM and published in the July issue of *Concrete-Cement Age* (v.3, 1913).

Hatt's second book – *Flood protection in Indiana* (of 1914) – was of more local interest and issued by the university printer W. B. Burford. He returned to concrete with his election as president of the ACI for the years 1917–1919 (only the third incumbent in the institute's history) – an involvement marked by further articles in the *Proceedings*:

- *Genesis of reinforced concrete construction*. February 1916 (v.12, pp.21–39)
- *Instruction in reinforced concrete construction*. February 1917 (v.13, pp.284–287)
- *Moment coefficients for flat-slab design with results of a test*. June 1918 (v.14, pp.145–189)

Results of Hatt's laboratory tests on the *Circumferential system of flat slab construction* appeared in Edward Smulski's article of the same June issue, for which Smulski [1925] gave full credit.

In 1921 he collaborated with Walter Voss in writing his first book on concrete aimed at a wider audience. *Concrete work: a book to aid the self development of workers in concrete* was published in two volumes by John Wiley & Sons and issued in the UK by Chapman & Hall. It was followed a year later by *Answers to concrete work*.

Also in 1921 Hatt was appointed as director of the advisory board on highway research of the National Research Council in Washington. The role was responsible for overseeing the expenditure of $1billion on transport research nationally. His appointment was reported at length in the *New York Times* of 31 July 1921.

His remaining titles on concrete in the years following were university publications, available to a more limited academic audience. *Researches in concrete* appeared as Bulletin No.24, (Engineering Experiment Station, 1925), and *Physical and mechanical properties of Portland cements and concretes* followed in 1928. Work at Purdue and by researchers such as A. N. Johnson, R. E. Davis and G. E. Troxell at other universities, was summarised in a paper for the World Engineering Congress in Tokyo, 1929. *Some recent researches in fundamental properties of concrete in the United States of America* described investigations into creep, volume change and the effects of moisture on fatigue – aspects of concrete that were widely studied at the time on both sides of the Atlantic.

He was made an honorary member of the ACI in 1932.

Hatt, W. K. & Voss, W. C. *Concrete work: a book to aid the self development of workers in concrete*. New York: John Wiley & Sons, 1921 [BL, C&CA 1959] and Chapman & Hall, 1921.

Hatt, W. K. & Voss, W. C. *Answers to concrete work*. Chapman & Hall, 1922 [Yrbk 1934].

1921/USA

Howe, Harrison Estell (1881–1942)

Son of William James Howe and his wife Mary (*née* Scott), Harrison was born in 1881 at Georgetown, Kentucky. He was educated at Earlham College across the Ohio River in Richmond, Indiana, where he was awarded his BSc in 1901. Postgraduate studies in chemistry followed at the universities of Michigan and Rochester, combined with employment at a series of chemical companies. Eventually in 1913, he gained his MSc. Howe's war service was undertaken as a consultant to the nitrate division of the US Army Ordnance. After the war (in 1919), he was appointed chairman of the division of research extension of the National Research Council – a post characteristic of one who was to make a name for himself in public service.

During this time at the NRC, Howe wrote an influential book for the mass market about the impact of the cement and concrete industries on wider society. *The new stone age* – whose title was adopted later for a promotional film by the Portland Cement Association – was well received, being reviewed in no less a paper than the *New York Times*. On 23 January 1921 the *Times* announced that:

> In *The new stone age*, Harrison E. Howe, Chairman of the division of research extension of the National Research Council, tells the romantic story of cement and concrete – tells it with scientific accuracy and fullness, but without the verbal encumbrances which make so much scholarly writing seem detached from all connection with every day life.

Very soon after the book-launch Howe was invited to take over the editorship of *Industrial and Engineering Chemistry* – a high-profile role in chemistry which he fulfilled from 1921 until his death in 1942. During that time he wrote extensively for the public too. *The new stone age* was followed by several books on chemistry:

- Profitable science in industry
- Chemistry in the world's work
- Chemistry in the home

Other titles included six 'nature and science readers' for school children (written in collaboration with E. M. Patch). He was also author of numerous articles in both scientific and general-interest journals.

Besides writing, Howe became chairman of a committee on work periods for the American Engineering

Council of the Federated Engineering Societies in 1922. He recommended adoption of an eight-hour, rather than the prevalent 12-hour day. He was a member of the Purdue Research Foundation and a trustee of the Science Service. Throughout his career he sought to ensure that scientists received a fair remuneration for their research, and later he was director of the American Institute of Chemical Engineers. Following on from his wartime service, he became a colonel in the Chemical Warfare Reserve of the US Army.

For his contribution to the advancement of chemistry, Harrison Howe was awarded many honours. He was presented with honorary degrees from Rochester, Southern College, Rose Polytechnic Institute, South Dakota State School of Mines, and various professional honours including the Chemical Industry Medal.

On the 10 December 1942 he died 'in harness' in his Washington home, suffering the effects of a heart condition. He was aged 60. An obituary by F. J. Van Antwerpen was published appropriately in the popular scientific magazine *Science*, on 22 January. An annual award is now made in his name.

Howe, H. E. *The new stone age*. New York: The Century Co., 1921 and London: University of London Press, 1921 [C&CA 1959]

1921/USA

Voss, Walter Charles (b.1887)

Beside his collaborations with W. K. Hatt (above) in 1921 and 1922, Walter Voss went on to write *Architectural construction*. Volume I (1925, with Ralph Henry), analysed the design and construction of recent American building, and Volume II (1926, with Edward Varney), addressed timber construction.

1921/UK

Warren, William Henry (1852–1926)

William Henry Warren
Courtesy of the University of Sydney

Son and namesake of a railway guard and his wife (Catherine Ann [*née* Abrahams]), William Henry Warren was born in Bristol on 2 February 1852. He was destined to expand the social and geographical boundaries of his birth to play a leading role in engineering education and the establishment of engineering as a discipline in his adopted country, Australia.

He was educated at the Reverend Trevelyan's school, and apprenticed at the Wolverton workshops of the London & North Western Railway Company for the years 1865 to 1872. He undertook a course of study at the Royal College of Science, Dublin, in 1872 to 1873, gaining the Whitworth scholarship in the latter year and a Society of Arts technological scholarship. He combined practical experience with further study in Manchester. This culminated with admission to the Institution of Civil Engineers in 1877.

On 27 July 1875 Warren married Albertine King, with whom he had two sons. The family took the momentous step of migrating to Australia in 1881.

On arrival Warren found employment in the Public Works Department, Sydney, commencing on 9 May 1881. Besides his work on roads, bridges and sewerage, he started to teach applied mechanics at Sydney Technical College in the evenings and in 1883 he was appointed lecturer in engineering at the University of Sydney. His career gathered momentum and he was made professor of a new department the following year, and then Challis professor in 1890. A tour of engineering establishments in Britain and the USA during 1895 influenced his direction and in 1900 the senate at Sydney approved the university's first four-year engineering course. In 1908 Warren became dean of the Faculty of Science. For 42 years he was professor and, in the words of the *Australian Dictionary of Biography*, 'built up a great engineering school' that remains to this day.

His international reputation was partly secured by his writing. In 1892 his *Australian timbers* was published, followed by another on the theme – *Properties of NSW hardwood timbers* – in 1911. More relevantly for our purpose, 1894 saw the publication of *Engineering construction in iron, steel and timber*, which like Sutcliffe's book of the same year (see Introduction), was subsequently revised to incorporate the new structural material: reinforced concrete. In Warren's case the revision was long-delayed, as concrete was added (in the form of a second volume) only in the third edition of 1921. However, the *Australian Dictionary of Biography* describes it as 'an important textbook'. Although concrete was included so late, Warren had in fact contributed to the early understanding of reinforcement with papers submitted to the Royal Society of New South Wales as early as 1902, 1904 and 1905 [ICE]. In all he contributed over 50 papers to learned societies, of which he was a member of several.

Coincidentally with his appointment at the university, he was elected to the Royal Society of New South Wales in 1883 and having served as honorary secretary, was twice president in 1892 and 1902. In 1887 he was president of Section J of the Australasian Association for the Advancement of Science and much later (in 1919), he was the inaugural president of the Institute of Engineers of Australia. Internationally he was the Australian representative at the Institution of Engineering in Great Britain and a council member of ISTM.

Besides teaching and writing, Warren served as consulting engineer to the New South Wales government and was appointed to several prominent positions, including Royal Commissions on railway bridges (1885) and the Adelaide and Melbourne Exhibitions (1887/8). He sat on the commission of inquiry on Baldwin locomotives (1892) and on the Sydney Harbour Bridge advisory board (1901). He was chairman of the electric tramways board, served on the automatic brakes board and in 1910 and 1912 was an advisor on dams to the Indian and Egyptian governments.

After a long and illustrious career, Warren finally retired at the end of 1925. He had little time to enjoy his private pastimes of singing, golf and prize-winning bulldogs as he died only days later, on 9 January 1926, at his home in Elizabeth Bay.

That he was highly regarded is implicit in his inclusion in the *Australian dictionary of biography*, which describes him as 'the acknowledged leader of his profession, with a reputation extending beyond Australia … One of his major achievements was to convince the engineering industry by his personal example that graduate engineers were a sound investment'. His life is also recorded in a university publication of 1999, by M. R. Gourlay, entitled *Australia's great engineers: William Henry Warren*. The University of Sydney's engineering school is named after him and the Institution of Engineers of Australia awards an annual Warren Medal in his memory.

Warren, W. H. *Engineering construction. Part II: in masonry and concrete*. London: Longmans Green, 1921 [Yrbk 1934]

1921/USA

Whitney, Charles S. (1892–1959)

Charles S. Whitney first appears in the literature for his collaboration with G. H. Hool – *The Concrete designers' manual* published in 1921. Hool [1912], as we have seen, was an academic at the University of Wisconsin,

Charles S. Whitney Courtesy of the American Concrete Institute

and Whitney was to base his career in Wisconsin, at Milwaukee. Though this was the only book explicitly on concrete to bear Whitney's name, his later books on bridges treat the material; and his involvement in the ACI (to the very highest level) was reflected in several significant papers in the professional press.

Originally from Pennsylvania, Whitney read civil engineering at Cornell and graduated in 1915. He moved to Milwaukee in 1919 where he acted as engineer for the architect, A.C. Eschwheiler. As a young consulting engineer Whitney earned a reputation for his bridges. In his later years he moved to New York and entered into partnership with Othmar Ammann – one of America's most highly regarded bridge builders – as Ammann & Whitney of New York and Milwaukee. Certainly by the late 1920s he was concentrating on bridges. His 6th Street Arch Bridge in Racine, Wisconsin, was built in 1928, and then in 1929 the New York publisher William Edwin Rudge issued his book *Bridges: a study in their art, science and evolution*. Other bridge commissions followed, with Range Line Road Bridge in River Hills, Milwaukee (built in 1935), and Highland Drive Bridge in Cedarburg, Wisconsin (built in 1939).

His first paper for the ACI's *Proceedings* reveals his interest in bridges, too. *Plain and reinforced concrete arches* (v28(3)) was published in March 1932. His next paper – *Design of reinforced concrete members under flexure or combined flexure and direct compression* (1937) – was a significant contribution to the literature, and to practical engineering. Many years later (in 2004), it was selected by the ACI as one of its 'landmark series' to commemorate the institute's centenary. According to the selectors:

> After carefully reviewing the literature and the available test data, Whitney proposed an equivalent rectangular stress block to represent the real variation of stress in the concrete above the neutral axis. After showing that the calculations made using the rectangular stress block provided ultimate strength that was nearly identical to that obtained from tests, he concluded that 'no further theoretical justification is necessary if the formulas derived therefrom accurately predict the ultimate strength of the member'. The code provision for flexure and combined flexure and axial load in ACI 318-02 are based on Whitney's proposal.

With his reputation secured, Whitney found his work published overseas. In Great Britain, *C&CE* published his *Eccentrically-loaded reinforced concrete columns* (v33(11)) in November 1938.

After the Second World War, Whitney's career and his involvement in the ACI reached its peak in the 1950s. The trajectory of both is charted by his papers to the *Proceedings*, written (in some cases) with the participation of Boyd Anderson and Edward Cohen:

- *1950 Cost of long-span concrete shell roofs, v.46(6)*
- *1951 Comprehensive numerical method for the analysis of earthquake resistant structures with B. G. Anderson and M. G. Salvadori, v.48(9)*
- *1953 Reinforced concrete thin shell structures, v.49(2)*
- *1955 Design of blast resistant construction for atomic explosions with B. G. Anderson and E. Cohen, v.51(3)*
- *1956 Guide for ultimate strength design of reinforced concrete with E. Cohen, v.53(11)*
- *1957 Ultimate shear strength of reinforced concrete flat slabs, footings, beams, and frame members without shear reinforcement, v.54(10)*

Projects at the stage included the maintenance hangars for Southwest Airlines at Midway Airport in Chicago, in 1953. He also undertook the engineering design for several projects led by the Japanese architect, Mr. Yamasaki.

When (in 1955) Whitney was elected president of the ACI, the decision to erect a new headquarters building for the institute was taken. It was this team of Yamasaki and Whitney that won the commission. On retiring at the 52nd annual convention in Philadelphia, Whitney was able to offer an optimistic review of progress during his term. His review was later published in *Proceedings* (52(4) April 1956), in which 1955 was introduced as a year of 'exceptional progress'. In this capacity Whitney was also the natural choice for describing American progress to a British audience, published as *USA* in *C&CE* (v.51(1) January 1956).

The headquarters building project took off after his presidency and Whitney was to describe it in the 1958 volume of

Proceedings as *Cantilevered folded plate roofs ACI headquarters*.

Not long afterwards, Whitney died and the ACI's 'Charles S. Whitney Medal for Engineering Development' was founded in 1961 to honour his memory. This award is open annually to noteworthy engineering development work that advances concrete design or construction. His memory is perpetuated too, by the continued availability of *Bridges of the world: their design and construction*, which was reissued by Dover Publications in 2003.

Hool, G. A. and Whitney, C. S. *Concrete designers' manual: tables and diagram for the design of reinforced concrete structures*. New York: McGraw-Hill, 1921 [Rev: *C&CE* v.16, p.746]. Second edition,1926 [BL, C&CA 1959, Yrbk 1934]

1922/USA

Dana, Richard Turner (1876–1928)

MASCE, MASEC, MAIM&EE, M. Yale Eng. Assoc.

Richard Dana is included for his *Concrete computation charts* of 1922, but this was not his first engineering book: he had written several titles before this date, two of them jointly with H. P. Gillette [1908] and one as a contributor to the *Cyclopedia of Civil Engineering*, edited by F. E. Turneaure [1907].

Son of Richard Dana and Florene (nee Turner), Richard was born in Lenox, Massachusetts of a long-established New York family. He became a civil engineer and for a long time – certainly at the date of each of his books between 1908 and 1922 – he was chief engineer at the Construction Service Co.

His first venture into print was Turneaure's eight-volume *Cyclopedia of Civil Engineering* (American Technical Society, 1908), where he was but one of several specialist contributors. Maybe this brought him to Gillette's notice, in his capacity as editor of *Engineering-Contracting*, or possibly it was through mutual membership of the American Societies of Civil Engineers and Engineering Contractors, and the American Institute of Mining Engineers, but within a year the two men had collaborated on *Cost keeping and management engineering*, published in 1909 by Myron C. Clarke for the American market and by Spon in London. Interestingly the book was dedicated to yet another of the authors in this survey, the management prophet, F. W. Taylor [1905].

Richard married and by 1910 Mary and he were living at 15 William Street, Manhattan. His next book was a *Handbook of construction plant: its cost and efficiency*, published in 1914, by (like the previous title) Clark and Spon in their respective territories. A revision of J. Kindelan's *The Trackman's helper* followed in 1917, jointly edited with A. F. Trimble. Next, with Gillette again, was a *Handbook of electrical and mechanical cost data*, published in 1918, this time by McGraw-Hill. Four years later Dana revised his first work with Gillette and prepared a new edition under the title *Construction cost keeping and management* (McGraw-Hill, 1922 and 1927).

1922 also saw him collaborate on his last book – *Concrete computation charts* – this time with J. W. Kingsley. A cost engineer, but never a concrete specialist,

Dana devised his charts as an estimating tool and drew on formulae suggested by the ASCE in 1917. The Codex Book Company of his native New York published them in a modest format.

In his later years Dana lived at 54 East 81st Street, New York. It was there he died in 1928, just months before his daughter Florine's engagement to William C. Kopper of Park Avenue was announced in the *New York Times*.

Dana, R. T. & Kingsley, J. W. *Concrete computation charts*. New York: Codex Book Co., Inc., 1922.

1922/UK

Fougner, Nicolay Knudtzon (1884–1969)

Although Frowde and Hodder & Stoughton published his work in the UK, Fougner was unusual as an author of a British book in that he was a Norwegian national. He was also closely associated with the USA and American government concreting schemes and (reasonably so) might have been expected to be published in New York. More distinctive than his cosmopolitan background, however, was his place in the literature as the author of the only mainstream book on the subject of shipbuilding in concrete – an application for which he will ever be associated.

Nicolay Fougner, younger brother of the engineer Hermann Fougner, graduated from the technical college at Trondheim at about the time of Norway's secession from Sweden. He moved to America to join his brother who, after an early career in the Russian Far East and South Africa, had taken a post with the Trussed Concrete Steel Co. at the company's New York office. Nicolay served as inspector for the firm at the Detroit River Tunnel from 1906–1908, but moved to the London office where he was soon made responsible for Truscon's activities east of Suez. Among his projects was the 156-ft dome of the Melbourne public library. He spent some time in Manila (he was certainly there in 1915), but during the war returned home to Norway via Manchuria and Siberia. Christiania (now Oslo) became his home for a while.

Norway's maritime losses during the war prompted Fougner to explore the

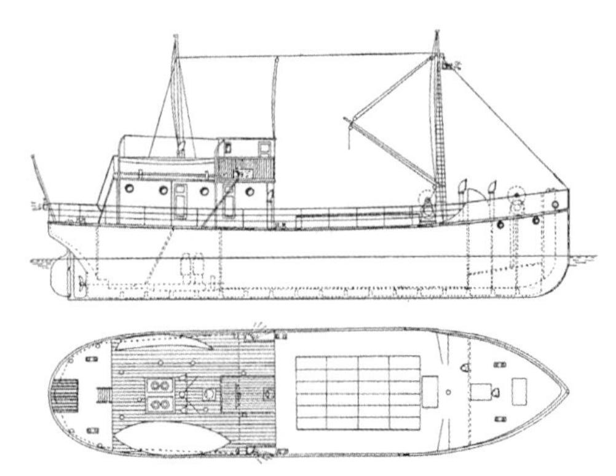

Fougner's Namsenfjord *in 1917, the world's first sea-going concrete ship Courtesy of The Concrete Society*

idea of building ships out of reinforced concrete – a vision that he had first expressed in experiments leading to a patent application in 1912 and developed through the construction of a concrete craft (a lighter called the *Buccaneer*) during his time in the Philippines. By 1916 his plan was to build a prototype sea going powered vessel. He applied to the Norwegian Department of Shipping and decided for expediency's sake, to build it at his own firm's expense. Fortunately, while seeking the department's permit, he secured an order from a Mr. Alizpan of Namsos for this first concrete ship. The permit was awarded on 31 May and the *Namsenfjord* was launched successfully on 2 August 1917 – the world's first powered sea-going concrete ship. It received its first cargo at the end of the month.

During this time Fougner was also negotiating with the Norwegian authorities (as far as the Prime Minister) to proceed with bigger ships without waiting for the full official reports into the *Namsenfjord*. This was finally granted, despite the prevailing scepticism, and Fougner's *Staal-Beton Skibsbygningscompani A.S.* of Jeloen, Moss, commenced construction of the flush-deck cargo carriers MS *Stier* (600t dw) and MS *Askelad* (1005t dw). These were also launched in 1918 and were followed by the tug *Staal-Beton 19*, built for the firm's private use. In all Fougner built 30 concrete vessels, including a floating dock, a lightship, a water carrier and sundry lighters and barges – all of which are described in his *Seagoing and other concrete ships*.

This book (published in Britain in 1922) contains an account of Fougner's own shipbuilding; technical details of the ships' specification and construction; an assessment of the merits and drawbacks of concrete as a shipbuilding material; a statement on costs; and a review of concrete ships elsewhere in the world. It also identifies specialist sources in the contemporary technical journals. By 1922 – although concrete ships had been built first in Norway and then in Britain, China, France, India, Italy and the Netherlands – the biggest government-sponsored programme of shipbuilding had been in the USA in 1918 and 1919. We have already touched on the research and design work of Slater and Hollister in this regard. It was to the USA that Fougner now turned.

Using his brother Hermann's position in the American concrete industry to present his case, Fougner argued for the benefits of concrete at the US Board of Shipping. In October 1917, with the *Namsenfjord* an approved reference ship, the Fougners agreed to prepare a shipyard in America in which to build ships for the newly announced national programme. By the time the First World War ended however, only 12 of the national total had been built, including just one – the 2,460t *Polias* – completed by the Fougner Shipbuilding Co. in late 1918. In addition they built the first ever concrete oil carriers in the Spring of 1918, for the Standard Oil Co. of New York.

At the end of the war much of the momentum for concrete ships was lost and, though plenty of ships that had been planned during hostilities were launched in 1919, the early 1920s saw a return to steel. Fougner's book of

1922 – while apparently a prospectus for concrete – actually represented a valedictory summing up of activity until war returned in 1939.

In 1923 Fougner took up a directorship of the Truscon Co. again, this time responsible for the South American market. He was based in Argentina, but travelled widely in south and central America before finally settling in New York.

Although Fougner's contribution to the technology of concrete shipbuilding is widely described in the literature, a rare biographical treatment appears in *Saga in steel and concrete: Norwegian engineers in America* by Kenneth Bjork (Norwegian-American Historical Association, 1947).

Fougner, N. K. *Seagoing and other concrete ships*. (Oxford Technical Publications). London: Henry Frowde and Hodder & Stoughton, 1922 [BL, C&CA 1959, ICE, Yrbk 1934. Copy: TCS]

1922/UK

Hudson, Richard John Harrington

BENG, AMICE, AMIMECHE, MCI, MACI

We have meagre evidence of Harrington Hudson's career, but what we have shows that he flourished in the 1920s. Towards the end of the First World War (on 13 September 1918), the young Hudson was commissioned as a 2nd lieutenant, and after it he joined the Concrete Institute and took an active part in its affairs. For instance, he became a member of the Science Committee, and so when his *Reinforced concrete* was published in 1922, it contained a foreword by E. Fiander Etchells, the then president of the institute.

Subtitled *A practical handbook for use in design and construction*, the book set out 'to furnish just those tables, charts, data and other items of information which are of the greatest practical value in designing and executing sound and economical reinforced concrete construction', in which 'every endeavour' was made 'to develop the practical side of the subject and present it in the simplest possible form' for those actually engaged in concrete work.

Hudson was keen to stress the monolithic character of reinforced concrete construction and to promote its economic design – an aspect that he had personally investigated. He had already written up his conclusions in *Concrete & Constructional Engineering* the previous year as *Secondary stresses in monolithic structures and how to calculate them* [v.16, pts.11–12]. He set out in Chapter XII what he believed to be a new method of analysing the bending stresses in pillars. On the subject of centring and formwork removal times however, he made no claim to originality, acknowledging the work of the Engineering Experiment Station at Illinois and of Oscar Faber writing in *C&CE*.

The book reproduced extracts from British Engineering Standards and the US Government's specifications for Portland Cement, but more notably it included the text of the London County Council Regulations for the first time in print, with the formulae written in the new standard notation.

On publication, Hudson's efforts and the book's novel approach were applauded by no less a reviewer than Oscar Faber, whose informed comments bear quoting at some length for what they reveal of the author's achievement:

> Mr. Hudson understands his subject well, has worked at it independently, and presents it in a manner which many will find useful and acceptable … Mr. Hudson belongs to the younger school of engineers who in the design treat the structure as monolithic, as in fact it is, who realise that a monolithic structure has many advantages, but imposes on a conscientious designer the obligation of investigating the effect of deflection of a member upon the members with which it is monolithic. He realises that there was growing up a tendency to claim all the advantages of a monolithic structure, and yet coolly ignore the stresses – sometimes very great stresses – which the monolithic character of the structure imposed on some of its members, and in fact, to treat it as a jointed structure in the design, knowing well that it would not be a jointed structure in execution. In his welcome attitude he has, of course, my fullest support and encouragement, for only along such lines will progress be made … By far the most valuable part is Part III (Monolithic Design), which is squeezed into twenty-two pages, and a little more of this and a little less standard notation would probably have been an improvement … Altogether the work is one which can be thoroughly recommended [*C&CE*, p.601].

Although Harrington Hudson wrote his preface in London, he is described as 'Executive Engineer, Public Works Department (India)' on the title page, and his acknowledgement of the assistance of the chief engineer to the Government of Behar and Orissa suggests the location of his professional employment. Presumably he returned to India after arranging the publication of his book with Chapman & Hall, and at the same time registering a patent application that illustrates his practical approach to concrete technology. British patent GB186561, dated 12 April 1922, is described as follows:

> Bars are bent to cranked form suitable for use as reinforcing-members for concrete, by a bell-crank lever provided with two bending formers and dies, a stop being linked to the lever in such a manner that the bent end of the bar remains parallel to its original direction while being displaced laterally.

Later in the 1920s – while living variously in Muzaffarpur in India, and Sidcup in Kent – he applied for further apparently unrelated patents for 'improvements in or appertaining to folding seats' and to plywood sheets. These were published on 21 August 1929 and 2 February 1931 respectively. Of his later career we know nothing more.

Harrington Hudson, R. J. *Reinforced concrete: a practical handbook for use in design and construction*. London: Chapman & Hall, 1922 [BL, C&CA 1959, Yrbk 1934. Copy: TCS. Rev: C&CA, 1922, p.601]

1923/UK

Badder, H. C.

Badder appears in 1923 as the author of *Impervious concrete*, issued by a publisher not normally associated with concrete: Educational Publishing Ltd. Three years later (in 1926) in an article entitled *British and American building* (*Building*, July 1926, pp.138–140), Badder addressed a more general construction theme and concluded that American tradesmen were paid more than their British counterparts, but earned it through greater productivity. After that Badder disappears from the record.

Badder, H. C. *Impervious concrete*. London: Educational Publishing, 1923 [BL, ICE]

1923/USA

Pond, Clinton de Witt

Pond was a lecturer at Columbia University, and in the year 1915/16, taught architectural engineering, in both the summer session and for extension programme. This was the subject of his book *Engineering for architects*, which was published in New York that year by Lemcke & Buechner, and in London by Humphrey Milford (sales agents for Columbia University Press). The purpose was 'to explain simply, and plainly, a few secrets of engineering, so that an architect can decide for himself the size of beams, girders and column sections necessary for construction'. He followed the theme with *Concrete construction for architects* eight years later.

Pond, Clinton de Witt. *Concrete construction for architects: a concise treatise on the design of reinforced concrete slabs, beams, girders, columns and footings, and a description of the actual design of a concrete building involving the use of flat slab construction*. New York: C. Scribner, 1923 [C&CA 1952]

1923/USA

Urquhart, Leonard Church (b.1886)

Like 'Taylor & Thompson' and 'Turneaure & Maurer' before it, 'Urquhart & O'Rourke' became a standard textbook in the authors' own day and has stood the test of time, with editions being updated right up to the 1990s. Both men were professors of structural engineering and wrote more widely than simply on reinforced concrete structures, but it is for their first collaboration – *Design of concrete structures* – that they are included here.

Urquhart was the professor in charge of structural engineering at Cornell University, and O'Rourke was his close colleague, so it should be no surprise to learn that *Design of concrete structures* was envisaged as a textbook for students on engineering courses:

> It is the aim of the authors to give sufficient development of the theory of concrete design with illustrative problems to insure the beginner a thorough understanding of the fundamentals. Complete designs of the more common concrete structures are given in order to furnish a vehicle for bringing together all the fundamental theory involved.

The work was derived from lecture notes based on teaching experience.

Writing in the *Structural Engineer*, E. S. Andrews applauded the high standard of presentation, and recommended the book as useful to Institution of Structural Engineering members, but was inclined to regard it as a little too advanced for the average elementary student. He noted differences between British and American practice, and regretted the absence of a bibliography and educational features such as problems set for solution. Nonetheless, the authors went on to other work whilst the book went on to several editions – the 11th (updated by Arthur H. Nilson and George Winter) was issued posthumously in 1991.

The immediate spin-off was *Tables and diagrams from design of concrete structures*, but several other titles followed:

- *Stresses in simple structures.* McGraw-Hill, 1926. Second edition, 1932
- *Design of steel structures.* McGraw-Hill, 1930
- *Elementary structural engineering.* McGraw-Hill, 1941

Besides their collaborative writing, Urquhart prepared an early 39-page pamphlet entitled *Elementary reinforced concrete building design* in 1915 (Ithaca: Carpenter & Co), and later edited the *Civil engineering handbook*, published by McGraw-Hill in 1932. Urquhart himself contributed the section on 'Stresses in framed structures', and the subject of concrete was addressed by S. C. Hollister [1918]. Hollister had already assisted Hool & Johnson in their *Concrete engineers' handbook* for McGraw-Hill, and in 1934 he transferred to Cornell where – as professor of civil engineering – he became Urquhart's colleague. Positive reviews (*SE*, October 1934 p.414) and satisfactory sales led in turn to further editions – the fourth was issued in 1959.

Urquhart, L. C. & O'Rourke, C. E. *Design of concrete structures*. New York: McGraw-Hill, 1923 [BL]. Second edition, 1926. Third edition, 1935. Fourth edition, 1940. Sixth edition, 1958.

1923/USA

O'Rourke, Charles Edward

Like Urquhart, O'Rourke was a professor of structural engineering at Cornell University, though not head of department. His collaborations with Urquhart have been described above while his solo work of the early 1930s was entitled *General engineering handbook* (also for McGraw-Hill). He was also a contributor to *Foundations and flat slabs* (published by the International Textbook Company), itself an amalgam of the earlier *Design of spread footings* (1932), *Flat-slab design* (1935) and *Foundations and piling* (1941).

1924/UK

Childe, Henry Langdon

MJI

It is difficult to dissociate any account of Childe's career from that of the development of the journal *Concrete & Constructional Engineering* and its publisher Concrete Publications Ltd – so closely were they linked. However, a fuller account of these enterprises already exists (Trout, E. A. R. *Concrete Publications Ltd and its legacy to the concrete industry*, *Construction History*. 2005) and it is largely from this that the following brief comments relating to Childe in the 1920s and 1930s – and specifically his role as an author – are drawn.

Henry Langdon Childe entered the world of concrete publishing as a young man of 29 years when in 1922 – having learned his trade at *The Builder* magazine – he was appointed 'to edit the journal and to manage and develop the business' at Concrete Publications Ltd. *C&CE* had been without a formal editor since the death of its founder Edwin Sachs, in 1919 – though Oscar Faber had acted as technical consultant in the interim – and Childe self-consciously declared his aim to 'maintain the high ideals' set by his predecessor. By 1956, when marking the 50th anniversary of *C&CE*, he was able to take pride in the fact that 'in the period of fifty years the journal has had two editors only'.

Childe was also the figurehead leading an expansion of the range of Concrete Publications Ltd's journals – *Concrete for the Builder & Concrete Products* (1926) and *Cement & Cement Manufacture* (1928) – the introduction of the first *Concrete Yearbook* in 1924 and a move into book publishing. In 1956, Childe looked back at this busy period:

> In the early 1920s some of the most popular textbooks in use in this country were of American origin. There were, of course, some excellent books by British authors, but more were needed if British students and British engineers were to have a sufficiently wide selection of British books. It was therefore decided to start the 'Concrete Series' books, and the guiding principle was to make available good books at low prices in the confident expectation that such a policy could not fail to be a financial success … It is a source of great gratification that some of these books, of which nearly fifty have been published, are now used as textbooks in American universities.

This new activity was not without precedent, as Concrete Publications Ltd had already issued two titles – one on concrete roads and another on houses – but these were occasional items, not part of a conscious drive to supply and sustain an eager market.

Early contributors to the new Concrete Series included Oscar Faber [1912] – Childe's collaborator and former technical adviser; A. C. Davis, representing APCM (the cement maker and Concrete Publications Ltd's financial backer); and Albert Lakeman [1913], a regular contributor to *C&CE*. Albert Wynn [1926] followed and soon Childe himself was adding to the list.

His first venture was *Manufacture and uses of concrete products and cast stone*, which as the 11th title in the Concrete Series, was published in 1927. This was only a year after the launch of his magazine *Concrete for the Builder & Concrete Products* and so Childe was presumably capitalising on recent journalistic activity, on his new contacts in the industry and a degree of lately acquired knowledge in this field. The book set out to deal with every phase of pre-cast concrete manufacture, from tiles to architectural cast stone, and its publisher claimed that it was the only such book to do so comprehensively. Given the scope of Childe's future publishing, it might be mentioned that he also covered a wide range of surface finishes and the design and production of moulds. Aimed at a practical audience, the content was expressed in simple direct language with line drawings and photographs on most pages.

The book was noticed by the *Structural Engineer* (1927, p.328) and was well received by manufacturers and builders alike, with comments such as 'this most interesting and helpful book' (Guildford Tile & Artificial Stone) and 'the finest technical book we have seen' (C. G. Powell, builder). Castle Concrete wrote: 'your book is of real practical value to us; the suggestions contained therein must prove useful to all interested in concrete.' Its timing was propitious, coinciding with an upsurge in manufacturing industry and preceding by months the establishment of the Cast Concrete Products Association in 1928. Even in the USA, *Concrete* described it as 'a book which any American pre-cast concrete maker will find of considerable value'. Its success can be measured by the rapid succession of new editions – by 1930 (a mere three years after publication) it had reached its fifth and was being 'thoroughly recommended' by the *Structural Engineer* (January 1931, p.31). After selling 16,000 copies this title reached its eighth edition in 1949 and was re-issued yet again as late as 1961, necessitated (as the publisher explained) by the 'phenomenal growth' of the pre-cast industry.

This success seems to have inspired Childe to issue another, related title – *Precast concrete factory operation* – two years later. His approach to his subject – a description of the production processes of twenty manufacturers – betrays him as a journalist and reporter. Nonetheless *The Builder* described the book as 'essentially practical in character' and noted 'the author gives scores of useful 'tips' which are the result of a close personal acquaintance with the inner workings of the industry'.

The third title in this period was a collaborative work: *Concrete Surface Finishes, Rendering and Terrazzo* was jointly prepared by W. S. Gray [1931] (a regular contributor to *C&CE*) and Childe in 1935. Two thousand copies were sold before revision in 1943 and reprinting in 1948. Finally (in 1964) it was re-issued by Childe alone under the modified title, *Concrete Surface Finishes and Decoration*.

Just outside our time range but closely associated in subject matter, was *Manufacture of Concrete Roofing Tiles* that Childe wrote in conjunction with **R.H. Baumgarten** in 1936. Baumgarten – a London-based equipment supplier and consultant to the precast industry (and later manager of the concrete products section at the Cement & Concrete Association) – had written the chapter on terrazzo for the anonymous *Concrete Surface Finishes* the year before, and had tried out his material on roofing tiles in a series of articles for Childe's magazine *Concrete Building & Building Products* (June 1935 to January 1936).

The Second World War brought its distinctive challenges to the distribution of journals and the publishing of books – the number of which had grown rapidly throughout the 1930s. But despite the shortages of paper, the requirements of the War Economy Standard, and the inevitable shift in subject matter to wartime applications, Childe was able to oversee revisions of his earlier books and to write a new basic textbook. His *Introduction to Concrete Work* came out in 1943 and ran to 10,000 copies. It was repeated again and again: in 1945; 1947; 1949; 1951; and 1961, by which date over 40,000 copies had been sold.

By the date of this last edition, Childe had celebrated the 50th anniversary of *Concrete & Constructional Engineering* and his introduction to the special issue (in January 1956) reveals something of Childe's success over much of this period.

Throughout the years, Childe had prefaced the journal with a couple of pages of 'Editorial Notes' and these in turn had generated many responses, both 'appreciative and vituperative' as he put it himself. So 'in view of the interest they have aroused', a selection of the notes was reprinted in 1958 in book form. Excerpts of letters to the editor were quoted in the foreword as 'some of the reasons' for publishing and with a becoming modesty, he cited as many negative as positive comments. Those below reflect upon the commendable qualities of the man:

> Your Editorial Notes have been the feature which has distinguished your journal in our eyes above all other technical publications. We have always admired their good sense, their penetrating thought, and the masterly quality of English expression.
>
> Over the years I have derived a lot of pleasure and information from your Editorial Notes. The standard of these has been uniformly high, and they have been particularly pleasing in that a broad approach to many engineering problems has been set out in them.
>
> Your last Editorial Note was so good that I had it copied and circulated throughout this organisation.

His choice of subjects reveal Childe's interest in the role of the engineer and the relationship between the engineering and architectural professions and wider society. He was an advocate of a rounded education, of a balance between science and the arts. He strove for logical thought expressed in clear and accurate English, and valued research, innovation and good literature. His personal choice of reading matter included classics by authors such as Aristotle, Bacon, the bishops Berkeley and Butler, Herodotus, Juvanal, Montaigne, Plutarch, Socrates and Adam Smith. Conversely he was hostile to the coining of ill-considered jargon, professional pomposity and self-promotion, and technical narrow mindedness. Each he condemned with some force.

Childe's time as Editor was drawing to a close however, as announced in November 1959:

> Retirement
>
> Mr. H. L. Childe, who has been Editor of this journal for thirty-eight years, is retiring on December 31 next. During this period he founded and edited the *Concrete Year Book*, the monthly journals *Concrete Building & Concrete Products* and *Cement & Lime Manufacture*, and the 'Concrete Series' books of which more than 600,000 have been sold, and has also been General Manager of the business of Concrete Publications Limited. [*Vol.54, No.11, November 1959*]

Even in retirement, Childe's enthusiasm and energy were not lost to the concrete industry. He maintained a correspondence in the pages of *C&CE*, and prepared new editions of his own books. A new title was *Practical concrete work* (1963), which he was asked by Evans Brothers to write for amateurs. In keeping with the intended audience, Childe himself made all the examples of garden-ware illustrated in the book, in his garden at Riverhead in Kent. It was also issued by Chemical Publishing Co. two years later. Ten years after retirement his final book – *Everyman's guide to concrete work* – was published by Godwin in 1969. It sets out to be 'an attempt to make available in one cover the information on concrete which the writer's experience has shown to be needed by many builders [etc] anxious to improve their knowledge'. Subtitled *better concrete at lower cost*, it acts as a worthy testament to a lifetime's achievement.

Childe, H. L. *Manufacture and uses of concrete products and cast stone*. (Concrete Series, No.11). London: Concrete Publications Ltd, 1927 [C&CA 1959. Copy: TCS. Rev: *SE* 1927, p.328]. Second edition, 1927 [BL] Third edition, 1927 [ICE]. Fifth edition, 1930 [BL, C&CA 1959. Rev: *C&CE* v.26, p.537; *SE* January 1931]. 1961 [TCS]

Childe, H. L. *Precast concrete factory operation*. London: Concrete Publications, 1929. [C&CA 1959, ICE, TCS]

W. S. Gray and Childe, H. L. *Concrete Surface Finishes, Rendering and Terrazzo*. London: Concrete Publications Ltd, 1935. [Copy: TCS]

1924/USA

Kinne, William Spaulding

William Kinne (whom we met as a collaborator of G. A. Hool), was a member of the University of Wisconsin, in whose yearbook he appears with the class of 1912. There, under Dean Turneaure [1907], he contributed articles to the *Wisconsin Engineer. Impact – the effect of moving loads on railway bridges* (v.22(8) May 1918) and *Deflection of structures by the method of elastic weights* (v.24(5) February 1920) are two examples. At the latter date he was associate professor of structural engineering.

By 1924 – at the time of his collaboration with Hool on *Reinforced concrete and masonry structures* – he had been acting as editor-in-chief of McGraw-Hill's Structural Engineers' Handbook Library for over a year. Besides *Masonry structures*, the other five volumes in the series were:

- *Foundations, abutments and footings*. (1923)
- *Stresses in framed structures*. (1923. Second ed, 1942)
- *Structural members and connections*. (1923. Second ed, 1943)
- *Movable and long-span bridges*. (1923. Second ed, 1943)
- *Steel and timber structures*. (1924, Second ed, 1942)

For an evaluation of these see the entry on Hool [1912].

Hool, G. A. and Kinne, W. S. (eds). *Reinforced concrete and masonry structures*. New York: McGraw-Hill, 1924 [BL, ICE, Yrbk 1934]

1924/UK

Manning, George Philip (b.1893)

MEng, FICE, MConsE

Born in Ashton-under-Lyne, Manning turned 18 in Rotherham and read for his degree at the University of Sheffield nearby. There he was awarded his B.Eng. with honours and was Mappin Medallist in Engineering for 1913. The medal

George Philip Manning (mounted) at Alexandria
Courtesy of The Concrete Society

– instituted by Sir Frederick Mappin, Mayor of Sheffield and 'father of the university' – was one of eight issued annually to recognise a student from each of the departments in the faculty of engineering. His early career as an engineer – 'in the halcyon days of pre-1914' (as he later described them) – was spent with L. G. Mouchel & Partners. Interestingly for one who was to become a significant author for British publishers, his view at the time was that 'there was no British textbook that I considered worth buying', though he admired the 'excellent textbook' by Turneaure & Maurer. His experience was soon diverted to military service, for which he was commissioned as an officer.

He served with the 53 Division against the Turks and after the Armistice (stationed in Alexandria while waiting for demobilisation), was asked to exploit his professional expertise by running a course on reinforced concrete. On returning to the UK he settled in Ealing Common, West London.

As a young engineer in 1913, Manning had taken the first steps in outlining his own method of arch design – a process he added to and refined over a 16-year period in practice from the end of the First World War to 1933. His early interest in reinforced concrete is indicated by the collection of glass slides and manuscripts dating from 1920, now held at the ICE.

This interest received published expression when in 1924 he finished his first book – *Reinforced concrete design* – issued by Longmans, Green & Co. Based largely on personal experience; it was intended to be useful as a practical guide for the design office. He also drew on American experience, referring to *Engineering* and Hool & Johnson for the material on flat slab construction. He again praised the quality of American literature and cited both Taylor & Thompson and Turneaure & Maurer as exemplars.

Ewart Andrews' review in the *Structural Engineer* (v.2, p.374) opens in arresting way: 'This is a remarkable book; it may be considered from two points of view – the technical treatment and the author's views on the principles of design which he sets out in an extremely interesting albeit rather an aggressive and heterodox manner.' The coverage was broad and the treatment of the subject in the early chapters 'exceedingly good'. Manning's section on the design of members included new material on the design of continuous beams. Part III covered the design of structures, which interestingly (given his future writing), included sections on water towers and bunkers. Andrews considered the book to be of value to readers familiar with the basics of design and to be written in a clear style with lots of 'pep'. Indeed he referred to the 'personality which is so strongly marked in Mr. Manning's work'. However, Andrews took issue with some of what he regarded as Manning's excesses: 'the author gives his views on the ethics of design; this part of the book contains some dangerous doctrine and the debonair manner in which the heresy is given is startling.' Andrews was an influential commentator but Manning did not rush to prepare a revision.

He did however, marry his fiancée Miss Kathleen Favell Everett – in the September quarter of 1924, just months after completing the book – and started writing for the leading British concrete journal *C&CE*. His articles were accepted for the next five decades:

- *Stresses in thin circular pipes, etc., under symmetrical loads*. 19(7/8) July/August 1924
- *Modern practice in reinforced concrete*. 21(6) June 1926
- *New factory for the Michelin Tyre Co., Ltd at Stoke-on-Trent*. 23(1) January 1928
- *Sand grading, water-cement ratio and practical concreting*. 23(10) October 1928
- *Bending moment envelopes*. 24(5) May 1929
- *New method of flat-slab construction*. 24(9) September 1929

The 1930s saw Manning pen two new books, published in both cases by Pitman. *Construction in reinforced concrete: an elementary book for designers and students* (1932) set out to treat concrete construction in a practical manner and appears to have been successful, particularly in its treatment of the otherwise neglected topic of timber formwork. *Building* described it as 'an excellent elementary book for those engaged on the practical side of concrete work'; an opinion shared by *Structural Engineer*, *Architects Journal* and *Municipal Engineering*.

It was followed rapidly by another title for Pitman: *Reinforced concrete arch design: a textbook for engineers and advanced students* (1933). This volume picked up Manning's early theoretical work and linked it to workaday examples from the 'general run of small and medium span bridges occurring in everyday practice'. Chapters 2 to 4 first appeared as a series of articles in *C&CE*, whose editor (H. L. Childe) Manning was happy to acknowledge. *A method of arch design* (as these ten articles were entitled) was published from February 1930 (v.25, no.2) to October 1931 (v.26 no.10).

Other articles followed, addressing tests, the strength of concrete and formwork joints:

- *Comparative pull-out tests*. 27(7) July 1932
- *Tests of bolted joints in shuttering*. 27(9) September 1932
- *A plain slab bridge*. 28(1) January 1933
- *Strength of nailed joints in shuttering*
- *What is the strength of concrete?* 29(7) July 1934
- *Asbestos-cement bar spacers*. 29(11) November 1934

In 1935 he concentrated on shrinkage, with the first of these foreshadowing his later interest in tanks: *Shrinkage cracks in reinforced concrete tanks*, 30(5) May 1935; *Shrinkage stresses in sliding shutter work* 31(7) July 1936; and *Shrinkage and the structural design of buildings*, 31(11) November 1936.

1936 saw the reissue of his *Reinforced concrete design*, in which he paid tribute to his late wife's 'encouragement and support'. Perhaps her loss had spurred him on to greater writing activity in the busy, productive years of the early to mid 1930s. His continuing interest in design is reflected in an article for the *Civil Engineering & Public Works Review*, entitled *Proportions of reinforced concrete cantilever walls* (v.28, 1933).

Articles continued to appear regularly in *C&CE*, picking up earlier themes of pipes and sliding shutters:

- *New pipe factory at West Thurrock*. 33(5) May 1938
- *Some developments in sliding shuttering in England*. 33(12) December 1938
- *Reinforced concrete silos at Welwyn Garden City*. 34(12) 1939

The Second World War brought an obvious change of emphasis to the war effort, exemplified by Manning's articles on air raid precautions published in July 1940 and March 1942. Other articles in 1942 included: *A spiral staircase in reinforced concrete*, 37(6) June; and *A culvert combined with a retaining wall*, 37(11) November.

Having revised *Reinforced concrete design* in 1936 he did the same for *Construction in reinforced concrete* ten years later and a second edition was published in 1947. By the time he turned to his *Arch design* in 1954, he had written a fourth new work – *The Displacement Method of Frame Analysis* (1952). Consequently the revision saw Chapters 7 and 13 brought into line with his displacement method as expounded in the new book. His writing in *C&CE* was similarly influenced: Manning starting the year with *Analysis of groups of piles by the displacement method* (47(1) January 1952). Other articles in the early 1950s included a rather general *Specification for concrete work* (46(1) January 1951) and a collaboration with W. Hamilton for the *Structural Engineer*: *Some structural uses of aluminium alloy with special reference to domes* (32(7) July 1954).

The Displacement Method of 1952 was published by Concrete Publications Ltd and was the first of several undertakings for this firm. Although he had reached retirement age, the 1960s represented a late flowering in Manning's career as an author and as an editor of the work of others. The decade started with Manning's revision of the late W. S. Gray's *Reinforced concrete reservoirs and tanks* – the fourth edition was issued in 1960 after his treatment of the *Design of slabs for rectangular tanks* in *C&CE* (August 1959). Next came his own *Design and Construction of Foundations* in 1961 (revised as a 2nd edition in 1972). A. E. Wynn's *Formwork for Concrete Structures* was reissued as a 5th edition in 1965 after Manning had updated it. Further updating work led to Manning's *Concrete reservoirs and tanks* of 1967, replacing an earlier work by Gray, whose *Water towers, bunkers, silos, etc* (1933) came out as a 5th edition by Gray & Manning in 1973. By this date, Concrete Publications Ltd had been absorbed by the Cement

& Concrete Association (whose own formation in 1935 marks the conclusion of this survey).

The subject matter of his *C&CE* articles during the 1960s reflected his work for the publisher. *Plain-slab water towers* appeared in three parts in the first three issues of 1962 (v.57); and *Design charts for timber formwork* in two parts at the end of 1963 (v.58).

Separately from his work in publishing and the trade press, and still in practice at the age of 77, Manning wrote a report for the Gas Council in 1970, entitled *Notes on frozen ground storage*.

Throughout the 1970s, despite his advancing age, he continued to play a role in the concrete industry – albeit from home in Acton – writing letters and reminiscences for publication in *Concrete* (the magazine of the recently formed Concrete Society).

Manning, G. P. *Reinforced concrete design*. London: Longmans, Green & Co., 1924 [BL, C&CA 1959, ICE, ISE]. Second edition, 1936 [C&CA 1959, ISE. Copy: TCS]. Third edition, 1966 [ISE]

Manning, G. P. *Construction in reinforced concrete: an elementary book for designers and students*. London: Pitman, 1932 [BL, C&CA 1959, ICE, ISE]. Second edition, 1947 [ISE. Copy: TCS]

Manning, G. P. *Reinforced concrete arch design: a textbook for engineers and advanced students*. London: Pitman, 1933 [BL, ICE, ISE. Copy: TCS]. Second edition, 1954 [ISE. Copy: TCS]

1925/UK

Scott, William Leslie (1889–1950)

MICE, MIStructE

Scott was born in 1889 and served a pupilage at the Thames Ironworks. In 1910 he was awarded the City & Guilds Institute's Bronze Medal on completing the structural engineering examination. He joined the concrete specialists Messrs Considere Constructions Ltd, who had set up in Britain only a couple of years before, as an assistant engineer. In 1914 he sailed to the Gold Coast (now Ghana) to serve as assistant engineer to the Public Works Department, but returned to Britain in 1917, to a post as surveyor at Lloyd's Register. In that capacity he developed an interest in the concrete ships that were then being designed and even visited America to study the vessels being constructed there.

Returning to Considere in 1919, Scott fulfilled the role of chief engineer until he left to establish his own consultancy in 1936. He soon made an impact in the professional press with a succession of articles for *Concrete & Constructional Engineering*. Some were on technical matters; others were project studies – presumably those on which his firm was engaged. These articles included:

- *1921* *Reinforced concrete suspended drain, v.16(8)*
- *1922* *Walkerburn hydroelectric scheme, v.17(1)*
  Mild v. high-tension steel for reinforced concrete work, v.17(5,7)
- *1923* *Reinforced concrete bowstring bridge at Nantes, v.18(7)*
- *1924* *Secondary stresses in reinforced concrete arched bridges, v.19(1)*
  Reinforced concrete substructure for Messrs John Barkers' new store at Kensington, v.19(9)

- 1925 *Open-air swimming pools at Addington and Bishop's Stortford*, v.20(1)
  Pit-head gears in reinforced concrete, v.20(6)
  Reconstruction of Pandy Bridge, v.20(12)

It will be seen that several of these related to bridges and it was this interest that was the subject of his first book – *Reinforced concrete bridges* – assisted by C. W. J. Spicer and published by Crosby Lockwood in 1925. It was a guide to practical design and was reissued in 1928. In the meantime he kept writing articles:

- *1927 Rapid-hardening concretes, v.22(6)*
  The New Kelvin Hall of Industries, Glasgow, v.22(8)
- *1930 Design of reinforced concrete slabs, v.22(3-5)*

In 1931 however, there was a concentration on bridge design. His *Recent developments in arch design*, v.26(3) was coincident with Manning's work of the same period, and his Chiswick and Twickenham bridges provided the practical outlet for work described on the pages of *C&CE*:

- *1931 Construction of Chiswick Bridge, v.26(7)*
- *1932 Construction of Chiswick Bridge, v.27(1)*
  Piles 93ft long at Refrew, v.27(2)
  Construction of Chiswick Bridge, v.27(6)
  Concrete test cubes, v.27(7)
  Twickenham Bridge, v.27(8)

The series ends in 1933 with *Sheffield City Hall* (v.28(1)), though his descriptions of Chiswick and Twickenham also featured in a *C&CE* supplement that year – *Three reinforced concrete bridges over the River Thames*. He was in illustrious company; the other contributors were: A. Dryland; the architect Maxwell Ayrton; and Mouchel's chief engineer, T. J. Gueritte. Perhaps it was this publication that suggested him as a contributor (along with Sir Henry Tanner [1907] and Oscar Faber [1912]) to the British Portland Cement Association's *Concrete Bridges* that followed soon after. Scott's piece was entitled *Large span bridges in reinforced concrete*.

His association with bridges was about to be eclipsed by his work on the new DSIR Code, serving as a member of the Reinforced Concrete Structures Committee. To guide the prospective user of the Code, Scott and W. H. Glanville [1934] (of the Building Research Station) wrote an *Explanatory handbook on the code of practice for reinforced concrete as recommended by the Reinforced Concrete Structures Committee of the Building Research Board*. Hearing that it was about to be published in 1934, the Reinforced Concrete Association wrote:

> The Association is profoundly interested in all matters which may lead to increased efficiency in design and construction, and the new Code – in the drafting of which it has been privileged to take an active part – is a very great step forward in that direction. A wide knowledge and understanding of the Code among those who are responsible for the design of modern buildings must lead to economy, and we are glad to hear that Messrs. W. L. Scott and W.

> H. Glanville are carrying out this very useful work. The Code owes much to their labours, and their comments and explanations cannot fail to be of great service.

Sir George Humphreys – chairman of the Reinforced Concrete Structures Committee – also drew attention to the intimate role of Scott and Glanville in drafting the Code and their suitability therefore, to write the *Handbook*:

> Of the competency of the authors to deal with the subject-matter there can be no question. The names of Mr. W. L. Scott (the chief engineer of Considere Constructions, Ltd.) and Dr. Glanville (the chief engineer at the Building Research Station) are both well known to all interested in the subject, and their pronouncements will carry the more weight inasmuch as they both served on a sub-committee entrusted with the task of drafting the Code of Practice.

The *Handbook* became a standard work and sold 16,000 copies. It was revised in 1950, and successively updated after his death in line with BS 114 and replacement Codes.

During the Second World War he served on a concrete committee for the Ministry of Supply and provided specialised consultancy for the DSIR with whom he had previously been involved through his work on the Code of Practice.

After the War he took Dr. Guthlac Wilson into partnership, founding Scott & Wilson in 1945. The practice was responsible for various bridges for the Ministry of Transport, the British Nylon Spinners' factory, the Pimlico housing scheme and several paving jobs in Nyasaland (now Malawi). Perhaps the most prestigious commission was for the Royal Festival Hall.

Scott died in November 1950 in Cheam, Surrey. In 1951 his practice merged with Sir Cyril Kirpatrick & Partners. Scott Wilson Kirkpatrick – now simply styled Scott Wilson – is still in business (as part of the URS group).

Scott, W. L. *Reinforced concrete bridges: the practical design of modern reinforced concrete bridges including notes on temperature and shrinkage effects.* London: Crosby Lockwood, 1925 [BL, C&CA 1959]. Second edition, 1928 [BL, ICE]. Third edition, London: Technical Press, 1931 [BL, ICE]

Scott, W. L. & Glanville, W. H. *Explanatory handbook on the code of practice for reinforced concrete as recommended by the Reinforced Concrete Structures Committee of the Building Research Board.* London: Concrete Publications Ltd, 1934 [BL, TCS]

1925/USA

Smulski, Edward (1884–1941)

We have met Smulski as a collaborator of S. E. Thompson and will refer back to the entries for both Taylor and Thompson, the co-authors of *Concrete plain and reinforced* (1905).

Smulski was a Polish consulting engineer from Austria – Hungary, who came to New York in 1907 and settled in Boston. He was naturalised as an American citizen in 1915. During these years he developed and patented a system of reinforcement called the 'Circumferential' or 'S.M.I. System'. According to a recent issue of *Structure* magazine, 'the system was unique in that the primary flexural reinforcement consisted of concentric rings of smooth reinforcing bars supplemented with diagonal and orthogonal trussed bars placed between the supporting columns and radial hairpin bars located at the column' (September 2007). The system began to be used from about 1915 and by 1918 was attracting professional attention with articles in *Concrete* and reference in the books by Taylor & Thompson (1919) and Hool & Kinne (1924).

Smulski obtained a patent on 21 September 1915, and in October the system's use for the Youths' Companion Building in Boston was described in the *Engineering Record*. Later, in February 1918, the system was explained in an article entitled *What S-M-I flat slab system means* for the magazine *Concrete* (v.12(2)). This was followed by his *Circular reinforcing the design feature of paper factory* (v.13(3)) March 1918) and *Tests of circumferentially reinforced floor* (v.13(9)) September 1919). Tests on the system – carried out by Professor W. K. Hatt on a slab at Purdue University – had already been published in the ACI's *Proceedings* as *A test of the S-M-I system of flat-slab construction* (v.14(6)) June 1918).

Unfortunately for Smulski, the patent was challenged in the courts by one Alfred E. Landau. Priority was awarded to Smulski, but Landau took the case to appeal in March 1922.

By this date Smulski had contributed to the revision of Taylor & Thompson's *Concrete plain and reinforced*. Taylor had died in 1915 with the work still in its second edition and Thompson undertook the revision. He asked Smulski to assist, though not (yet) as co-author. The two men had collaborated previously in December 1913, on a paper for the ACI *Proceedings*: *Design of rigid frames in steel and reinforced concrete* (v.9(12)) – while Smulski had gone on to write *Design of wall columns and end beams* (*Proc* v.11(7) July 1915).

The book was already a standard work and continued its success. Ewart S. Andrews' extended review for *C&CE* (quoted above) was generally very complimentary, though notably he took issue with Smulski's contribution on double reinforcement.

The fourth edition was rewritten and issued as three volumes in 1925, representing quite a change in approach: 'a complete departure has been made from the old Taylor and Thompson we have known'. This time Smulski was credited as a co-author and presumably responsible for aspects of the change. The emphasis on flat-slab floors – of which of course Smulski was a pioneer – was remarked on by a British reviewer as an example to follow in the UK.

A fifth edition followed in 1932 and, continuing their association, Smulski and (the by now elderly) Thompson jointly wrote *Reinforced concrete bridges*, published by Wiley in 1939.

Taylor, F. W., Thompson, S. E. & Smulski, E. (*A treatise on*) *concrete plain and reinforced*. Fourth edition. New York: John Wiley, 1925 [BL, C&CA 1959, ICE, IStructE, TCS, Yrbk 1934. Copy: TCS. Rev: *SE*, 1926, p.67]. Fifth edition, 1932. (Originally published in 1905)

1925/UK

Spicer, C. W. J.

C. W. J. Spicer assisted W. L. Scott [1925] in preparing the latter's *Reinforced concrete bridges* (published by Crosby Lockwood in 1925) and is acknowledged on the title page. It is probable that Spicer was, like Scott, an employee of Considere Constructions Ltd at this time. Certainly he replaced Scott as Chief Engineer at the firm, after the latter's departure in 1936, and was still in post in 1939.

Scott, W. L. *Reinforced concrete bridges: the practical design of modern reinforced concrete bridges including notes on temperature and shrinkage effects*. London: Crosby Lockwood, 1925 [BL, C&CA 1959]. Second edition, 1928 [BL, ICE]. Third edition, London: Technical Press, 1931 [BL, ICE]

1925/UK

Travers Morgan, Reginald (1891–1940)

MEng, MICE, AMIMechE

Born on 19 May 1891, the son of a Liverpool doctor, Reginald Travers Morgan was educated at Lodge Lane College, Liverpool and Liverpool University. Graduating in 1912, he headed west to work on the Canadian Pacific Railway. He returned in 1914 and enlisted in the Royal Artillery for the duration of the First World War. As the war progressed Travers Morgan undertook a course in military engineering at Woolwich and on qualifying, taught there for a while.

Reginald Travers Morgan Courtesy of the IStructE

In August 1919 he returned to Liverpool as assistant building surveyor to Mr. Swan, the city building surveyor. During his eight years in this role he wrote his book *Tables for reinforced concrete floors and roofs*, which was published by Chapman & Hall in 1925. Later, while working on St Martin's Bank in Liverpool, he met J. R. Sharman, a consulting engineer practicing in London. Sharman, who was due to retire, offered him a position with a view to taking over the practice.

Travers Morgan joined Sharman in 1927, attending to the firm's interests in Liverpool before moving to London as junior partner. From 1929 the practice traded as Sharman & Travers Morgan at 7 Victoria Street in SW1, and then as R. Travers Morgan on Sharman's retirement in 1931.

Throughout the 1930s he was active in the Institution of Structural Engineers, for whose journal he was photographed with other office holders in the special 1933/34 Charter Year journal issue.

As an engineer he was notable for his work on Thames House, which featured in the *Structural Engineer* (v.9(1) January 1931), and later for the new Westminster Hospital. He wrote up these projects for the professional press, with articles appearing in *C&CE* and the *Structural Engineer*: *Foundation work at the new* nurses' *home and medical school, Westminster Hospital* (*C&CE* 31(10) October 1936) and *New Westminster Hospital* (*SE* v.15(12) 1937).

Sadly Travers Morgan died in September 1940 following an operation, and his obituary appeared in the *Structural Engineer* for November 1940 (v.18(11) p.5). At the end of the month his premises were bombed and his partner, C. F. Pike, was taken on by Scott White –

a fellow engineer with whom the practice had cordial relations. R. Travers Morgan & Partners survived the war under Scott White's ownership and prospered for many years after, though in recent times it merged to form part of the Capita Symons group.

Travers Morgan, R. *Tables for reinforced concrete floors and roofs*. London: Chapman & Hall, 1925 [BL, C&CA 1959, Yrbk 1934]

1926/USA

Sutherland, Hale

As one might expect from a professor of structural engineering at a leading technical institute, Sutherland's book *Introduction to reinforced concrete design* (co-authored originally by Walter Clifford [below]), was intended for students. Its British reviewer, writing in the *Structural Engineer*, suggested that the compression of theoretical formulae into time-saving diagrams and practical tables also made the book a useful tool for the architect or engineer for whom concrete was but a minor aspect of his work. The second edition of 1943, published in the UK by Chapman & Hall, was co-authored with Raymond C. Reese in place of Clifford.

Sutherland, H. & Clifford, W. W. *Introduction to reinforced concrete design*. New York: Wiley, 1926 [BL, ICE. Rev: *SE*, 1927].

1926/USA

Clifford, Walter Woodbridge

A 'well-known' consulting structural engineer from Boston, W. W. Clifford was Sutherland's collaborator for the first edition of *Introduction to reinforced concrete design*.

1926/USA

Pulver, Harry E.

BS, CE

As described in the entry for G. A. Hool [1912], Harry Pulver joined Hool for the latter's final collaborative work. The two men were colleagues and like Hool, Pulver had contributed to the University of Wisconsin Extension Division's series; his book was entitled *Materials of construction*. By 1926 Pulver had taken Hool's former role as associate professor, on his colleague's promotion to professor of structural engineering. Together they wrote *Concrete practice: a textbook for vocational and trade schools* – the title of which is probably sufficiently explicit to indicate its intention. The book provided a resource from which instructors could select assignments for students on vocational course to undertake, varied according to suitability or the availability of equipment.

Hool, G. A. and Pulver, H. E. *Concrete practice: a textbook for vocational and trade schools*. New York: McGraw-Hill, 1926 [C&CA 1959, ICE, TCS, Yrbk 1934. Copy: TCS]

1926/USA

Wynn, Albert Edward (1888–1956)

BSc, AMASCE

Born in the West Midlands during the third quarter of 1888, Wynn was awarded his BSc by the local University of Birmingham. At the age of 22 he was employed as an engineer, based in West Bromwich, and continued to practice in the UK throughout the 1910s. The 1920s however, were spent in the USA, where he developed a 'flourishing business' designing and constructing concrete buildings, and joined the American Society of Civil Engineers. It was in this decade that his presence began to be felt on the literature, as a regular contributor to the British journal *Concrete & Constructional Engineering*. His business base was reflected in the many titles on American and Canadian practice:

- *1920 The construction of the best concrete highway in Canada. 15(3) March 1920*
- *1920 The causes of cracks in concrete pavements and how to prevent them. 15(4) April 1920*
- *1920 The results of American experience in the use of hydrated lime concrete roads. 15(9) September 1920*
- *1921 The American 'flat slab' type of building. 16(2/3), February/March 1921*
- *1922 The tallest reinforced concrete building in the United States. 17(6) June 1922*
- *1922 A method of erecting foundations and buildings simultaneously: pretest piles and the bulb of pressure. 17(10) October 1922*
- *1924 Building a reinforced concrete bridge around an old steel structure. 19(6) June 1924*
- *1924 Exterior treatment of concrete industrial building. 19(8) August 1924*
- *1924 Design of formwork for concrete construction [in seventeen parts]. 19(9) September 1924–21(3) March 1926*
- *1926 Concrete in the United States during the past twenty-one years. 21(1) January 1926*
- *1926 Controlling the manufacture of concrete to obtain uniformity of strength. 21(6–8) June–August 1926*

1926 saw a change in the pattern of his writing, with the publication of the first of his four books. *Modern methods of concrete making* was written jointly with E. S. Andrews [1912] – remarkable, when we recall that the authors resided on either side of the Atlantic. However, both were frequent contributors to *C&CE* and no doubt the editor (H. L. Childe) played a role in co-ordinating their efforts to produce one of the early volumes in Concrete Publications Ltd's Concrete Series. *Modern methods* was a simple guide explaining the variables inherent in mixing concrete and publicising the recently accepted American method of controlling strength by the water/cement ratio – the subjects of his two *C&CE* articles that year. The book sold out in six months and was reissued in 1928. A third edition followed in 1935, at the end of our survey period, with further editions beyond.

Design and construction of formwork for concrete structures (also published in 1926 by Concrete Publications Ltd) was a much more substantial work of 320 pages. The publisher claimed it was 'the only book published in the English language dealing solely and exhaustively with the subject of formwork'. It carried complete designs for an extensive range of plain and reinforced concrete structures, along with sizes and tables of quantities of timber. The book was an immediate success and the numerous 'appreciations' from readers that the publisher reproduced in publicity material repeatedly and independently described it as 'useful', 'comprehensive', and 'valuable'. One went as far as regarding it 'absolutely essential'. It was reviewed in the *Structural Engineer* and the reviewer – after expounding the benefits of a knowledge of economic formwork to reinforced concrete construction, and presenting an outline of the book's contents – opined that it 'deserves to be considered the standard work on a subject which no structural engineer can afford to ignore'. The book was not ignored; new editions being required in 1930, 1933 and on to 1956, after which updating was undertaking by G. P. Manning [1924]. Yet it is a measure of Wynn's thoroughness and clarity that in 1965 – nearly 40 years after the first imprint – the announcement of Manning's revision of the fifth edition was able to state: 'Nevertheless, it is surprising that it has been found that so much of the original matter could be retained and not appear to be dated' (*C&CE*, April 1965, p.159).

C&CE continued to receive articles by Wynn, along with so many book reviews that we gain the impression of him being retained for the purpose. Of the articles submitted in the late 1920s, the first, last and penultimate hint at Wynn's major preoccupations and the subject of his two books of 1930:

- *1928* *Cost keeping and estimating for reinforced concrete contractors [in fifteen parts]. 23(4) April 1928 – 25(3) March 1930*
- *1929* *Forms for dams, piers and heavy walls. 24(8) August 1929*
- *1929* *Preventing crazing by steam curing. 24(10) October 1929*
- *1930* *The design of concrete mixtures: details of economies effected by use of water-cement-ration proportioning [reviewing North American practice]. 25 (1/2) January/February 1930*
- *1930* *Typical quantity and cost estimate for a building. 25(5) May 1930*

Estimating and cost keeping for concrete structures (published in 1930) was clearly a reworking of the series of 15 articles he had contributed to *C&CE* over the previous two years. Wynn acknowledged this, but 'as now published in book form it covers the whole field of estimating, cost keeping and book-keeping'. His intentions were to promote a systematic and methodical approach to an essential aspect of contracting all to often left to 'guesswork and gambling'. The *Structural Engineer* regarded it as 'very readable' and 'most valuable', containing a quantity of information not found in other textbooks of the time. It was duly revised to reflect changes in practice after the Second World War.

The other book of 1930 was *Making precast concrete for profit.* This was a much slighter work of 56 pages, in a cheap binding, but again concerned with practical economics. It also reflected Concrete Publications Ltd's contemporary concern for the booming pre-cast industry, with several titles on the subject by Childe, Burren and others issued that year. The book described a system of cost-keeping adapted to the concrete industry and simple enough for 'the practical man'. It reproduced all the various forms required by this system.

Wynn returned to the UK in 1930, and was most closely associated with the Earl's Court Exhibition Centre. His articles for *C&CE* paradoxically, were fewer, but seven of them described the design and progress of Earl's Court:

- *1935 Reinforced concrete construction in Madrid. 30(7/11) July & November 1935*
- *1936 Carpark at Earl's Court Exhibition. 31(12) December 1936*
- *1936 The new Earl's Court Exhibition building [in six parts]. 31(9) September 1936–32(9) September 1937*
- *1938 The use of tubular steel. 33(4) April 1938*

This last one was coincidental with Wynn's decision to leave these shores again, and emigrate to the Transvaal in order to found the Concrete Association of South Africa (later the Portland Cement Institute and now the Cement & Concrete Institute), in which he served as director until his retirement. Under his leadership (1938–1951) the association was to 'collect and disseminate information', including the 'publication of pamphlets, books, papers, *etc*'.

Wynn died in February 1956 after a short illness. His obituary notice referred to his many friends 'to whom the benefit of his great experience was unfailingly given' and described his books on formwork and estimating as 'still the standard works on the subjects'. His life was not only marked by the continued presence of his books, but by the Institution of Structural Engineer's commencement of the Wynn Prize in 1958; and the Portland Cement Institute's erection of a memorial in the library at its new premises in Johannesburg in 1960. The 'Wynn Collection' constituted a case containing his books and articles along with other books purchased with money presented by his widow.

Wynn, A. E. & Andrews, E. S. *Modern methods of concrete making*. London: Concrete Publications Ltd, 1926 [ICE, IStructE]. Second edition, 1928. Third edition, 1935 [C&CA 1959]. 1944 [TCS]

Wynn, A. E. *Design and construction of formwork for concrete structures*. London: Concrete Publications Ltd, 1926 [BL, ICE. Rev: *SE*, 1927]. Second edition, 1930 [ICE]. Third edition,1933 [C&CA 1959], revised 1939 [IStructE]. Fourth edition, 1951 [IStructE], revised 1956 [IStructE]

Wynn, A. E. *Making precast concrete for profit: how to keep costs and determine profits*. London: Concrete Publications Ltd, 1930 [C&CA 1959. Copy: TCS]

Wynn, A. E. *Estimating and cost keeping for concrete structures*. London: Concrete Publications Ltd., 1930 [BL, C&CA 1959, ICE, IStructE. Rev: *SE* v.9, February 1931]. Second edition, 1946, revised, 1949 [IStructE]

1927/UK

Bennett, Sir Thomas Penberthy (1887–1980)

KBE, FRIBA, HON FIOB

Sir Thomas Penberthy Bennett in 1939, painted by his wife Mary Courtesy of TP Bennett LLP

Born in 1887, Bennett trained part-time as an architect at the Regent Street Polytechnic whilst employed in the drawing office of the London & North Western Railway. He continued his studies at the Royal Academy School, joining the Office of Works in 1911. In 1921 he left government service to establish his own practice and a year later became a fellow of the RIBA.

Born on 14 August 1887 Bennett trained to be an architect from the age of 14. He studied building and architecture in the evenings, at the Polytechnics, whilst employed in the drawing office of the London & North Western Railway. He continued his studies at the City & Guilds and Royal Academy Schools, at which he won silver medals for Builders Quantities and Sculpture. Leaving the railway he joined the Office of Works in 1911 where he managed construction projects during the Great War, and developed a parallel career teaching part-time. His academic interests led him to publish the *Relation of sculpture to architecture* in 1916, an interest he pursed practically by incorporating sculpture and friezes into many of his buildings. In 1920 he left government service to take up the appointment as Head of Architecture at the Northern Polytechnic Institute in Holloway (1920–1929), establishing his own practice, TP Bennett & Son, in 1921. A year later he became a Fellow of the RIBA.

During the interwar years his practice was responsible for several landmark building in and around London: the Saville Theatre; Dorset House, at 170-172 Marylebone Road (both Grade II listed); Esso House; the John Barnes department store; Hampstead Westminster Hospital; one of BOAC's air terminals; and the London Mormon Temple. A compilation of commissions, entitled *The work of TP Bennett FRIBA*, described some of the innovatory work with which the practice was making its name.

Bennett was still relatively recently established when he wrote his seminal work, *Architectural design in concrete* in 1927. In the words of the *Structural Engineer* at the time, 'this volume expresses the possibilities of designs in concrete from the aesthetic point of view, is profusely illustrated by photographs showing the development of concrete building during the last twenty years, and a number of architects in various parts of the world have attempted to handle designs which logically develop the constructional features of a concrete building.' In preparing the book he was assisted by F. R. Yerbury (the architectural photographer), who assembled a hundred or so illustrations to supplement the text. As with their contemporary Francis Onderdonk in the USA and three continental authors whose work was published over the following months, Bennett and Yerbury demonstrated that concrete had arrived as an architectural medium in its own right. In this country, the architectural profession's emerging interest in concrete – of which this book was a manifestation – can be seen to have been sparked by the success of Maxwell Ayrton's and Owen Williams's pioneering work on the highly publicised Empire Exhibition at Wembley just two years previously.

Bennett appears to have refrained from any other writing in the years immediately following publication of *Architectural design in concrete*. However, he started to contribute to the professional periodicals in the mid-1930s, addressing topical issues with his *Points in planning flats*. This appeared in *Building* (August 1934) and *Architectural Design & Construction* (May and June 1935). *Marsham Court, Westminster* was published in both *Building* and *Architect & Building News* (April 1937). Other papers on flats were carried by *Architect & Building News* (February 1939) and *Architectural Design & Construction* (April 1939).

Once the Second World War started Bennett was appointed Controller of Bricks and Director of Works at the Ministry of Works and was awarded the CBE while in post. His writing reveals the wartime preoccupations of planning and the preparation for post-war building. A paper at the Architectural Association in 1943 on the organisation of building operations, found its way into *Architect & Building News* (April), *Architects' Journal* (April) and *Official Architect* (May). *Post-war building industry* came out in *Architect & Building News* in August, while another paper presented the following year (this time at the RIBA) covered *War experience in the organisation of building contracts*. This was published in the *Builder* (February 1944), *Architectural Design & Construction* (March) and *RIBA Journal* (March). The pointers to his future career become clearer still later that year and in 1945: *The post-war prospect: prefabrication, pro and con* in *Architect & Building News* (September 1944), related to yet another of his government roles as Controller of Temporary Housing, was followed by *Engineering production and its relation to the building industry* in *Builder* and *Architect & Building News*.

He returned to private practice straight after the ending of hostilities and was knighted for his wartime services in 1946. The practice was to become a pioneer of the post-war New Town Movement, closely associated with the development of Crawley and Stevenage.

In 1947 he was appointed as the chairman of the Development Corporation of Crawley New Town – a post he held until 1960. He quickly discarded the existing plans and commissioned a new master plan and Crawley represents perhaps his greatest legacy. In 1958, with much of the town built, a comprehensive school was named after him.

In 1960 Sir Thomas returned to concentrate on his private practice and undertook a string of prestigious commissions. One of the last projects with which he was personally involved was Smithfield Poultry Market, developed with Ove Arup in 1961-1963, which featured a pioneering concrete shell roof. This also is now Grade II listed.

In 1967 the practice passed to Philip Bennett (the original '& Son') and from him through two more generations of partners to become TP Bennett LLP, the award-winning, 170-strong practice of the present time.

The National Portrait Gallery owns portraits that commemorate his knighthood and eventual retirement: a bromide print of 1946 by Walter Stoneman; and another of 1967 by Walter Bird (1903–1969). A brief biography has been prepared by the practice librarian, Jo French.

Bennett, T. P. *Architectural design in concrete*. London: Ernest Benn, 1927 [C&CA 1959, ICE. Copy: TCS]

1927/UK

Yerbury, Frank R. (1885–1970)

HON ARIBA

As a photographer Yerbury is not an obvious candidate for inclusion in a review of authors, but his contribution to Bennett's book (above) was as integral as W. S. Gilbert's libretti to the music of Sir Arthur Sullivan. He was also responsible for a great many books in the 1920s and 1930s, on aspects of architecture if not especially on concrete.

Between 1925 and 1933 (in the company of Howard Robertson) Yerbury travelled throughout Europe and the USA in search of new architecture. Their accounts were covered extensively in the architectural press and Yerbury's photographs featured in many of his and other books of this period. It was this output and experience that brought him to Bennett's attention and admirably qualified him to illustrate his book.

Yerbury started to appear in print in the mid 1920s, initially as a photographer to others' work on *Nicholas Hawskmoor* (Benn, 1924) and *Sir William Chambers* (Benn, 1924). His own *Human form and its use in art* (Batsford, 1924) was not typical of his later work, but *Small houses for the community* (Crosby Lockwood, 1924) with C. H. James, registered an ongoing interest in domestic architecture. He collaborated with James again the following year with *Modern English houses and interiors* (Benn, 1925) and followed it with *Georgian details of domestic architecture* (Benn, 1926) and *Small modern English houses* (Gollancz, 1929).

1924 saw the first of a series of books exploring European architecture (mostly modern architecture) that paralleled his travels with Howard Robertson. His titles included:

- *Old domestic architecture of Holland*, edited by Yerbury (Architectural Press, 1924)
- *Swedish architecture of the Twentieth Century* (Benn, 1925)
- *Lesser known architecture of Spain* (Benn, 1925)
- *Lesser known architecture of Spain, II* (Benn, 1926)
- *Dutch architecture of the XXth Century*, edited by J. P. Miras & Yerbury (Benn, 1926)

- *Modern Danish architecture*, edited by Kay Fisker & Yerbury (Benn, 1927)
- *Modern European buildings* (Gollancz, 1928)
- *Modern Dutch buildings* (Benn, 1931)

These consisted largely of photographs, taken and selected by Yerbury in an editorial capacity, with supporting text by either himself or others. His work with Bennett appeared midway through the series – a natural accompaniment to *Modern Danish architecture* of the same year. This highly productive period established Yerbury's reputation and has been the subject of at least two retrospective anthologies:

- *Frank Yerbury: itinerant cameraman – architectural photographs 1920–35.* (Architectural Association, 1987)
- *Howard Robertson and Frank Yerbury: Travels in modern architecture* (Architectural Association, 1990)

Later books treated more conventional architectural subjects, such as *Roedean School* (Benn, 1927) with L. Cope Cornford; and *The old Bank of England* (Benn, 1930) with H. Rooksby Steele. His *One hundred photographs* (Jordan Gaskell, 1935) perfectly sums up Yerbury's *oeuvre* in the closing year of our survey.

Besides the enduring appeal of his photographs Yerbury has found a place in architectural history as the secretary of the Architectural Association and co-founder of the Building Centre in London.

1927/USA

Crane, Theodore (b.1886)

Crane was associate professor of building construction at the Sheffield Scientific School, Yale University. His book – *Concrete building construction* – was a straightforward textbook, summarising current practice. It was intended as a simple practical guide to the 'accepted principles' of concrete design and construction, keeping the theoretical treatment to a minimum. Crane was keen to credit the role of the late Professor **Thomas Nolan** (formerly at the University of Pennsylvania) in editing the text. Much of the information was derived from 'the practice of the more successful building organisations', for which the Turner Construction Company in particular was acknowledged. Likewise many firms contributed illustrations derived from their own activities. Distributed in Great Britain by Chapman & Hall, the book was reviewed briefly by the *Structural Engineer* with the conclusion: 'altogether a very useful work.' In later life Crane was to write *Architectural construction: the choice of structural design*, published by Wiley in 1947.

Crane, T. *Concrete building construction*. New York: Wiley, 1927 [BL, Yrbk 1934, C&CA 1952. Copy: TCS. Rev: *SE* v6 1928]

1927/USA

Harrison, J. L.

Harrison occupied the post of highway engineer for the US Bureau of Public Roads, based in Washington DC. His book *Management and methods in concrete highway construction*, described the conclusions of the bureau's investigations into the management of cost and efficiency in road building made under Harrison's immediate direction. It supplemented and consolidated the sporadic publication of his findings in the bureau's own journal, *Public Roads*. Harrison addressed the practical needs of the contractor while demonstrating an awareness of the increasingly scientific approach to production management then being developed.

Harrison, J. L. *Management and methods in concrete highway construction*. New York: McGraw-Hill, 1927 [ICE, Yrbk 1934]

1927/USA

Moore, Prof. Herbert Fisher (b.1875)

MASTM

Co-author of a book peripheral to our review, Moore is included briefly for *Fatigue of metals*, which he wrote with J. B. Kommers in 1927, one chapter of which specifically treats concrete.

Moore was closely involved with the Engineering Experiment Station that was established by the University of Illinois in 1903 and so was instrumental in researching the fundamentals of concrete. Moore's speciality, however, lay in metals. As assistant professor in theoretical & applied Mechanics he wrote several Experiment Station bulletins for the university:

Bulletin 42 The effect of keyways on the strength of shafts 1909

Bulletin 52 An investigation of the strength of rolled zinc 1911

Bulletin 68 The strength of I-beams in flexure 1913

He wrote another Bulletin, *An investigation of build-up columns under load*, in collaboration with Arthur Talbot [1917].

His Bulletin 98, *Tests of oxyacetylene welded joints in steel plates* of 1917, was written after his elevation as research professor of Engineering Materials. It was in this role that he wrote his *Text book of the materials of engineering* (McGraw Hill, 1917). Although it didn't specify concrete in the title, this book aimed to be 'a concise presentation of the physical properties of the common materials used in structures and machines'. Intended for use in technical schools the text was 'distinctly elementary in character'. The second edition of 1920, did have a chapter on concrete; it was contributed by his colleague, Harrison Frederick Gonnerman, who had replaced Moore in his earlier position as assistant professor in theoretical & applied mechanics.

It seems that although he was the author of two books on engineering materials, concrete was not more than of peripheral interest to him, the chapters treating that subject being the work of others.

Moore, H. F. & Kommers, J. B. *The fatigue of metals: with chapters on the fatigue of wood and concrete*. New York: McGraw-Hill, 1927 [BL]

1927/USA

Kommers, Jesse Benjamin (1884–1966)

BSc, MASTM

Born on 11 March 1884 in Sheboygan, Wisconsin, Jesse B. Kommers was the son of Jacob Kommers and Bina Van Der Wall.

In 1919 Kommers was appointed special research associate professor of engineering materials at the University of Illinois, alongside his future collaborator Professor Herbert Moore. Later he was associated with the University of Wisconsin's Department of Mechanics where he was Associate Professor and wrote on billet steel (1924), riveted joists (1925), machine design (1929) and brass (1933). His *Fatigue of metals*, written with H. F. Moore in 1927 – and based on the earlier University of Illinois Bulletin *Investigation of the fatigue of metals* – included a chapter specifically on concrete.

He died at the age of 82 in October 1966.

1928/USA

Bauer, Edward Ezra (1896–1964)

MSc, CE, MASCE, MACI, MASTM

Son of William John and Christina Bauer of Rocky Ridge, Ohio, Edward was born on 6 June 1896. He was educated at that bastion of concrete development, the University of Illinois, graduating in 1919 with a BSc in civil engineering after a period of wartime service with the Student Naval Training Corps. He married his fiancée Amy Weir and commenced teaching at the university in 1919. In the years following, the Bauers had two children – Paul and Virginia – in the years following and in 1927 Edward was awarded the degree of civil engineer. He was thus a young man, recently qualified, when he wrote the book that justifies his inclusion in this survey.

As assistant professor of civil engineering, Bauer was concerned for the education of engineers and so his *Plain concrete* was conceived as an educational tool for students. It was primarily to meet the needs of undergraduates on courses in plain concrete that 'a number of engineering colleges' had started to offer in the 1920s. It took the form of 'a combination textbook and laboratory manual', the latter including instructions for all the usual tests on cement and aggregates, along with a number of research problems on concrete suitable for class use. Included too were standard forms for recording laboratory data. Perhaps misled by his own preconceptions, the reviewer W. S. Gray was reserved in his judgement. He was disappointed to note that with such an emphasis on materials and proportioning only 46 pages of the entire 340 considered the making, placing, finishing and curing of concrete. Nearly a half of the book was devoted to specification and testing of aggregates, about which Gray was pleased to declare would 'provide a valuable handbook for the laboratory where these tests are specified'.

Shortly after publication Bauer was awarded his Master of Science. He was to remain teaching at Illinois for the rest of his life: 45 years of continuous service. During that time he belonged to various professional societies, not least the American Society for Concrete Education, and (from 1925) the American Society of Civil Engineers. He had several other books to his name, such as *Highway materials*, and with T. H. Thornburn, *Introductory soil and bituminous testing*. His *Laboratory manual for plain concrete* was written with G. W. Hollon.

A memoir of his life is published in the *Transactions* of the ASCE for 1965, p.813.

Bauer, E. E. *Plain concrete*. New York: McGraw-Hill, 1928 [BL, ICE, Yrbk 1934. Rev: *C&CE* v24 p.116]. Second edition, 1936 [Copy: TCS]

Frontispiece to Onderdonk's book, The Ferro-concrete style, *featuring the Shrine of the Sacred Heart in Washington DC Courtesy of The Concrete Society*

1928/USA

Onderdonk, Francis S.

Born in New York, the son of Francis Onderdonk Sr., the young Francis spent 20 years in Europe where he received his education and training as an architect. His thesis – the basis for the work by which he was later known – developed the idea that a distinct style of architecture was emerging as a result of the adoption of reinforced concrete. In 1918 he was awarded the degree of Doctor of Technical Sciences for this thesis from the Vienna Imperial and Royal Technical University. Returning to the USA after the First World War, he was appointed instructor at the college of architecture of the University of Michigan, and continued to survey concrete buildings in his native country. Building on his initial thesis and ten years further research, Onderdonk prepared his book – the purpose of which was to 'trace the beginning of the Ferro-Concrete Style and to indicate the developments which are to be expected in the future'. *The Ferro-concrete style* was published in 1928, in the wake of T. P. Bennett's book of 1927 and coincident with three European books that indicated the seriousness with which architecture now regarded concrete. International in its coverage, drawing upon a multitude of European and American examples over a ten-year period, Onderdonk's book demonstrated the validity of his theories of figural concrete tracery and the parabolic arch as features of a new style of concrete architecture.

Onderdonk, F. S. *The ferro-concrete style: reinforced concrete in modern architecture*. New York: Architectural Book Publishing Co., 1928 [BL, C&CA 1959, ICE. Copy: TCS]

1929–1935: Consolidation and Codification

1929/USA

McMillan, Franklin R. (b.1882)

Franklin McMillan was appointed assistant professor of structural engineering at the University of Minnesota with effect from 1 July 1919 – after earlier research at the university that resulted in his *Shrinkage and time effects in reinforced concrete* (*Studies in Engineering* 3, 1915), and a period of government service during the First World War. He served as a research engineer for the Emergency Fleet Corporation of the US Shipping Board, and from the title of his first published paper in the ACI *Proceedings* – *The strainagraph and its application to concrete ships* (v.15) – this would appear to have been in the concrete shipbuilding programme that was such a feature of 1918 and 1919. In this paper McMillan described how he, assisted by Henry S. Loeffler, was charged with developing the instruments needed for a programme of testing the stresses that concrete ships were subject to in service. Another paper followed in 1921, entitled *A study of column test data*, for which the author was cited as assistant engineer at the Turner Construction Co. of New York. This paper demonstrated 'the important effect of shrinkage and time yield of concrete upon the deformations and the resulting stresses in the reinforcement' (Richart & Brown, 1934) and was to shape McMillan's research interests in the early 1930s as we will see below.

Franklin R. McMillan
Courtesy of the American Concrete Institute

It was through his involvement with the ACI that he came to national prominence in 1928. The *Concrete Primer* – a summary of concrete knowledge expressed in question and answer format, aimed at those responsible for, but not active in, concrete construction – came out in the *Proceedings* of 1928. It was a great success and was repeatedly reprinted; over 100,000 copies in English were issued, and the text translated into seven languages. After the passage of time it was revised and expanded in 1958 (and edited at later dates by L. H. Tuthill). It is likely that the nature and success of the *Concrete Primer* suggested that an equivalent be issued by a commercial publisher, and *Basic principles of concrete making* was published the following year by McGraw-Hill. It was reviewed in *Engineering News Record* in 1929 (v.103(25)).

By this time McMillan had risen to the role of director of research at the Portland Cement Association, taking the place formerly occupied by Duff Abrams [1918]. In 1930 he reported on testing work at the PCA's research laboratory on behalf of Committee C-9 of the American Society for Testing Materials. *Suggested procedure for testing concrete in which the aggregate is more than one-fourth the diameter of the cyclinders* appeared in ASTM's *Proceedings* for 1930.

Prompted by McMillan's paper of 1921, a comprehensive investigation into columns was planned in 1929 and placed under the direction of the ACI's Committee 105. McMillan was a member of the committee. With laboratory work undertaken by Lehigh and the University of Illinois, the project was written up in the University of Illinois Bulletin of June 1934 by R. E. Richart and R. L. Brown.

During this period McMillan wrote for the *ACI Journal*, namely: *Some permeability studies of concrete* with Inge Lyse (v.26-7, December 1929); *Study of defective concrete* (v.27–36, May 1931); *A method of evaluating admixtures* with T. C. Powers (v.30–32, March/April 1934). By 1936 he was elected as president of the ACI, having completed a term as one of the directors. At the end of his period of office he wrote about the difficulties faced by the ACI during times of economic hardship and promoted his views on areas of progress. It was entitled *The Institute carries on* (v.33–19, March/April 1937). Other papers followed immediately:

- *Some comparisons of European and American concrete practices.* (v.33–21, March/April 1937)
- *Classification of admixtures as to pozzolanic effect by means of compressive strength of concrete,* with T. C. Powers. (v.34–9, November/December 1937)
- *Field survey of mass concrete.* (v.34–32, May/June 1938)

The 1940s saw McMillan's career come to its climax, during which (at the age of 61), he was made an honorary member of the ACI in 1943 and his 'long-time' studies of concrete were written up:

- *Progress in the long-time study of cement performance in concrete.* (v.38–27, April 1942)
- *Long-time study of cement performance in concrete: 1,* with I. L.Tyler. (v.44–21, February 1948)
- *Long-time study of cement performance in concrete: 2,* with W. C. Hansen. (v.44–26, March 1948)

Long after retirement age, McMillan was prevailed upon to revise his *Concrete Primer*, which was reissued in early 1958. For all his contribution to research and our understanding of concrete, this slim volume stands as his most widely read legacy.

McMillan, F. R. *Basic principles of concrete making.* New York: McGraw-Hill, 1929 [BL, ICE]

1929/UK

Turner, Leslie (1891–1971)

BSc, MICE, MIStructE

Leslie Turner, president of the Institution of Structural Engineers, 1949-50 Courtesy of IStructE

Born in Birmingham on 28 October 1891 and educated at George Dixon School, Leslie Turner attended Birmingham University from which he graduated in civil engineering in 1912. His first position was as assistant engineer at Southend-on-Sea Corporation, at a time marked by seafront improvements, before starting with the Indented Bar & Concrete Engineering Co., Ltd of Westminster. Clearly this early experience directed his expertise in concrete. For a short while he returned to Birmingham where he worked for Harry Jackson, consulting engineer, before serving the war effort.

He had already undertaken some military training before the outbreak of the First World War and so when he enlisted he was commissioned as an officer in the North Midland Divisional Royal Engineers. At first he was an instructor and then saw two and a half years of active service in France, including the Somme. He was a captain on demobilisation in 1919.

After the war he rejoined Indented Bar, where he worked for the next nine years. His role took him abroad and the articles he wrote at the time for *Concrete & Constructional Engineering* reveal something of his travels, intermixed with more generic topics:

- *Reinforced concrete bridge at Elorn, Brest*, v.21(12) December 1926
- *Reinforced concrete bridge at Brest*, v.22(1) January 1927
- *Low retaining walls*, v.22(7) July 1927
- *Floor and roof slabs*, v.22(8) August 1927
- *Reinforced concrete bridge over the Elorn, Brest*, v.23(6) June 1928
- *Pont de la tournelle, Paris*, v.23(1/2) January/Febraury 1928
- *The Elorn Bridge, Brest*, v.23(1&3) January & March 1929

It was probably at the end of this phase of his career that he collaborated with Albert Lakeman to write *Concrete construction made easy* (discussed under Lakeman [1913]). Published in 1929, it set out the necessary information – designs, tables and illustrations – for the builder to carry out reinforced concrete work for which no designs had been prepared.

By the time the book was published, Turner had accepted a position with the Salonika Plain Reclamation Works in Greece. From 1928 and for the next four years, he was both bridge engineer and deputy director. Back in London in 1932, he started with another concrete specialist – the Trussed Concrete Steel Co., Ltd. Here, he was able to indulge his interest in research. He specialised in the performance of concrete in tension and torsion, and established a basis for design in torsional problems. His work with V. C. Davies on torsion was published in 1934 as No.165 of the ICE's Selected Engineering Papers: *Plain and reinforced concrete in torsion, with particular reference to reinforced-concrete beams*.

In 1934 he was elected to membership of the Institution of Structural Engineers and went on to serve on a number of committees, acting as chairman of the Science, Foundation and Research Committees. His published output reflects this involvement. *Some bridge and foundation problems* for instance, appeared in the institutions' journal the *Structural Engineer* (v.13(11) November 1935). His research activity continued, with Turner undertaking about a thousand tests to establish a quantitative law for the autogenous healing of cracked mortars and concretes. This was written up as *The healing of cement and concrete* in *C&CE* (v.32(2) February 1937). He was also an examiner for the Institution of Structural Engineers.

Under the name Leslie Turner & Partners he commenced private practice in 1935, acting as consulting engineer to a series of projects; including frame buildings, reservoirs, water towers, silos, bridges, subways. He specialised in foundations, as is suggested by his article in *Structural Engineer – Some bridge and foundation problems* (v.13(11) November 1935). Presumably his *Reconstruction of Bonchester Bridge* (published the month before in *C&CE*), indicates a recent project. Besides consultancy he also accepted legal work as a technical witness.

Having written a series of five articles on the theme for *C&CE* in 1938, Turner collaborated with Percy Taylor in writing his second book for Concrete Publications Ltd: *Reinforced concrete chimneys*; which was published in 1940. The year 1940 also saw *Examples of modern buildings: with particular reference to the application of reinforced concrete* in the *Structural Engineer*, v.18(9).

He acted as a consultant to the Government during the Second World War, but also was elected to the council of the Institution of Structural Engineers in 1941 and served as honorary secretary, honorary treasurer and vice-president. After the Second World War his ongoing interests in bridges and foundations were apparent in his *Reconstruction of the Longeray Viaduct* (*C&CE* v.41(3) March 1946), and a paper presented to the institution on 17 January 1947 entitled: *Foundation problems in relation to soil mechanics*.

In 1949 he was elected president of the Institution of Structural Engineers for the year 1949–1950. A fulsome introduction in the *Structural Engineer* indicates his active participation in the professional bodies. He was a member of the Association of Consulting Engineers, Gold Medallist of the Junior Institution of Engineers, one of the Court of Governors of Birmingham University, a member of IABSE and also of the British section of the *Societe des Ingenieurs Civils de France* in London. Indeed he was to act as president of the latter in 1957; its diamond jubilee year.

In the final months of his career Turner turned his mind again to chimneys. *Design of chimneys for resistance to weight and wind* was published in *C&CE* (v.54(12) December 1959), as a prelude to a new edition of Turner & Taylor's *Reinforced concrete chimneys* the following year. This was to be his last published work as a practising engineer.

In 1960 he retired to Dorset and Leslie Turner & Partners, consulting structural engineers, became Kenneth Severn & Partners, as his younger associate took on the business. Turner died on 16 April 1971 and a notice of his passing appeared in the June issue of the *Structural Engineer*, v.49(6).

Turner, L. *Concrete construction made easy*. London: Concrete Publications Ltd, 1929 [ICE, Yrbk 1934]

1930/UK

Burren, F.

Moulds for cast stone and concrete – written with **G. R. Gregory** – was aimed at 'the practical man', the manufacturer or builder. It comprised a series of working drawings for an extensive range of architectural elements and garden features – balustrades, bird baths and the like – many of which had previously been featured in the magazine *Concrete Building & Concrete Products*. Burren and Gregory wrote the accompanying passages of text separately, initialled appropriately to indicate respective authorship. Additional design sketches were contributed by the distinguished architectural illustrator T. Raffles Davison.

Burren, F. & Gregory, G. R. *Moulds for cast stone and concrete*. London: Concrete Publications Ltd, 1930 [C&CA 1959. Copy: TSC]

1930/UK

Davison, Thomas Raffles (1853–1937)

HON ARIBA

Like the photographer Frank Yerbury [1927], the architect T. Raffles Davison was principally an architectural illustrator, a draughtsman and painter, known mostly for his views of buildings designed by others. Pevsner described him as 'the well-known architectural illustrator', while others have gone as far as calling him 'the doyen of architectural illustrators' and a 'key arts and crafts movement figure'. Again like Yerbury, he wrote a little and edited anthologies, but it is for his association with authors on concrete that he is included here.

He was born in Stockton on Tees in 1853, the son of a Congregational minister, and was educated privately at Shrewsbury. Showing an aptitude for drawing he was articled to W.H. Spaull, an architect based in Oswestry. His time served, he worked in architectural offices in Nottingham and Manchester. By the 1870s, while still in his twenties, he was making a name for himself as a purveyor of 'sketches' for the architectural press, and in 1878 was appointed editor of the recently founded *British Architect & Northern Engineer*. He remained in this post until 1916 when the *British Architect* merged with *The Builder*. Free to travel, Raffles drew assiduously. His *Rambling sketches* (London & Manchester: The British Architect, 1883), drawn by Davison with notes by William E.A. Axon, brought him to a wider public. He moved from Cross St., Manchester, to 114 Fleet Street, the first of several addresses in London. *Gleanings from the past and memorials for the present: being a series of sketches at the Health Exhibition, 1884* followed. Also that year was *A visit to the Architectural Museum* 'written for students by J.P. Sellers, FRIBA, with sketches from the museum by T. Raffles Davison', published by the Architectural Museum, 1884. This strand of work during the 1880s concluded with *Pen and ink notes at the Glasgow Exhibition: a series of illustrations by T.R. Davison with an account of the exhibition by R. Walker,* published by Virtue of London in 1888.

As the century wore on, the contemporary Arts and Crafts Movement influenced his approach and he became linked with architects such as C. F. A. Voysey, M. H. Baillie Scott and Edwin Lutyens. These was all renowned for their residential commissions in the years either side of 1900, and such an interest led naturally to a wider concern for house design and the Garden City movement that can be seen in his published work. 1909 marked a highpoint: he edited a volume entitled *The arts connected with building: lectures on craftsmanship and design delivered at Carpenters' Hall, 1909*, to which contributors included Voysey and Baillie Scott. Of wider coverage was his *Modern houses: selected examples of dwelling houses*, described and illustrated by Davison and published by Graham Bell & Sons. Drawing on his periodical press background, Davison reproduced articles he had originally written for *The Idler* and *Magazine of Art* with the purpose of demonstrating 'that that houses may be designed with individual artistic character whatever their size and cost'. He deprecated any appearance of dogmatism in matters of taste, but hoped that an 'individual expression of feeling resulting from a long study' would stimulate interest in the lay reader. It is in keeping with his interest in the design of housing that his next production should be *Port Sunlight: a record of its artist and pictorial aspect* (London: Batsford, 1916).

Before then however, he was embroiled in a controversy about a very different kind of design. Davison and Barclay Niven designed a new bridge to cross the Thames at Charing Cross. Conflicting views were reported in *The Times* in 1914 and an article appeared in *The Builder* on 28 April 1916.

Other more characteristic work in *The Builder* formed the basis of his next book: *Wren's city churches: a series of pencil illustrations reprinted from The Builder* (London: Builder, 1923).

By the 1920s Davison had moved to Woldingham in Surrey and retired from practice. His life to this point was described in a biographical anthology edited by Maurice Everett Webb and Herbert Hardy Wigglesworth, entitled *Raffles Davison: a record of his life and work from 1870 to 1926* (London: Batsford).

However it was in retirement that his connection with concrete was made. In 1930 he was asked by H. L. Childe – himself formerly a journalist for *The Builder* – if he would contribute some sketches to Burren & Gregory's *Moulds for cast stone and concrete*; the latest volume in Concrete Publications Ltd's Concrete Series. The presence of illustrations by a man of Davison's reputation could add nothing but lustre to an otherwise modest book. He continued supplying sketches for associated articles by Burren in *Concrete Buildings & Concrete Products* until 1936.

Then – perhaps more in keeping with Davison's own background in domestic architecture – he illustrated the revised 1932 edition of Albert Lakeman's *Concrete cottages, bungalows and garages*; the original of which was the very first title in the Concrete Series.

Thomas Raffles Davison died on 5 May 1937 at his home in Woldingham. An obituary appeared in *The Builder* on 14 May 1937 (p.1024), and details of his life are held at the British Architectural Library.

1930/UK

Corkhill, T.

MIStructE

Corkhill's role in the literature of concrete was as editor of *Brickwork, concrete and masonry*; a set of eight volumes in the series of craft titles published by Sir Isaac Pitman and originally issued as 33 weekly parts between 11 October 1930 and 23 May 1931. As the lengthy subtitle makes clear, the subject coverage was extensive, the approach practical and the authorship divided between specialists. Two reviewers at the time – each writing in the *Structural Engineer* (the journal of the institution of which Corkhill was a member) – were generally complimentary. To summarise their comments, the volumes were useful and of a handy size, attractively presented. 'E. G. W.' criticised a couple of the contributors but declared the editing to have been 'thorough throughout the work', while 'H. R. C.' congratulated Corkhill and the authors on the 'results of their joint labours'.

Corkhill added considerably to his corpus of building books after the Second World War, concentrating on carpentry and joinery, and acting variously as editor and author. Among these titles were: *A concise building encyclopaedia* (Pitman, 1945); *A glossary of wood* (Nema Press, 1948); *Definitions and formulae for students: building* (Pitman, 1952); and *Joinery and carpentry*, Vol.I (New Era Publishing). His *Complete dictionary of wood* was still in print in the 1980s.

Corkhill, T. (ed). *Brickwork, concrete and masonry: a complete work by practical specialists describing modern practice in the work of the bricklayer, concreter, mason, plasterer and slater*. (8 vols). London: Pitman, 1930 [ICE. Rev: *SE* v.9 July 1931, p.260; *SE* v.10 April 1932, p.198]

1930/UK

Scott, Ernest A.

Scott, a structural engineer from Glasgow, flourished in the early 1930s, known principally for his authorship of *Arrol's reinforced concrete reference book*. This popular book balanced the twin roles of textbook with company promotion; in effect a manual of company practice. It provided practical formulae, diagrams tables and graphs, and reproduced the text of regulations and standard specifications, while dressing the content with a selection of photographs to illustrate the achievements of Sir William Arrol & Co. This venerable contracting firm had established a reinforced concrete department ten years previously and with the move to standardisation in prospect, the management presumably felt there was more to be gained by demonstrating the company's competence than by maintaining commercial secrecy. Scott expressed the firm's intentions:

> It is hoped, therefore, that this book will in some measure bring to public notice the firm's resources for both technical and construction work of this class. At the same time no pains have been spared to make this publication of value in assisting both designers and constructors in the application of reinforced concrete work to structures suitable for its use ... throughout it has been the aim to keep technically correct, and at the same to give practical and economic guidance.

In doing so, Scott drew on the experience of the staff; notably the chief engineer Adam Hunter, and interestingly in view of his subsequent ubiquity as an author himself, C. E. Reynolds [1932]. Scott was evidently not on the staff himself.

His other published work for which we have a record was an article in the *Structural Engineer* of September 1934, entitled *Tables for slabs designed to Ministry of Transport load and stress requirements* (*SE* v.12(9)).

Scott, E. A. *Arrol's reinforced concrete reference book of examples, general formulae, model regulations, and designing table as used by the reinforced concrete department of Sir William Arrol and Company, Ltd. Glasgow*. London: E.& F.N. Spon, 1930 [BL, C&CA 1959, ICE. Copy: TCS. Rev: *C&CE* v26, p.244; *SE* December 1930, p.435]

1930/UK

Singleton-Green, John (b.1897)

MSc, MICE, AMIMechE, MIStructE, MACI, MIQ, FIHE

John Singleton-Green
Courtesy of IStructE

John Singleton-Green was born in the latter part of 1897 at Haslingden in the northwest of England, and attended the local grammar school. Studies at the University of Manchester, between 1917 and 1920 led to the award of a BSc, and an MSc followed in 1925. His first job was as a teacher of mechanical engineering at Warrington Technical College, probably while studying for his MSc. He also served as deputy engineer for the Borough of Hythe.

His considerable contribution to the literature began in these early years of his career. His paper, *Concrete & fire resistance* – presented to the Manchester & District branch of the ICE on 22 February 1922 – formed the basis of a series of seven articles on that subject in *Concrete & Constructional Engineering. Part 1: insurance and inspection* appeared in v.17(9) September 1922, concluding with *additional data and general conclusions* in v.19(3) March 1924.

His interest in concrete was manifest in his first submission to the *Structural Engineer – The strength of concrete, its relation to the cement, aggregate and water.* It was written in two parts, concluding in March 1924 (*SE* v.2(3) March 1924, pp.117–119). In the subsequent weeks his private life was transformed when he married his fiancée Rosina M. Leach.

Recently married and with his newly earned MSc, Singleton-Green took a post as engineer at the cement maker G. & T. Earle Ltd of Wilmington, Hull. This well-regarded firm – which was part of the Associated Portland Cement Manufacturers while retaining its autonomy and a separate identity for marketing purposes – published *The making & testing of Portland cement and concrete* in 1925. It is tempting to see this as indicative of common interest between the firm and its new employee. The book – besides describing the cement-making process and promoting its corporate presence – gives comparable weight to the mixing and testing of concrete, including identification of aggregates for fireproof concrete and discussion of the strength of concrete – each a subject of Singleton-Green's early writing.

His next few articles in the technical journals reflect his work on concrete materials at G. & T. Earle. The year 1927 saw two pieces for *C&CE*: *Polishing Portland cement concrete* came out in March (v.22(3)); and *Curing concrete* in December (v.22(12)). In the latter he described the setting and hardening of Portland cements and summarised the 16 conclusions of H. J. Gilkey's work on the effect of curing conditions on compressive strength (published by ACI the previous year). He continued with his interest in effects on strength in the *Structural Engineer*, addressing *The practical application of the water-cement ratio theory* in the September issue of 1928 (*SE* v.6(9)).

Painting Portland cement concrete surfaces was highly topical when published in the *National Builder* that same year (1928, v.8(2)). Maximilian Toch [1931] covered much the same ground in America, as did S. F. Ross for the ACI. Singleton-Green was concerned in his article to stress the deleterious effects of lime upon oils and vegetable pigments and the precautions available to reduce its alkalinity.

Two years later Singleton-Green tackled the opposite problem of acidity, in the same journal. *Acids and concrete* was published in *National Builder* (v.9(9) 1930). In it he assessed the potency of various acids, described the preparation of concrete to enhance resistance to acids, and identified the various treatments and admixtures available to protect concrete against acid attack.

His standing with the Institution of Structural Engineers at this time was indicated by his acceptance of the role of chairman at the Lancashire & Cheshire Branch (the institution's first and largest), for the year 1929–1930. In this he was the sixth incumbent – R. Travers Morgan [1925] preceded him as the fourth.

In 1930, his writing took a significant turn. His first book *Reinforced concrete roads* (published by the Contractor's Record, Ltd), was on an application quite different from those that he had tackled previously. His aim was to present, 'in compact form, to the practising engineer, the wealth of information which has resulted from the enormous amount of investigation carried out in recent years in the subject of concrete for roadwork'. In that it covered the practical aspects of producing concrete for road as well as the theoretical, Singleton-Green was continuing his usual approach and, according to his reviewer in the *Structural Engineer*, 'in this direction the author has succeeded'. This book however, was clearly not derived from experimental work at G. & T. Earle, for it drew extensively on American experience in the various areas of materials, specification, production and testing. The references were extensive, but the index limited.

Florescence (published in *Municipal Engineering* of 25 February 1932) was a more familiar theme, but it was only a prelude to a bigger preoccupation that year. *Limestone as an aggregate for concrete* was read to the 27th annual general meeting of the Central Association of the Lime & Limestone Industry of Great Britain, held on 23 May 1932 at the Grand Hotel, Harrogate. It subsequently was published in *Good Roads* v.8(12) 1932. It was a polemical work, urging greater recognition of the role of limestone as an aggregate for concrete. Citing American experience and tests for the concrete used in Edale in Derbyshire, Singleton-Green examined the strength and fire-resistance of limestone. Perhaps it was no coincidence that G. & T. Earle had recently built a new cement works near Edale.

The early 1930s were a productive period for Singleton-Green, for at the same time he embarked on a series of articles for *Cement, Lime & Gravel*, entitled *The manufacture of concrete products and cast stone* (Volumes 6–7, 1932–1933):

1. *Simple solid blocks* (April & May 1932)
2. *Increasing output* (June 1932)
3. *Special shapes* (September 1932)
4. *Choice of constituent materials* (November 1932)
5. *Surface treatment of products* (December 1932)
6. *Efflorescence and crazing* (January 1933)
7. *Curing of products* (February 1933)

Testing concrete in flexure also appeared in the November issue of the same journal (*CLG* v.7, November 1932).

At the same time he was working on his next book: a summary of his work to date. He completed *Concrete engineering:* Vol. I, *practical concrete* in 1932, and it was published by Charles Griffin & Co. in the New Year. He was then described as 'consulting engineer', but his debt to former employer G. & T. Earle was evident throughout the book. It was dedicated to Mr. F. Gibson (the magnificently moustachioed secretary & director of the firm), and acknowledged the writer's indebtedness to 'the directors of Messrs. G. & T. Earle, Ltd., Hull, for permission to use the results of laboratory tests'. Both the chief chemist and head tester were thanked. Much of the text was drawn from Singleton-Green's journal articles, including reference to American practice. Research from the new Building Research Establishment was beginning to become important too. The author's emphasis was on currency, concision and practicality; the book acting as a textbook for students and a reference book for professional and practical men. It was well referenced and authoritative.

Lightweight concrete construction, in the *Structural Engineer* of April 1934 (v.12(4)), described the various methods of reducing weight in concrete construction by the use of clay pots, hollow blocks, high strength cements, lightweight aggregates and entrapped air. The more distinctive feature of 1934 was the award of British patent 409323. This was for: 'Improvements in and relating to reinforcing mesh of concrete asphalt, rubber and like materials'.

This burst of literary activity came to a climax in 1935, at the end of our survey, with Volume II of *Concrete engineering,* subtitled *properties of concrete*. Singleton-Green was still resident in Hull in June (at an address in Endike Lane), and his acknowledgements followed the pattern set in Volume I. Where the book differed was in its emphasis on the properties of concrete in its hardened state: creep; strength; permeability and fire resistance; along with the various processes of deterioration. A chapter on lightweight concrete picked up the author's interest of the previous year. Again, it was aimed at both the student and the practitioner.

What Singleton-Green undertook during the Second World War is not known, but his article in *Surveyor & Municipal & County Engineer* (8 November 1940) indicated a concern for the war effort. It was entitled *Camouflage with coloured cement*.

After the war he was back with the cement industry, representing APCM on the Cement & Concrete Association's newly formed advisory committees for research and education in 1948. These committees met to prioritise topics for the association's research programme and to comment on the drafts of all new C&CA publications before production.

In the 1950s he became chairman and managing director of Stressed Concrete Design, Ltd, London, and it was in this capacity that he contributed to the debate arising from *Applications of prestressed concrete to water supply and drainage* by Professor Rhydwyn Harding Evans, published in the *Proceedings* of the ICE on 3 May 1955. He also took an interest in the Pavings Development Group, and chaired its 'Concrete Roads and Runways' conference on 17 October 1956. This interest in roads saw him elected as the president of the Chartered Institution of Highways and Transportation in 1961 – the 16th to hold office.

He was a vice-president of the Institution of Structural Engineers at the time of the Golden Jubilee in 1958 – a date at which he was acting as advisory editor to 'The Concrete Library' books published by the Contractor's Record Ltd in London and John Wiley & Sons in New York. The series included titles such as *Concrete technology* by D. F. Orchard (1958), *Joints and cracks in concrete* by P. L. Critchell (1958) and *Prestressed concrete cylindrical tanks* by L. R. Creasy (1961). Both as author and editor, his lifetime's output was considerable.

Singleton-Green, J. *Reinforced concrete roads*. Contractor's Record, 1930 [BL, ICE]

Singleton-Green, J. *Concrete engineering:* Vol.I, *practical concrete*. Charles Griffin & Co., 1933 [BL, C&CA 1959, ICE. Rev: *SE* v.10, p299]

Singleton-Green, J. *Concrete engineering:* Vol.II, *properties of concrete*. Charles Griffin & Co., 1935 [BL, C&CA 1959, ICE]

1931/USA

Baker, Samuel

CE

The International Textbook Company of Scranton,

Pennsylvania, published a range of compact textbooks for the use of students following the curriculum of the International Correspondence Schools. Samuel Baker was the dean of the schools of technology at the ICS, and with his colleague A. DeGroot (below), wrote several of the textbooks prepared for this purpose. *Cements and aggregates* was published in 1931 and *Concrete and mortar* in 1935. Together with their *Concrete Products* of 1941, these were issued later as a compilation volume; *Cements, concrete and mortar*. Similarly the two men worked with J. J. Wickham (principal of civil engineering) and Professor C. E. O'Rourke [1923] of Cornell University on: *Design of spread footings* (1932); *Flat-slab design* (1935); and *Foundations and piling* (1941), which were later amalgamated into a single volume – *Foundations and flat slabs*.

Baker, S. & DeGroot, A. *Cements and aggregates*. Scranton, PA: International Textbook Co., 1931 [Copy: TCS]

Baker, S. & DeGroot, A. *Concrete and mortar*. Scranton, PA: International Textbook Co., 1935 [Copy: TCS]

1931/USA

DeGroot, A.

BS

Director of civil engineering schools, DeGroot was a colleague of Samuel Baker (above) and collaborator on the various works that comprised *Cements, concrete and mortar* and *Foundations and flat slabs*.

1931/UK

Gray, William Samuel

BA, MAI (Dublin), AMICEI

An Irishman – to judge from his qualifications and affiliations – William Gray was closely associated with Concrete Publications Ltd in London, as an author both of books and of articles in *Concrete & Concrete Engineering*, as well as a regular reviewer of new titles.

He first appears in *C&CE* in the mid 1920s with a couple of articles on matters of concrete technology – *Influence lines applied to continuous beams* (v.20(6), June 1925) and *The rigidity of joints in reinforced concrete* (v.22(5), May 1927) – but he was soon to start on the subject matter for which he is now associated through his well-known books: tanks; bunkers; gantries and related industrial applications:

- *Reinforced concrete bunkers*, v.23(2–7), February-July 1928
- *Coal bulkheads*, v.23(11), November 1928
- *Shallow circular tanks of large diameter*, v.24(1) January 1929
- *Reinforced concrete gantries*, v.24(2) February 1929
- *Conveyor pits*, v.24(4) April 1929
- *Shallow rectangular bunkers of large dimensions*, v.24(9) September 1929
- *Reinforced concrete tanks*, v.25(4) April 1930

Consolidating this work – and perhaps encouraged by *C&CE*'s editor, H. L. Childe – Gray wrote his first book for Concrete Publications Ltd in 1931.

It was entitled *Reinforced concrete reservoirs and tanks*. In defining its coverage, Gray included swimming pools and public baths, but (at this stage) decided against water towers and dams. The book was not confined to broad principles, but treated its subject in detail with the aim of providing practical guidance to engineers unfamiliar with designing these particular structures. Gray offered his own opinions and presented the fruits of continental practice to a British audience. The performance of tanks in areas of subsidence and water-logged ground were treated for the first time in a British textbook. Sales of the book were respectable, given the nature of the subject matter, with the first edition running to 3,200 copies. Revisions in 1924 and 1948 generated 3,250 and 5,000 respectively, and a third edition appeared posthumously in 1954.

Extending the subject matter to various forms of elevated tanks, Gray followed this with his second book – *Reinforced concrete water towers, bunkers, silos and gantries* – in 1933. Although he had touched on these topics before writing *Reinforced concrete reservoirs*, his *C&CE* articles of 1931 and 1932 provide evidence of further work for incorporation in the new book:

- *Circular culverts*, v.26(3) March 1931
- *Design of silos*, v.26(5-6) May-June 1931
- *Shallow trough bunkers*, v.26(10) October 1931
- *Roofs and floors of elevated circular tanks*, v.27(4) April 1932

In its review, the *Structural Engineer* considered that: 'there is a real need for books enabling engineers who possess a general knowledge of some branch such as reinforced concrete to apply it in specialised directions with the assurance that they are benefiting by the best and most recent work of others. Here that need is filled in a work which is clear, authoritative and practical … (full of the kind of knowledge accumulated by an experienced contractor)' (v.11, August 1933, p.361).

Although the subject matter of these two books dominated this period, it would be wrong to think of it excluding anything else, and indeed his writing for *C&CE* included several articles on more general aspects of reinforced concrete:

- *Influence lines for continuous beams*, v.25(5) May 1930
- *Reinforced concrete beams of non-uniform section*, v.25(7) July 1930
- *Continuous slabs reinforced in two directions*, v.25(12) December 1930
- *Maximum moments in multi-storey frames*, v.27(6) June 1932
- *Design calculations for estimating*, v.27(9) September 1932
- *Wind stresses on portal frames*, v.27(11) November 1932

After such a productive period in which his distinctive output represented pioneering work, his last book that falls within our survey was relatively lightweight; an exercise in technical collaboration with his editor Childe, to produce a practical guide to the options for decorative finishes currently available.

Concrete Surface Finishes, Rendering and Terrazzo was jointly prepared by Gray and Childe in 1935. Two thousand copies were sold before its revision in 1943 and further reprinting in 1948. Long after Gray's active involvement, it was re-issued in 1964 by Childe alone under the modified title, *Concrete Surface Finishes and Decoration*.

At the very end of our review period, in December 1935, Gray and F. E. Wentworth-Shields – that stalwart of the original Concrete Institute – collaborated to write *Reinforced concrete piling* for *C&CE* (v.30(12)). Like so much of Gray's journal output, this too found its way into print as a book by Concrete Publications Ltd. It carried the same title – *Reinforced concrete piling* – and was issued in 1938. Further editions followed in 1948 and 1960; the latter revised by H. W. Evans.

Three other articles appeared after 1935 – *Shallow foundations* (v.31(5) May 1936); *Bending moments in columns* (v.32(12) December 1937); and *Small arched roofs* (v.26(5) May 1941) – but from wartime onward Gray was silent.

Gray, W. S. *Reinforced concrete reservoirs and tanks*. London: Concrete Publications Ltd, 1931 [BL, Yrbk 1936. Rev: *SE*, v.9, November 1931, p.382; *C&CE*, v.26, 1931, p.676]

Gray, W. S. *Reinforced concrete water towers, bunkers, silos and gantries*. London: Concrete Publications Ltd, 1933 [BL, C&CA 1959, ICE, Yrbk 1936. Rev: *SE*, v.11 August 1933, p.361]

Gray, W. S. and Childe, H. L. *Concrete Surface Finishes, Rendering and Terrazzo*. London: Concrete Publications Ltd, 1935. [Copy: TCS]

1931/USA

Toch, Maximillian (1864–1946)

Principally being a specialist in paints and pigments, Dr. Maximillian Toch was far from the average writer on concrete subjects, despite the promise of his *Protection and decoration of concrete*, which was published by the mainstream Van Nostrand in 1931.

Born on 17 July 1864 to German immigrants, Toch spent most of his life in New York. He was educated at New York University and after graduating studied at Columbia. Specialising in Chemistry, Toch also gained a degree in Law. Later he taught Chemistry, Industrial chemistry and Chemical engineering at various institutions: Cooper Union, Peking University, Columbia, and City College, New York

In business he was the President of Toch Bros, Inc., as well as Head of Research, and VP of Standard Varnish Works in New York. His academic and commercial interests combined in a series of publications on paint:

- The Chemistry and technology of mixed paints (1907)
- Materials for permanent painting (1911)
- The chemistry and technology of paints (1916)
- How to paint permanent pictures (1927)
- Paint, painting and restoration (1931)

In the early years of the Great War, before America entered the conflict,

Toch developed a tone of 'battleship grey' and a 'dazzle' camouflage scheme for painting naval ships which in 1917 was one of six approved by the US government.

His interest in the colour effects of concrete was manifest when, in 1925 he asked: *Shall anything be added to Portland cement?* in the ACI's *Proceedings*, v.21(2). Here he challenged the cement manufacturers' limit of 2% additional material, claiming there was a valid place for pigments to modify the 'sombre' monotony of grey. He followed this with a more prescriptive article in *Concrete* (v.33(3), 1928), entitled *Decorating concrete*. The book of 1931 took his argument further, but other references to the literature point to fine art as his main field of activity in later years. He appears as an expert on the sources of pigment for historical paints, and in the year after his book on concrete, he was writing in support of an attribution to Karel van Mander of an early 17th Century painting of Shakespeare and Ben Jonson. In Tracy Kingman's book *An authenticated contemporary portrait of Shakespeare* (New York: William Edwin Rudge, 1932), Toch provides the final two chapters, reporting on his x-ray and fluoroscope investigations and providing his conclusions on the provenance of the paint.

He died on 20 May 1946 and his obituary appeared in the *New York Times* of 31 May.

Toch, M. *The protection and decoration of concrete*. New York: Van Nostrand, 1931 [ICE]

1932/USA

Cross, Hardy (1885–1959)

Hardy Cross
Courtesy of the University of Illinois

Largely forgotten by practitioners today, Hardy Cross developed the moment distribution method for structural calculations in large buildings – a method widely used from the mid 1930s to the 1960s, during which time 'his reputation was absolutely world wide' (Professor Robert Davas). Professor William J. Hall at the University of Illinois has described his technical achievements as having 'significantly changed the field of structural analysis and the understanding of structural behavior.'

Cross was born on 10 February 1885 in Nansemond Co., Virginia, near the Great Dismal Swamp and was educated at Hampden-Sydney College. His first degree was in English, awarded in 1902 when he was just 17, and he began his teaching career in that department. A year later (in 1903), he received a BSc, and moving to Norfolk Academy, continued to teach. He took a further BSc in civil engineering, at the Massachusetts Institute of Technology in 1908. After a brief spell in the bridge department of the Missouri Pacific Railroad in St Louis, he returned to the Norfolk Academy in 1909. Then, to complete his education, he attended Harvard for a year, becoming a Master of Civil Engineering in 1911.

Moving to Brown University he served as assistant professor for seven years, and spent some time in general practice. In 1921 he married Edith Hopwood Fenner and was appointed professor of structural engineering at the University of Illinois. The 16 years spent at this institution were his 'most creative years', during which he wrote 43 papers and developed his 'moment distribution method' of analysing statically indeterminate structures – the method for which he became famous.

Its genesis lay partly in the personal animosity between Cross and his dean at Illinois – Milo Ketchum [1908]. According to L. K. Eaton, Ketchum was resentful of Cross's reputation as a great teacher, and critical of his limited output of research papers. In response Cross first produced his *Dependability of the theory of concrete arches* in the *Bulletin of the University of Illinois Engineering Experiment Station, No.203* (1930). Then, weeks later, he submitted his ten-page *Analysis of continuous frames by distributing fixed end moments* for publication in the *Proceedings* of the American Society of Civil Engineers. Published in the May 1930 issue, the paper had immediate impact: Cross 'was immediately hailed as a man who had solved one of the knottiest problems in structural analysis' in a way that could be accepted by the practical engineer with only a functional grasp of mathematics. Between publication and the closure of discussion, 38 commentators responded to the paper, occupying 146 pages of comment. Such a record response is testimony to the importance of the problem and the ingenuity of Cross' solution. In 1949 he summarised his method thus:

> Moment Distribution. The method of moment distribution is this: (a) Imagine all joints in the structure held so that they cannot rotate and compute the moments at the ends of the members for this condition; (b) at each joint distribute the unbalanced fixed-end moment among the connecting members in proportion to the constant for each member defined as 'stiffness'; (c) multiply the moment distributed to each member at a joint by the carry-over factor at the end of the member and set this product at the other end of the member; (d) distribute these moments just 'carried over'; (e) repeat the process until the moments to be carried over are small enough to be neglected; and (f) add all moments – fixed-end moments, distributed moments, moments carried over – at each end of each member to obtain the true moment at the end.

The method was not exact, but close enough in practice and had the advantage of speed. With N. D. Morgan, Cross developed his ideas in the 1932 book, *Continuous frames of reinforced concrete*, and again in the ASCE paper, *The relation of analysis to structural design* (*Proceedings*, v.61(8)1935). The 'Hardy Cross' method was rapidly adopted and by 1935 – the year in which he was awarded the ACI's Wason Medal – was being taught in most American schools of engineering.

Having written a further important paper in 1936, *Analysis of flow in networks of conduits or conductors* (Experiment Station Bulletin No.286) he left Urbana-Champagne in 1937 to become Strathcona professor and chairman of the department of civil engineering at Yale. There, his teaching was recognised by the American Society for Engineering

Education with the award of the Lamme Medal in 1944. Other lifetime awards included honorary doctorates from Yale, Lehigh and Hampden-Sydney; the ASCE's Norman Medal; and the Gold Medal of the Benjamin Franklin Institute of Philadelphia.

He retired in 1951 and published *Engineers and ivory towers* (1952) the following year. Shortly before his death on 11 February 1959, at Virginia Beach, he was made an honorary member of the ACI and awarded the Gold Medal of the Institution of Structural Engineers. At the time Cross was the first American to be so honoured and only the fifth recipient of the medal's fifty-year existence. His final work (published posthumously) was *Arches, continuous frames, columns, and conduits: selected papers of Hardy Cross* (1963).

As methods of analysis changed, facilitated by the use of computers, Cross and his moment distribution method fell into obscurity. In recent years however, his reputation has been resurrected by the efforts of Leonard K. Eaton, whose work has culminated in a biography entitled *Hardy Cross: American engineer* (University of Illinois Press, 2006) – from which the quotations herein are taken. Hardy Cross is now recognised as 'one of the first internationally important, American-educated engineering theoreticians' (*Technology and Culture*), and one whose publication made a significant difference to structural engineering practice for a generation.

Cross, H. & Morgan, N. D. *Continuous frames of reinforced concrete*. New York: John Wiley & Sons, 1932 [BL, C&CA 1959, ICE]. Rev edition, 1954 [BL]

1932/USA

Morgan, Newlin Dolbey

Joint author of *Continuous frames of reinforced concrete* with Hardy Cross. Like Cross he was an academic at the University of Illinois, appointed in 1925 to the position of assistant professor of architectural engineering in the Department of Architecture.

1932/UK

Dowsett, John Frederick

AIStructE

John Dowsett – a member of the Institution of Structural Engineers – had written two earlier publications before he turned his hand to addressing concrete. His first was *Advanced constructive geometry* (published by Oxford University Press in 1927), followed by *Stair-building and handrailing* (by the Library Press of London, 1929). The latter was presumably a commercial success for the Library Press undertook the publication of his next title – a collaborative work with Eric Bartle (below), called *Practical formwork and shuttering*. It was welcomed by Dowsett's institutional journal, the *Structural Engineer*:

> Both authors and publishers are to be congratulated on the production of this book. Perhaps one its greatest values is that it impresses the reader with the importance of the subject, which for so long was considered the Cinderella of the craftsman's activities. The book is beautifully produced and well illustrated, and the authors evidently understand the subject. They claim that

> each example has been designed from the standpoint of both cost and efficiency, and taken from actual practice" (*SE*, April 1932, v.10, p.198)

Dowsett, J. F. & Bartle, E. G. *Practical formwork and shuttering*. London: Library Press, 1932 [IStructE. Rev: *SE*, April 1932, v.10, p.198]

1932/UK

Bartle, Eric Godfrey (b.1899)

Collaborator of J. F. Dowsett (above), Eric Bartle also wrote *Building construction: Book II. A course for third year students* in 1946, as part of Pitman's technical building series.

Dowsett, J. F. & Bartle, E. G. *Practical formwork and shuttering*. London: Library Press, 1932 [IStructE. Rev: *SE* April 1932, v.10, p.198]

1932/UK

Reynolds, Charles Edward (1900–1971)

BSc, MICE

We have already encountered Reynolds as one of the concrete specialists at Sir William Arroll that was acknowledged by Ernest Scott [1930]. Reynolds' own book – *Reinforced concrete designers' handbook* – was to become the standard work of reference on its subject in Britain, continuously in print from 1932 to the present (2009), and with nearly 200,000 copies having been sold.

Charles Edward Reynolds was born at the dawn of the 20th Century and educated at Tiffin Boys' School in Kingston-on-Thames and at Battersea Polytechnic. He joined Sir William Arrol shortly after the inception of the firm's reinforced concrete department in the early 1920s. His first essay in the literature of concrete was an article for *Concrete & Constructional Engineering*: *Combined bending and direct force: a practical method for the design and analysis of rectangular sections* (v.20(12), December 1925). This was followed by *Concrete road construction in the Argentine* (v.25(12), December 1930) – in the year that Scott's book appeared – and *Combined bending and direct forces* (v.26(7)) in July 1931.

These articles brought Reynolds into contact with H. L. Childe – editor of Concrete Publications Ltd – and when Childe decided to publish a handbook on concrete design (a generic equivalent to Scott's book), Reynolds was his choice of author. *Reinforced concrete designers' handbook* was published in 1932.

In drawing attention to the book and complimenting its layout, a review in the *Structural Engineer* (v.10, p.446, October 1932) paraphrased the author's intentions:

> The object of this book is to present, in one volume, complete data for the design of all classes of reinforced concrete structures. It is claimed that it will enable the reinforced concrete designer to dispense with out-of-date reference books, to possess hitherto unpublished information, to dispense with loose graphs and notes, to execute designs and detail drawings rapidly, and to use one book instead of a dozen.

These were grand claims, but the book-buying engineer found them supported by the content, and the *Reinforced concrete designers' handbook* sold in significant numbers. Not only was it to become the standard work on design, but it also cemented a relationship with Concrete Publications Ltd that was to see Reynolds eventually succeed Childe as managing editor. The book also established Reynolds' reputation, which over the years helped secure employment at BRC, Simon Carves, Leslie Turner & Partners and C. W. Glover & Partners, before moving into private practice.

Within a couple of years, the practice of structural design in concrete was standardised by the new DSIR Code of Practice. Reynolds' response to the code can be seen in the articles he wrote for professional press: *Some observations on the new code of practice for reinforced concrete* (*SE* v.12(12), December 1934); and a series entitled *The new code applied to design* (*C&CE* v.29(3–6, 8, 10–12) 1934 and 30(2) 1935), in which he undertook the design of a hypothetical building.

With very few exceptions, the many articles he wrote in *C&CE* over the following years describe problems of design – notably the calculation of bending moments – and in 1950 and 1957 they comment on the revised Codes of Practice:

- *Design and cost of shuttering.* v.31(8), August 1934
- *Measurement of quantities.* v.34(2), February 1939
- *Design of slabs in liquid-containing structures.* v.35(2), February 1940
- *Design of pre-cast reinforced concrete frames: frame profiles.* v.41(11), November 1946
- *Direct calculation of bending moments in continuous beams.* v.43(12), December 1948
- *Bending moments in continuous beams.* v.44(1), January 1949
- *Bending moments on continuous beams with haunches.* v.44(5), May 1949
- *Design of reinforced concrete members in accordance with the British standard code.* v.45(1/ 5), January/May 1950
- *Loads on groups of free-standing piles.* v.48(12) December 1953
- *Groups of symmetrically-arranged free-standing piles.* v.49(3), March 1954
- *Bending moments on continuous beams in accordance with the revised Code.* v.52(10), October 1957
- *Beams and slabs designed by the load-factor method: in accordance with BS code of 1957.* v.53(5/ 6), May/June 1958

He also wrote in *C&CE*'s sister periodical – *Concrete Building & Concrete Products* – though these articles were on more practical subjects. *Proportioning concrete mixtures* (November 1936), is an example.

In parallel to the articles, Reynolds produced a number of books to follow on from his first title, all issued by Concrete Publications Ltd. *Practical examples of reinforced concrete design* was the first, appearing in May 1938 to reflect the new London County Council Building By-laws that came into force in the previous January, as well as the earlier DSIR Code of 1934.

When these codes were replaced in the post-war period, by the British Standard Codes, Reynolds wrote a book to illustrate them with examples: *Examples of the design of reinforced concrete buildings in accordance with the British Standard Codes of Practice*; published first in 1952 (and revised in 1959), interpreted CP3 and CP114, with further reference to CP101, 111 and 123.

The changing codes also affected the *Handbook*, which was revised successively in 1939, 1946, 1948, 1951, 1954 and 1957.

At some point (probably during the 1950s), Reynolds became the technical editor of Concrete Publications Ltd in addition to his private practice, and then following the retirement of Childe at the end of 1959, managing editor. James Steedman was appointed to replace Reynolds as technical editor in 1961. With Reynolds in charge, this period saw an initial burst of activity from his own hand. Firstly, the *Reinforced concrete designers' handbook* was revised to its 6th edition in 1961 and then, a year later, Reynolds introduced a new title: *Basic reinforced concrete design: a text-book for students and engineers*. This book – published in two volumes – aimed to introduce the specifics of concrete design to those who had an understanding of general engineering principles and enable the reader to achieve the stage of designing simple structures. Part III of the text extended its coverage to review the many forms of advanced structure through examples and short bibliographies for further reading.

During the 1960s the basis of reinforced concrete design was revised and the codes reissued to reflect the changes. Reynolds consequently prepared a 7th edition of his *Handbook* and planned an 8th edition to cover the proposed Unified Code, CP110. After a period of illness and two operations however, Reynolds died on Christmas Day, 1971. A tribute to Reynolds and his life's work appeared in *Concrete* (February 1972). His work was unfinished, but in due course was faithfully completed by Steedmen. Besides his consulting and publishing work, he was also a member of the council for the Junior Institution of Engineers and was honorary editor of its journal at the time of his death.

A travelling bursary for students was awarded in his memory for several years and Reynolds lives on – as is the way with authors – through his books. The remarkable fact is that his first book, published in 1932, is still in print. James Steedman completed the 7th revised edition and undertook the planned 8th edition that incorporated CP110. After further revisions in 1981 and 1988, the baton was passed to Tony Threlfall whose 11th edition – still bearing Reynolds' name – was published in 2008.

Reynolds, C. E. *Reinforced concrete designers' handbook*. London: Concrete Publications Ltd, 1932 [BL, C&CA 1959. Copy: TCS. Rev: *SE* v.10, p.446, October 1932]

1933/UK

Chettoe, Cyril Stanley (1893–1963)

BSc (Eng.) London, MICE

With an engineering degree from London, and membership of the Institution of Civil Engineers, Chettoe was a colleague of – and collaborator with – H. C. Adams at the time of the one book for which he is now remembered: *Reinforced concrete bridge design*. The two men were employed in the same department of the Ministry of Transport, headed by Sir Henry Maybury, and had enjoyed recent experience in their chosen discipline. Indeed their role in checking the plans and calculations submitted for the majority of road bridges, qualified them to observe official requirements and to compare the best in contemporary British practice.

The stated aim of the joint authors was to reconcile the overlooked art of bridge building with the rapidly developing science of this subject. The book concentrated on the arch bridge – the form that embodies the greatest variety and beauty according to the authors, who claimed in their preface to have made a special study of this type. Whilst combining art with science throughout, the book also contained a chapter specifically on aesthetics. The second edition of 1938, incidentally, paid tribute to the work of the Royal Fine Art Commission in this respect.

The author's approach was thought by a reviewer at the time (*Structural Engineer*, June 1933) to be particularly appropriate for students or junior engineers as it concentrated on the basics, leaving more difficult problems of design to specialist sources. Indeed, he identified several topics that would benefit from more detailed treatment in any subsequent edition – and the 1938 revision did contain new material. However, the section on widening and strengthening existing bridges was welcomed as a previously neglected aspect of bridge engineering.

During and after the Second World War, Chettoe's continued professional interest in bridges was demonstrated by his output for the Institution of Civil Engineers:

- *The effects of modern road layout on bridge design*, 1941/42
- *The strength of cast-iron girder-bridges*, October 1944
- *The military classification of bridges in Great Britain*, 1948

In 1942 he was awarded the Institution's Telford Premium.

As an engineer, Chettoe had an active interest in the conservation of the finer qualities of building. According to Haddon Adams, 'he had a sound instinct for knowing what was worthy to be saved in old buildings of all kinds, and always showed intense interest in find suitable ways and means of safeguarding any fine old bridge and in strengthening and widening it, when necessary, without destroying its character'. His interest extended to supporting the work of the National Trust, the Council for the Preservation of Rural England and the Society of the Protection of Ancient Buildings.

After retiring from his position as head of the bridges branch of the Ministry of Transport in 1954, he undertook consultancy work as a bridge and road engineer and was chairman of the British section of the International association for Bridge and Structural Engineering.

He died suddenly on 8 November 1963 at the age of 70, having lately been elected by IABSE as a 'Member of Honour'. In an obituary published in *C&CE* in January 1964, Adams described Chettoe as 'a man of quiet personal charm with many friends', who would be 'widely missed'.

Chettoe, C. S. and Adams, Haddon C. *Reinforced concrete bridge design*. London: Chapman & Hall, 1933 [BL, ICE, Yrbk 1934]. Second edition, 1938 [C&CA 1959]

1933/UK

Adams, Haddon Clifford (b.1899)

MC, MA (Cantab), AMICE

Born in Salisbury, Haddon Adams spent part of his youth in Ipswich. After military honours (presumably in the First World War) and an MA at Cambridge, he embarked on a career in civil engineering. He was for some time assistant highway engineer for the State of Illinois in the USA, and duly became an associate member of the Institution of Civil Engineers. Away from work he appears to have been associated with the Ruskin Collection at Bembridge (the progressive independent school), now held at the University of Lancaster – an artistic interest that seems to have found expression in his appreciation of the aesthetics of bridge design. His two works on reinforced concrete both date from 1933: *Reinforced concrete bridge design* (with C. S. Chettoe); and *Elements of reinforced concrete design.*

The former – described under Chettoe – was the more specific of the two, but like the latter, tackled its subject with simplicity of approach directed as much to the student as to the bridge engineer. It encapsulated much of the complicated mathematics in the form of tables and design charts, and thus appears to be a natural companion to the volume following.

Elements of reinforced concrete design (issued only months later), was a short book of 146 pages, aimed at the student, and was both promoted and praised in the *Structural Engineer* (January 1934) for its simplicity and ease of reading: 'the author has a happy style which carries the student with him from first to last.' It covers elementary design in a practical way, avoiding 'mathematical gymnastics' and including worked examples of detailed designs. Despite Adams' highlighted experience in America, the publisher explicitly advertised that it was by 'an author fully conversant with the practice of British specialist designers' and that it presented 'students and junior designers with a text-book based on the practice of British firms specialising in reinforced concrete design'.

His close association with Chettoe made him the natural choice to write the latter's obituary in 1964. Much earlier in his career, his assistance to W. S. Gray in checking calculations was acknowledged in *Reinforced concrete reservoirs and tanks* (1931), and in 1948 he was appointed to represent the Ministry of Transport on the Cement & Concrete Association's committee for co-operative research.

Chettoe, C. S. and Adams, Haddon C. *Reinforced concrete bridge design*. London: Chapman & Hall, 1933 [BL, ICE, Yrbk 1934]. Second edition, 1938 [C&CA 1959]

Adams, Haddon C. *Elements of reinforced concrete design*. London: Concrete Publications, 1933 [ICE, TCS]

1934/UK

Glanville, Sir William Henry (1900–1976)

CB, CBE, DSc, PhD., FRS, FICE, FIStructE., Hon MACI

Sir William Henry Glanville Courtesy of The Concrete Society

We have already encountered Glanville in our discussion of W. L. Scott [1925] – the two men worked together on the DSIR Code of Practice and its explanatory *Handbook* described above. Although the *Handbook* prompts his inclusion in this survey, Glanville's association with concrete and his place in the literature is however, considerably more extensive and constituted only part of a distinguished career in engineering research.

William Henry Glanville was born on 1 February 1900 in Willesden, Middlesex, the second of William and Amelia Glanville's three children. His father (propitiously) was a builder. He received his secondary education at Kilburn Grammar School and enlisted in the army during the final year of the First World War. After demobilisation he read civil engineering at East London College, graduating in 1922 with first class honours.

At the age of 22 he embarked on what would become an illustrious career in research, taking employment as an engineering assistant at the newly founded Building Research Station (later the BRE), then based in East Acton. In fact his was only the third appointment in the life of the station, which had been established by the Department of Scientific and Industrial Research in April 1921. His first investigation was into the permeability of concrete and he was one of the early supporters of Duff Abrams' discovery of the importance of the water/cement ratio. He also promoted curing by immersion rather than air curing. He was awarded his PhD in 1925 and his work was published as *The permeability of Portland cement concrete* (Building Research Technical Paper No.3) in 1926 – the year in which the enlarged BRS was established at Watford.

Glanville was promoted to the post of chief engineer in 1928 and continued to specialise in concrete research. 'By the early 1930s, a great deal had been discovered about the basic properties of concrete. Permeability, curing, temperature development during setting and hardening, creep, shrinkage and workability had all been studied (Glanville. *Research on concrete* in *Concrete*, 1967). This work culminated in a series of reports in 1930, entitled *Studies in reinforced concrete*. They were issued as Building Research Technical Papers, 10–12. In the first – *Bond resistance* – he examined the chemical nature of adhesion, and the frictional resistance between concrete and reinforcing steel. He explored the distribution of bond stresses in relation to the amount of slip and the composition (water and cement content) and condition of the concrete. *Shrinkage stresses* described the effect of concrete shrinkage along the reinforcing bar over time. The third paper – *The creep or flow of concrete under load* – extended his study of movement to creep and the modulus of elasticity for different types of concrete. He summarised his conclusions in *C&CE* as *Small movements in concrete*, later in the year (v.25(11) November 1930). His *Creep of concrete under load* (written with F. G. Thomas), extended this work and was presented as a paper to the Institution of Structural Engineers before publication in the *Structural Engineer*, v.11(2) February 1933.

Beside his seminal work on concrete, 1930 was a pivotal year in his private life, for on 20 June he married. His bride was Millicent Carr with whom he was to have a son and a daughter.

In 1931 he was consulted by the London County Council on devising of a code of practice for reinforced concrete construction and was appointed technical officer to the subsequent Reinforced Concrete Structures Committee. Its report – drawing heavily on Glanville's research – was published in 1933 and was accepted as the new DSIR Code. To guide the prospective user of this code, Glanville joined W. L. Scott the following year in writing an *Explanatory handbook on the code of practice for reinforced concrete as recommended by the Reinforced Concrete Structures Committee of the Building Research Board* – a book that is described at length above.

Glanville summarised BRS research after publication of the Code of Practice: 'Study of the use of higher stresses and of high-tensile and deformed bars, of limited crack widths, of concrete slabs, of fire resistance, of lightweight aggregate, of shear in beams, and of special problems such as the impact strength of beams and of reinforced concrete piles during driving' (*Concrete*, 1967). Much of this work found its way into print.

A BRS colleague – D. A. G. Reid – joined Glanville in preparing *Mortar tests as a guide to the strength of concrete* (*SE* v.12(5) May 1934), which described the development of a small-scale mortar test for service as an index to the strength of a concrete made from the same cement. The technique achieved accuracy to within 10 per cent, and prompted discussion in a subsequent issue of the *Structural Engineer* (v.13(2), pp.114–123). The year 1934 also saw publication of Glanville & P. W. Barnett's *Mechanical properties of bricks and brickwork masonry* (Building Research Special Report 22) – the fruition of a testing programme that had started in 1927.

With G. Grime and W. W. Davies, Glanville reported the results of the investigation into concrete piles in a paper entitled *The behaviour of reinforced-concrete piles during driving*, published in the ICE's *Journal* in 1935. A development of this programme was subsequently published by HMSO in 1938 as *An investigation of the stresses in reinforced concrete piles during driving* (Technical Paper No.20) by Glanville et al.

A conspicuous feature of early work at the BRS was the extent to which researchers had to develop their own methods and equipment. In Glanville's own words: 'we had to mix our own concrete, and design and make our apparatus and most of our testing facilities ... There were constant temperature and humidity chambers to construct, testing machines and testing methods to develop, methods of producing uniform materials to introduce and special apparatus to design' (*Concrete*, 1967). This necessity led in Glanville's case to the award in of a British patent (No.429942) in 1935, entitled *Method and means for adjusting the water content of concretes or mixtures of granular materials*. He was also awarded his DSc that year.

In 1936 Glanville was asked to take on the role of deputy director at the Road Research Laboratory (RRL) in Crowthorne. Here, work on concrete continued, and with both BRS and RRL led by the same director, the two organisations acted as one in this field of research. Glanville's own output in the late 1930s included:

- Glanville. *Strength tests for cement. SE* v.15(2) February 1937.
- Glanville, Collins, A. R. & Matthews, D. D, (DSIR). *The grading of aggregates and workability of concrete* (MoT Road Research Technical Paper, 5). London: HMSO, 1938 and 1947.
- Glanville and Thomas, F. G. *Studies in reinforced concrete: IV, further investigations on the creep or flow of concrete under load*. London: HMSO, 1939.

However, Glanville's role was becoming more administrative than practical, and in 1939 he was appointed to the role of director. With the outbreak of war, he was put in charge of the research and experiments department of the Ministry of Home Security, initially at Princess Risborough and later back at Crowthorne. As advisor to the Air Ministry he was responsible for work on airstrips and applying a new understanding of soil mechanics to the war effort. His interest in explosives led among other applications to his participation in the bouncing bomb project.

This, and Glanville's later work on roads and traffic at a much enlarged RRL (which is described in a biography by G. Charlesworth), lies beyond the scope of this study, but it is a reflection of his importance to concrete research that among the various elected offices and accolades of the post war years – CBE in 1944, president of the ICE in 1950, CB in 1953 and knighthood in 1960 among them – he was awarded Gold Medals of both the IStructE (1961) and ICE (1962) and was made an honorary member of the American Concrete Institute in 1966 (the year after his retirement from the RRL). He was also an honorary member of the newly formed Concrete Society (established at the end of that year), for which he wrote a personal account of the development of research into concrete throughout his lifetime (*Research on concrete*, *Concrete*, January 1967).

After a period of private consultancy and eventual retirement, he died of a stroke on 30 June 1976 at his home in Northwood, Middlesex. His career had been full of achievement and honours – his name perpetuated by buildings at Crowthorne and Imperial College London – and his research recorded in 115 articles, papers and books. One of these might perhaps, be selected as a fitting memorial – the four-volume set he edited for the Caxton Publishing Co. in 1939: *Modern concrete construction: a comprehensive and practical treatise dealing with every phase of concrete, materials, and modern construction*.

Scott, W. L. & Glanville, W. H. *Explanatory handbook on the code of practice for reinforced concrete as recommended by the Reinforced Concrete Structures Committee of the Building Research Board*. London: Concrete Publications Ltd, 1934 [BL, TCS].

1934/UK

Lynam, C. G.

C. G. Lynam's *Growth and movement in Portland cement concrete* was indicative of the topical interests of its day. Following on from Glanville's work in this field at the BRS, it reviewed previous research and summarised existing knowledge of the setting and hardening of cement and the consequent elasticity, creep and strength of concrete. Lynam's particular contribution to concrete technology – according to commentators such as P. C. Aitcin – lay in identifying the phenomenon of 'autogenous shrinkage'. Like Glanville, he also looked at the nature of bond between concrete and reinforcing steel. The *Structural Engineer* described his book as 'clear', 'interesting' and 'a useful addition to concrete literature'. Other interests that exercised him at the time lay in decorative finishes to concrete. He was resident in London and an active member of the Institution of Structural Engineers.

Lynam, C. G. *Growth and movement in Portland cement concrete*. London: Oxford University Press, 1934 [C&CA 1959, ICE, IStructE. Rev: *SE* December 1934, p.512]

1934/UK

Smith, Major R. A. B.

MC, ACGI, AMICE, MIStructE, MIT, AITP

R. A. B. Smith was a member of the Institution of Structural Engineers and co-designer (with W. R. Manning) of the Chelsea Embankment's reinforced concrete section. By the 1930s he was working for the British Portland Cement Association. Colleagues of his from here were acknowledged for their assistance in the book for which he is remembered: *Design and construction of concrete roads*.

Concrete Publications Ltd (which had close connections with the BPCA) had already issued one book on roads – *Concrete roads and their construction* – initially for the Concrete Utilities Bureau in 1920, and revised in its own imprint in 1923. This consisted of a series of descriptions of road projects in Great Britain, the Dominions and the USA – disparate accounts melded by its anonymous editor in a journalist style, advocating and promoting the idea of concrete construction as demonstrated by the collected examples. In this it followed the early approach of its sister publication, *Concrete & Constructional Engineering*.

Smith's book replaced the propagandist *Concrete roads*, but was of quite a different character: not polemical with case studies, but offering technical guidance – the product one assumes of research at the BPCA. It does however, draw upon the experience of Smith's contacts with local authority engineers – evidence of the extent to which concrete road construction had expanded in practice.

Major Smith was the ideal choice to write this book: qualified; connected and well known to the managing editor of Concrete Publications Ltd – H. L. Childe [1924]. Throughout the 1920s he had contributed articles on the subject of roads to *C&CE*:

- *1924 Concrete roads in hot climates, v.19(4)*
- *1926 Concrete roads in America, v.21(1)*

American concrete road construction, v.21(2)

- *1927 Concrete roads: faults and their lessons, v.22(6–7)*
  Joints in concrete roads, v.22(8–9)
  Materials for concrete road construction, v.22(10–11)
- *1928 Reinforcement for concrete roads, v.23(1–2)*
- *1930 Organisation of plant and labour on concrete road construction, v.25(1)*
  Cement-bound roads, v.25(3)
- *1931 Organisation in concrete road construction, v.26(3)*

In the *Structural Engineer* of February 1927 (v.4 (2)), he described a study tour of the USA where he was 'generously' received by the Bureau of Public Roads and the [American] Portland Cement Association. The purpose of the visit was to identify the differences in practice that could lead to improvements at home. He was back in the USA again in October 1930, as one of Britain's delegates (representing the British Portland Cement Association) at the sixth International Roads Congress in Washington DC. There, the British delegation presented their report on progress – the text of which was published by the BPCA as *Some recent finding in connection with concrete road construction*.

Design and construction of concrete roads was a natural next step for Smith, consolidating his work at the BPCA. His expertise transferred with him to the Cement & Concrete Association when the BPCA was incorporated into the new organisation a year later, and Smith was appointed as its first director. The book clearly filled a niche as a new edition, revised with T. R. Grigson (1884–1968), was published in 1946.

In post-war life Major Smith was resident at 11 Haroldslea Drive in Horley, Surrey, and by 1952 (having resigned from the C&CA in January 1941), he was associated with the National Road Transport Federation.

Smith R. A. B. *Design and construction of concrete roads*. London: Concrete Publications Ltd., 1934 [BL, C&CA 1959, ICE, IStructE]

1934/UK

Stroyer-Nielsen, (Jens Peter) Rudolf

Rudolf Stroyer's origins – presumably Scandinavian – remain obscure, but his business papers in the ICE archives date from 1913, and he emerged in the British technical press in the years directly after the First World War. He wrote extensively for *Concrete & Constructional Engineering*, contributing individual articles on a variety of subjects:

- *The largest pile-driver in the world* v.15(8) August 1920
- *Arches with partially fixed ends* v.17(8) August 1922
- *Tests on hoped cast iron concrete* v.21(2) February 1926
- *The Brigitta Bridge, Vienna* v.21(9) September 1926
- *Some special types of retaining walls* v.22 1927
- *Concrete sleepers* v.23(3) March 1928

His work appeared in the institutional press, too, at much the same time, with *Hooped cast iron and long span arches* in *Structural Engineer* (v.5(11) November 1927) and papers in the ICE's *Minutes of Proceedings*, in January 1928 and June 1932. Contributions to *C&CE* culminated in a series on concrete in marine engineering at the end of the decade:

- *Marine structures in reinforced concrete* v.23, May 1928–v.25, May 1930
- *Marine embankments in reinforced concrete* v.23(8–9) and v.24(2) 1928–1929
- *Vertical embankments in reinforced concrete*v.23 (11/12) November/ December 1928
- *Piers, jetties and dolphins in reinforced concrete* v.24(4–6) April–June 1929
- *Anchored caisson launch* v.25(10) October 1930

Compiling the content of these articles into book form, *Concrete structures in marine work* was published by Knapp, Drewett & Son in 1934.

Stroyer practised as a consulting engineer in partnership with James Beattie and Alan Robert William Adcock, based at 547-9, Victoria Street, and trading as 'R. Stroyer'. Beattie left in 1941 and the partnership was reformed as Stroyer & Adcock.

Besides his bibliographical presence, Stroyer's contribution to new thinking is recalled in Charles Reynolds' *Reinforced concrete designer's handbook* with the passing comment, 'Stroyer suggested a formula applicable to reinforced concrete sheet-pile walls with ties'.

Stroyer, R. N. *Concrete structures in marine work*. London: Knapp, Drewett & Son Ltd, 1934 [BL, ICE, IStructE]

1935/UK

Desch, Cecil Henry (1874–1958)

DSc (Lond.), PhD (Wurzb.), FRS, FIC

Cecil Henry Desch by Walter Stoneman bromide print, 1924 © National Portrait Gallery, London

Desch was a chemist, specialising in metallurgy, and so is an unlikely choice for inclusion in a volume on concrete. He did however, write about the chemistry of cement, and whilst therefore may be deemed more suited to Spackman's tome on that subject, his later work with Lea in 1935 addressed concrete too.

Born in London on 7 September 1874, Cecil was the son of Henry Thomas Desch and his wife Harriet. The elder Desch was chief surveyor and eventually a director, of the construction firm W. Cubitt & Co. of Grays Inn Road. During his 52 years with the firm, he was involved with the construction of the Cunard Building in Liverpool, and prestigious offices in Cornhill and Lombard Street. It may be recalled that Cubitt & Co. commissioned William Dunn [1904] to undertake testing for the original RIBA report into reinforced concrete, and Henry Desch turned his hand to writing about concrete himself in November 1908 in the quaintly entitled article, *Don'ts on concrete*, in *C&CE*, v.3(5). He died in 1923.

The younger Desch was educated at Wurzburg University and University College London, where he was awarded his PhD. His academic career started in the metallurgy department at King's College (1902–1907) and continued with his appointment as Graham Young lecturer in metallurgical chemistry at the University of Glasgow, from 1909 to 1918. Metallurgy was his speciality and he sealed his reputation in this discipline with his *magnum opus*: *Metallography*; first published in1910, but revised in successive editions to a fourth in 1937.

Metallurgy may have been his speciality, but it was not his exclusive interest. Early in the 20th Century he had gathered sufficient authority to be among the first writers on cement included in the new journal *Concrete & Constructional Engineering*. *The setting of Portland cement* appeared in Parts 2 to 4 of the first volume, and was followed by *Portland cement: the compounds of silicon and lime* in 1908 (v.3(2)). Just two months later (v.3(4)), the first of a six-part series on *Testing laboratories for concrete and cement* was issued, concluding in 1910 (v.5(6)).

This phase of interest in cement culminated in his book *The chemistry and testing of cement*, published by Edward Arnold in 1911 and by Longmans, Green & Co. in New York. Transferring the investigative methods of metallurgy and drawing on the practical experience of H. K. G. Bamber (of Associated Portland Cement Manufacturers), and H. Earle (of its subsidiary, G. & T. Earle), it placed Desch in the league of cement chemists inhabited by the likes of Henry Faija, Harry Stanger, Henri Le Chatelier, Bertram Blount and eventually, F. M. Lea – his co-author of 1935.

He dabbled with chemical aspects of concrete in the years following – *The electrolytic corrosion of steel reinforcement* (*C&CE*, v.6(11)) in 1911; *Microscopical examination of clinker* (*C-CA*, v.3, p.27) in 1913; and *The mechanism of the setting process in plaster and cement* (*C&CE*, v.13(2)) in 1918 – but his principal focus had shifted back to metallurgy. His *Intermetallic compounds* was published by Longman, Green & Co. in 1914.

The post-war years were quiet ones; Desch appearing only to revise earlier work. He transferred to the chair of metallurgy at the Royal Technical College, Glasgow in 1918 and then to a professorship at Sheffield in 1920. In 1928 however, after two years as president of the Faraday Society, he was writing again – contributing to the first issue of the new journal *Cement & Cement Manufacture*, with a report on *Progress in the cement industry*. He followed it two years later with *The setting and hardening of cements* in *Cement*, v.3(1).

Perhaps this revived his interest in cement, for only a year after *The chemistry of solids* (published by Oxford University Press in 1934), he joined F. M. Lea as co-author of *The chemistry of cement and concrete*, published – like his *Chemistry and testing* of 1911 – by Edward Arnold. We will look at this classic work under Lea's entry (below). By this date, Desch was superintendent of the metallurgy department at the National Physical Laboratory – a post he occupied from 1932–1939.

During 1930s and 1940s, Desch became a vice-president of the Royal Institution, and president of both the Institute of Metals (1938–1940) and the Iron & Steel Institute (1946–1948); and for the Royal Institution of International Affairs wrote *Science and the social order* in 1946. He died on 19 June 1958.

Lea, F. M. & Desch, C. H. *The chemistry of cement and concrete*. London: Edward Arnold, 1935 [BL, ICE]

1935/UK

Lea, Sir Frederick Measham (1900–1984)

CB, CBE, DSc, FRIC, Hon RIBA

Sir Frederick Measham Lea Courtesy of BRE Press

It seems entirely appropriate to end this survey with a writer who was to be prolific for years to come, to play a key role in developing concrete technology, and whose book of 1935 remains in print to this day.

Like his contemporary W. H. Glanville, Lea was born in 1900 (on 10 February) and – after studying at the University of Birmingham and a stint at the Admiralty – was an early recruit to the Building Research Station during its formative years in East Action. His first article for *Concrete & Constructional Engineering* in October 1922 was *Adhesion of steel and concrete, and preservation of bright-drawn steel embedded in slag and other concretes* (v.17(10)), written with R. E. Stradling (first director of the BRS).

His next two published articles (written with S. R. Carter) appear to have no direct relation to concrete and may reflect his work at the Admiralty: *The diffusion-potential and transport number of hydrochloric acid in concentrated solution*; and *The influence of acid concentration on the oxidation-reduction potential of cuprous and cupric chlorides* (*J. Chem. Soc., Trans.*, 1925, v.127, pp.487 and 499).

Lea's research at the BRS in the 1920s appears to have followed two strands: the chemistry of cement and the use of lightweight aggregates. With F. L. Brady he addressed the latter in 1927 with *Slag, coke breeze and clinker as aggregates*. This he followed up two years later with his report, *Investigations on breeze and clinker aggregates* (HMSO, 1929). It was a theme that was to culminate in 1936 with his Building Research Bulletins, numbers 5 and 15:

- Lea, F. M. *The properties of breeze and clinker aggregates and methods of testing their soundness*. London: HMSO, 1936. (Building Research Bulletin No.5)
- Lea, F. M. *Lightweight concrete aggregates*. London: HMSO, 1936. (Building Research Bulletin No.15)

By this date he had already established his reputation as one of the leading specialists on the chemistry of cement, and by extension, concrete. In the year before the publication of his *magnum opus* – *The chemistry of cement and concrete*, written with Cecil Desch in 1935 – he wrote extensively on this theme in the scientific press.

Modern developments in the chemistry of Portland cement – published in the *Society of Chemical Industry Journal* (v.53(25) 1934) – summarised the more important contributions of the previous 25 years to the knowledge of chemistry, hydration and setting heat of Portland cement. It also discussed the behaviour of cement on exposure to chemical attack by seawater, ground water containing sulphate salts and acid waters. *The rate of hydration of Portland cement and its relation to the rate of development of strength* by Lea and F. E. Jones appeared in the same journal the following year (v.54(10) 1935).

Hydraulic cements in *Science Progress* (v.30(117) 1935), explored recent developments in the understanding of the cementing properties of Portland, high alumina and calcium sulpho-aluminate cements. Possibly his most important contribution at this stage was his work with F. W. Parker on the 'quaternary system'. This started as *Investigations on a portion of the quaternary system* $CaO\text{-}Al_2O_3\text{-}SiO_2\text{-}Fe_2O_3$, published in the *Philosophical Transactions of the Royal Society*. This, according to his obituary, 'laid the foundations of the cement and concrete chemistry group that flourished at BRS'. More immediately it led to Building Research Technical Paper No.16 – *The quaternary system* $CaO\text{-}Al_2O_2\text{-}SiO_2\text{-}Fe_2O_3$ *in relation to cement technology* (HMSO, 1935), also written with T. W. Parker.

Such diverse and rapid progress in the subject led to his collaboration with Desch in writing *The chemistry of cement and concrete*, linking Lea's work to the achievements of the older man. The book encapsulated the then-current understanding of cement chemistry, and was a great success, but this was a time of rapid progress and new editions followed in turn. Besides the recognition brought about by his book, Lea received a doctorate from Birmingham University and later (in 1940), the Beilby Memorial Award for his work on the physical chemistry of cement.

With such developments to his name, it is not surprising that he participated in the International Symposium on the Chemistry of Cements held at Stockholm in 1938, contributing *The chemistry of pozzolanas* to the *Proceedings*. Continuing research in this field – building on earlier work by a former colleague F. T. Meehan, and supported by the cement industry – culminated in Technical Paper No.27: *Investigations on pozzolanas* (HMSO, 1940).

The last year of peace saw a broadening of interests again, in both subject matter and choice of publisher:

- *Comparative tests on the fineness of cements*, *J. SCI*, April 1939
- *The solubility of cements* (Technical Paper No.26). London: HMSO, May 1939
- *Floors for industrial purposes.* Fitzmaurice, R. and Lea. *SE*, v.17(5), May 1939
- *The action of waste waters from the milk industry on cement products*, *C&CE*, v.34(11), November 1939

During the Second World War he played a leading role in the experimental work of the BRS into fire resistance and protection against incendiary bombs. He continued to specialise in cements and high temperature chemistry, however, directing his expertise to the war effort. With J. R. Hawes he studied the 'Rhenania method' of producing the plant food 'silicophospate fertiliser' from phosphate rock by calcining it with soda ash and sand in converted cement kilns. He also studied (with T. W. Parker) the possibilities of blast-furnace slag as a substitute material for aggregate, and as a constituent of blast-furnace cement. In 1944 he was awarded the OBE.

His interest chemistry was reflected by membership of the Royal Institute of Chemistry – of which he was by then a fellow, and on whose council he served during the war (1943–1946). On 19 December 1944 he delivered a lecture on cement and concrete, published as a pamphlet in the New Year.

At the end of the war he was appointed director of building research, and remained so until his retirement in 1965. He continued to publish; the two examples below suggesting a new emphasis on durability:

- *The deterioration of concrete in structures*. Lea & Davey. *J. ICE*, v.74, 1949
- *The durability of reinforced concrete in seawater*. Lea & Watkins, C. M. (National Building Studies, 30). London: HMSO, 1960

One of the more important achievements of these years was his involvement in setting up the international research body RILEM (of which he was an early president), and the International Council for Building Research, Studies and Documentation (CIB) – of which he was president for two terms. He also chaired the concrete committee of the International Commission on Large Dams between 1953 and 1959, building on earlier involvement with the British Committee on Large Dams before the war.

As director of the BRS, the nature and purpose of building research were clearly preoccupations and various statements made in his final years in his post (or published just after retirement) reflect his concerns:

- *The pattern of building research*. Watford: BRS, 1967 (from *Report of the Director of Building Research*, 1962)
- *The users of building research*. Watford: BRS, 1967 (from *Report of the Director of Building Research*, 1963)
- *Research for a changing industry*. Watford: BRS, 1966 {*from Proceedings of the Building Research Congress, Pretoria*, July 1964)
- *New Paths in building*. Watford: BRS, 1967 (from the *National Provincial Bank Review*, 1965)

Retirement in 1965 brought many accolades and recognition of an important contribution to scientific knowledge. Adding to his CBE (1952) and CB (1960), he received a knighthood. Like his contemporary Glanville a year earlier, he was made an honorary member of the ACI in 1967, adding to his honorary fellowships of the RIBA and IOB. His Walter C. Voss Award from ASTM in 1964 was the first to be conferred outside North America.

He enjoyed the recognition of his long role at BRS during the Golden Jubilee Concrete in June 1971 – for which he wrote an historical volume to accompany the *Proceedings – Science and building: a history of the Building Research Station* (HMSO, 1971). Similarly he took part in the Concrete Society's Aspdin anniversary celebrations in May 1974, giving *Cement & building: from Roman times to the twentieth century* as a public lecture at the University of Leeds.

After a long retirement he died on 7 July 1984, and an obituary appeared in the *RILEM Bulletin*. Following his death, the Institute of Concrete Technology instituted the Sir Frederick Lea Memorial Lecture, seven of which were given between 1986 and 2004.

Lea, F. M. & Desch, C. H. *The chemistry of cement and concrete*. London: Edward Arnold, 1935 [BL, ICE]

Appendix A – Sources

Primary Sources

Library catalogues

The following library catalogues were searched systematically for 'books' and/or 'monographs', combined with the keyword 'concrete' in the title, across the date range 1897 to 1935. Further title words were included *ad hoc* when browsing, and specific author searches across all media undertaken for individual biographies.

- British Library
- Cement & Concrete Association's printed catalogue of textbooks, November 1959
- Concrete Society (the library formerly at the Cement & Concrete Association)
- Institution of Structural Engineers
- Institution of Civil Engineers
- *Catalogue of books, periodicals and pamphlets in the library of the Portland Cement Association*, 1918

The British Architectural Library at the Royal Institution of British Architects was consulted for those authors who were architects.

Those books identified from such sources and that were available in the Concrete Society's library have all been physically examined.

Bibliographies

The extensive bibliography published in *The Concrete Yearbook* fulfilled a similarly descriptive function, and was examined in its entirety. The 1934 edition was the last to attempt a comprehensive coverage of books on the subject; that of 1935 was limited to those titles issued by Concrete Publications Ltd, complete with detailed descriptions.

Building Science Abstracts from the Building Research Establishment, provides a comprehensive and detailed description of the scientific literature – books, pamphlets and journal articles – on all aspects of construction, including concrete. Appropriate subject sections from the volumes for the period 1928 to 1935 were examined systematically.

A review of the literature of reinforced concrete by Leon S. Moisseiff and *Engineering and metallurgical books, 1907–1911* by R. A. Peddie treat the early years of our period of survey, while Purdue University's *Comprehensive bibliography of cement and concrete 1925–1947* by Floyd O. Slate provides a thorough listing for the later phase. Moisseiff's annotated bibliography provided a contemporary crosscheck for the entries in this survey, though Slate's was simply too extensive to be of practical use in this case. The C&CA's *Bibliography of cement and concrete* (1952) provides a survey of books that were available in selected London libraries at the time of compilation. The C&CA also published a 'Library Bibliography' in 1969, entitled *Historical and rare books on cement and concrete and allied subjects held by the C&CA library* (Ref: Ch 58).

JOURNAL INDICES

Two of the key journals of the period have electronic indices, and these were checked for items associated with individual authors. These journals are:

- *Concrete & Constructional Engineering* (indexed by the IStructE)
- *Proceedings of the American Concrete Institute* (indexed by the ACI)

A printed subject and author index is also available for publications of the Institution of Civil Engineers (1917–1948) and a volume of synopses for the ACI (1929–1935) in the latter part of our period.

JOURNAL RUNS

Complete runs of the following journals were physically consulted at the Concrete Society's library – systematically for reviews of new books during the years to 1935, and *ad hoc* for obituaries and articles by appropriate authors:

- *Cement Age* (USA)
- *Cement, Lime & Gravel* (UK)
- *Concrete & Constructional Engineering* (UK)
- *Proceedings of the American Concrete Institute* (USA)
- *Transactions of the Concrete Institute* (UK)
- *Structural Engineer* (UK)

Internet bibliographical resources

The Internet has a wealth of second-hand and antiquarian book search and sale sites, of which AbeBooks, Alibris and similar were repeatedly checked for corroborative bibliographical details and to identify non-concrete titles. Amazon.com too, has a surprisingly good selection of out-of-copyright reprints advertised for sale.

Full text access to early books – complete or in part – is available on Google Books, Internet Archive, J Store and Open Library. These resources proved especially valuable for consulting the text of American works in the absence of physical copies.

Also on the Web are numerous alumni magazines, archived collections of professional papers, reports, *etc*, which in this survey are cited informally under individual author entries.

Secondary sources

Biographical sources

Standard biographical dictionaries were checked for those authors who were public figures:

- *Dictionary of National Biography*
- *The Times Obituaries*
- *Who was Who*

More specialised sources consulted where appropriate included the *Biographical dictionary of Scottish architects* and the *Biographical database of the British chemical community, 1880–1970*.

British authors whose lifespan suggested a worthwhile search were checked on the 1881 and 1901 Census websites, and FreeBMD for dates of birth, marriage and death.

Histories

The following are of general application to the history of the concrete industry and its literature – all of which have been consulted for this present work. Sources relevant solely to individual authors are cited informally under their own biographical entries.

Addis, B. & Bussell M. *Key developments in the history of concrete construction and the implications for remediation and repair* in Macdonald, Susan (ed). *Concrete building pathology*. Oxford: Blackwell Science, 2003.

American Concrete Institute. *Concrete – a century of progress* [a supplement to *Concrete International*]. Farmington Hills: ACI, 2004.

Barton, Barry. *Water Towers of Britain*. London: Newcomen Society, 2003.

Bewsey, Ken. *The history of Mouchel's of West Byfleet*. [Precis of a talk to the Woking History Society]. Woking History Society website.

Booth, L. G. Discussion of *The development of reinforced concrete: design theory and practice*, in *Proceedings ICE, Structures and Buildings*, v.128, November 1998, pp.397–400.

Bussell, M. N. *The development of reinforced concrete: design theory and practice* in *Proceedings ICE, Structures and Buildings*, v.116, August 1996, pp.317–344.

Bussell, M. N. *The era of the proprietary reinforcing systems*, in *Proceedings ICE, Structures and Buildings*, v.116, August 1996, pp.295–316; and in Sutherland, J., Humm, D. and Chrimes, M. (eds). *Historic Concrete*. London: Thomas Telford, 2001.

Campbell, P. *1890-1910*, in Collins A. R. (ed). *Structural engineering – two centuries of British achievement*. London: Tarot Print Ltd, 1983.

Concrete Publications Ltd. *Concrete & Constructional Engineering* [21st anniversary issue], January 1926, v.21(1).

Cusack, Patricia. *Architects and the reinforced concrete specialist in Britain 1905–08* in *Architectural History*, v.29, 1986, pp.183–196.

Cusack, Patricia. *Agents of change: Hennebique, Mouchel and ferroconcrete in Britain, 1897–1908*, in *Construction History*, 3, 1987, pp.61–74.

Draffin, Jasper O. *A brief history of lime, cement, concrete and reinforced concrete*, in *Journal of the Western Society of Engineers*, v.48(1), March 1943, pp.14–47.

Gasparini, D. A. *Contributions of C.A.P. Turner to development of reinforced concrete flat slabs 1905–1909*, in *Journal of Structural Engineering*, v.128(20), October 2002, pp.1243–1252.

Gueritte, T. J. *The first decade of reinforced concrete in the UK (1897–1906)*, in *Concrete & Constructional Engineering*, v.21, 1926, pp.89–92.

Hamilton, S. B. *The history of the Institution of Structural Engineers*, in *The Structural Engineer*, Jubilee issue, July 1958, pp.16–20.

Harley-Haddow, T. *1910–1930*, in Collins A. R. (ed). *Structural engineering – two centuries of British achievement*. London: Tarot Print Ltd, 1983.

Hollister, S. C. *Sixty-two years of concrete engineering*, in *Concrete International*, October 1979, pp.63–66.

Hurst, L. *Edwin O Sachs – engineer & fireman*, in Wilmore, D. (ed). *Edwin O. Sachs: architect, stagehand, engineer & fireman*. Summerbridge: Theatresearch, 1998.

Institution of Structural Engineers. *The Structural Engineer*, Royal Charter issue, 1934.

Macdonald, Susan (ed). *Concrete building pathology*. Oxford: Blackwell Science, 2003.

McBeth, D. C. *Francois Hennebique (1842–1921): reinforced concrete pioneer*, in *Proceedings, ICE*, 1998, pp.86–95.

McWilliam, Robert C. *BSI: the first hundred years*. London: Thanet Press, 2001.

Moisseiff, L. S. *A review of the literature of reinforced concrete*. New York: Engineering News Publishing Co., 1909.

Morley, Jane. *Frank Bunker Gilbreth's concrete system*. *Concrete International*, November 1990, pp.57–62.

Newby, F. *The innovative uses of concrete by engineers and architects*, in *Proceedings ICE, Structures and Buildings*, v.116, August 1996, pp.264–282.

Newlon, Haward (ed). *A selection of historic American papers on concrete, 1876–1926*. (SP-52). Detroit: ACI, 1976.

Sachs, E. *My father – that remarkable man Edwin O. Sachs*, in Wilmore, D. (ed). *Edwin O. Sachs: architect, stagehand, engineer & fireman*. Summerbridge: Theatresearch, 1998.

Spackman, Charles. *Some writers on cement: from Cato to the present time*. Cambridge: W. Heffer & Sons, 1929.

Sutherland, J., Humm, D. and Chrimes, M. (eds). *Historic Concrete*. London: Thomas Telford, 2001.

Titford, R. M. (ed). *The golden age of concrete*. Dorothy Henry Publications, 1964.

Trout, E. A. R. *Concrete Publications Ltd and its legacy to the construction industry*, in *Construction History*, 2005, v.19, pp.65–86.

Wilde, Robert E. *75 years of progress: the ACI saga*, in *Concrete International*, October 1979, pp.8–42.

Witten, A. *The Concrete Institute 1908–1923, precursor to the Institution of Structural Engineers*, in Sutherland, J., Humm, D. and Chrimes, M. (eds). *Historic Concrete*. London: Thomas Telford, 2001.

Woodward, C. Douglas. *BSI: the story of standards*. London: British Standards Institution, 1972.

Yeomans, David and Cottam, David. *Owen Williams: the engineer's contribution to contemporary architecture*. London: Thomas Telford Ltd., 2001.

Yeomans, David. *Engineering design books in the inter-war years*. Paradigm, v.3(2), May 2006, pp.28–32. http://faculty.ed.uiuc.edu/westbury/paradigm/documents/yeomans.pdf.

Appendix B – Bibliography

The books considered in the foregoing survey are listed here according to author's surname, arranged in alphabetical order.

Abrams, Duff Andrew

Abrams, D. A. *Design of concrete mixtures*. (Bulletin No.1). Portland Cement Association, 1918.

Adams, Haddon Clifford

Adams, H. C. *Elements of reinforced concrete design*. London: Concrete Publications, 1933. [Yrbk 1934, C&CA 1949, ICE. Rev: *SE* January 34]

Chettoe, C. S. and Adams, H. C. *Reinforced concrete bridge design*. London: Chapman & Hall, 1933. Second edition, 1938. [Yrbk 1934, C&CA 1959, BL, ICE. Copy: TCS Second edition Rev: *SE* June 33]

Adams, Henry

Adams, H. & Matthews, E. R. *Reinforced concrete construction in theory and practice: an elementary manual for students and others*. London and New York: Longmans, Green & Co., 1911 [C&CA 1959, ICE, Peddie]. Second edition, 1920. [C&CA 1959, BL, ICE, Yrbk 1934. Copy: TCS. Rev: *C&CE* 1912]

Andrews, Ewart Sigmund

Andrews, E. S. *Elementary principles of reinforced concrete construction: a text-book for the use of students, engineers, architects and builders*. London: Scott, Greenwood & Son, 1912 [C&CA 1959, ICE. Copy: TCS. Rev: *C&CE* v. 7, p. 391; *Trans C. I.* v. 5, p. xii], and New York: Van Nostrand [Rev: *C-CA*, v. 2, p. 248]. Second edition, 1918 [BL]. Third edition, 1924 [BL, ICE, Yrbk 1934]

Andrews, E. S. *Regulations of the London County Council relating to reinforced concrete (London Building Act, 1909, Amendment) and steel framed buildings (L.C.C. General Powers Act, 1909): a handy guide, containing the full text, with explanatory notes, diagrams and worked examples*. London: B. T. Batsford Ltd, 1915 [ICE]. Second edition, 1924 [BL, Yrbk 1934. Copy: TCS. Rev: *C&CE* v. 11, p. 267]

Andrews, E. S. *Detail design in reinforced concrete: with special reference to the requirements of the reinforced concrete regulations of the L.C.C.* London: Pitman, 1921 [BL, ICE, Yrbk 1934]

Andrews, E. S. and Wynn, A. E. *Modern methods of concrete making*. Second edition. London: Concrete Publications Ltd, 1928 [Yrbk 1934]

Andrews, Hiram Bertrand

Andrews, H. B. *Practical reinforced concrete standards for the designing of reinforced concrete buildings*. Boston: Simpson Bros., 1908 [Moisseiff. Rev: *CA*, v. 7, September 1908]

Andrews, H. B. *Design of reinforced concrete slabs, beams and columns, confirming to the recommendations of the Joint Committee on Concrete and Reinforced Concrete*. Boston: Andrews, 1910 [Peddie]

Badder, H. C.

Badder, H. C. *Impervious concrete*. London: Educational Publishing, 1923 [BL, ICE]

Baker, Samuel

Baker, S. & DeGroot, A. *Cements and aggregates*. Scranton, PA: International Textbook Co., 1931 [Copy: TCS]

Baker, S. & DeGroot, A. *Concrete and mortar*. Scranton, PA: International Textbook Co., 1935 [Copy: TCS]

Balet, Joseph W.

Balet, J. W. *Analysis of elastic arches: three-hinged, two-hinged and hingeless, of steel, masonry and reinforced concrete*. New York: Engineering News Publishing, 1908 [ICE].

Ball, James Dudley Ward

Ball, J. D. W. *Reinforced concrete railway structures*. (Glasgow Text Books of Civil Engineering). London: Constable & Co., 1913 [BL, C&CA 1959, ICE, Yrbk 1934. Rev: *C&CE* v. 9, pp. 69–70)

Ballard, Fred

Ballard, F. *Concrete for house, farm and estate*. London: Crosby Lockwood, 1919 [Copy: TCS. Rev: *C&CE* v. 15]. Second edition, 1921 [Rev: *C&CE*]. Third edition, 1925 [BL, Yrbk 1934]

Ballinger, Walter Francis

Ballinger, W. F. & Perrot, E. G. Inspector's handbook of reinforced concrete. New York: Engineering News Publishing Co., 1909 [PCA], McGraw-Hill, 1909 [ICE. Rev: *CA*, v. 10], and London: Constable, 1909 [Peddie]

Bartle, Eric Godfrey

Dowsett, J. F. & Bartle, E. G. *Practical formwork and shuttering*. London: Library Press, 1932 [IStructE. Rev: *SE* April 1932, v. 10, p. 198]

Bauer, Edward Ezra

Bauer, E. E. *Plain concrete*. McGraw-Hill, 1928 [BL, ICE, Yrbk 1934. Rev: *C&CE* v24 p. 116]. Second edition, 1936 [Copy: TCS]

Baxter, Leon H.

Baxter, L. H. *Elementary concrete construction*. Milwaukee: Bruce Publishing Co., 1921 and London: Batsford, 1921[Yrbk 1934]

Becher, Heinrich

Becher, Heinrich. *Reinforced concrete in sub- and superstructure*. (The Deinhardt-Schlomann series of technical dictionaries in six languages, 8). London: Constable, 1910 and New York: McGraw, 1910. Translated chiefly by Dr. A. B. Searle from *Der Eistenbeton im Hoch-und Tiefbau* (Schlomann-Oldenbourg illustrierte technishe Worterbucher, 8. Munich: R. Oldenbourg, 1910. [BL, Peddie. Rev: *C&CE* v. 5, 1910, p. 454]

Bennett, Sir Thomas Penberthy

Bennett, T. P. *Architectural design in concrete*. London: Ernest Benn, 1927 [C&CA 1959, ICE. Copy: TCS]

Beyer, Walter F.

Campbell, H. C. and Beyer, W. F. *Practical concrete work for the school and home*. Peoria, Illinois: Manual Arts Press, 1917 and London: Batsford, 1917 [Yrbk 1934]

Bowie, Percy George

Faber, O. and Bowie, P. G. *Reinforced concrete design* (later Vol. I – theory). London: London: Edward Arnold, 1912 [BL, ICE. Rev: *C&CE* v. 7, p. 390], and New York: Longman, Green & Co., 1912

[PCA. Rev: *CA*, v. 14, p. 314]. Second edition, 1919 [BL, C&CA 1959, ICE. Copy: TCS] and 1924 [BL]

Boynton, Walter Channing

Boynton, W. C. & Marshall, R. (comp). *How to use concrete*. Detroit: Concrete Publishing, 1910. [PCA. Rev: *CA*, v. 10; *C&CE* v. 5, pp. 612–13]

Brayton, Louis F.

Brayton, L. F. *Brayton standards: a pocket companion for the uniform design of reinforced concrete*. Minneapolis, 1906 [Moisseiff, Peddie. Rev: *CA* v. 3, p. 129]; Second edition, 1907]

Brett, Allen

Brett, A. (ed). *Reinforced concrete field handbook*. Cleveland: Technical Publishing Co., 1909 [Peddie]

Brooks, John Pascal

Brooks, J. P. *Reinforced concrete: mechanics and elementary design*. New York: McGraw, 1911, and London: Hill, 1911 [ICE; Peddie. Rev: *C&CE*, v. 7, 1912]

Buel, A. W.

Buel, A. W. and Hill, C. S. *Reinforced concrete construction*. New York: Engineering News Publishing, and London: Archibald Constable & Co., 1904 [ICE]. Second ed., 1906 [C&CA 1959, BL, ICE, IStructE. Copy: TCS. Rev: *EN*, 15 November 1906, *C&CE* May 1907, p.166]

Burren, F.

Burren, F. & Gregory, G. R. *Moulds for cast stone and concrete*. London: Concrete Publications Ltd, 1930 [C&CA 1959. Copy: TSC]

Cain, William

Cain, W. *Theory of steel-concrete arches and of vaulted arches*. Second edition. New York: Van Nostrand, 1902.

Campbell, Henry Colin

Campbell, H. C. and Beyer, W. F. *Practical concrete work for the school and home*. Peoria, Illinois: Manual Arts Press, 1917 and London: Batsford, 1917 [Yrbk 1934]

Campbell, H. C. *How to use cement for concrete construction for town and farm*. Stanton and Van Vliet, 1920.

Cantell, Mark Taylor

Cantell, M. T. *Reinforced concrete construction: elementary course*. London and New York: Spon, 1911[Peddie. Rev: *C&CE* v. 7]

Cantell, M. T. *Reinforced concrete construction: advanced course*. London: Spon, 1912 [BL, C&CA 1959. Copy: TCS. Rev: *C&CE* v. 7, p. 942]

Cantell, M. T. *Reinforced concrete construction: part I*. London: Spon, 1918? Second edition, 1920 [BL, Yrbk 1934]

Cantell, M. T. *Reinforced concrete construction: part II*. Second edition London: Spon, 1921 [BL, C&CA 1959, ICE. Copy: TCS. Rev: *C&CE*, v. 16, p. 745]

Cantell, M. T. *Practical designing in reinforced concrete: part I*. London: Spon, 1928 [BL, C&CA 1959, ICE, Yrbk 1934. Copy: TCS]

Cantell, M. T. *Practical designing in reinforced concrete: part II*. London: Spon, 1933 [BL, C&CA 1959, ICE. Rev: *SE*, v. 12, p. 114]

Cantell, M. T. *Practical designing in reinforced concrete: part III*. London: Spon, 1935 [BL, C&CA 1959, ICE]

Chandler, Albert H.

Lewis, M. H. & Chandler, A. H. *Popular handbook for cement and concrete users*. Norman W. Henley Publishing Co., 1911 [C&CA 1952, *CA*, v. 12, p. 220. Rev: *C&CE* v. 6, pp. 633–634]. Second edition. Henry Frowde and Hodder & Stoughton, 1921 [BL. Rev: *C&CE* pp. 53–56]

Chettoe, Cyril Stanley

Chettoe, C. S. and Adams, H. C. *Reinforced concrete bridge design*. London: Chapman & Hall, 1933 [BL, ICE, Yrbk 1934]. Second edition, 1938 [C&CA 1959. Copy: TCS 2ed ed. Rev: *SE* June 1933]

Childe, Henry Langdon

Childe, H. L. *Manufacture and uses of concrete products and cast stone*. (Concrete Series, No.11). London: Concrete Publications Ltd, 1927 [C&CA 1959. Copy: TCS. Rev: *SE* 1927, p. 328]. Second edition, 1927 [BL] Third edition, 1927 [ICE]. Fifth edition, 1930 [BL, C&CA 1959. Rev: *C&CE* v. 26, p. 537; *SE* January 1931]. 1961 [TCS]

Childe, H. L. *Precast concrete factory operation*. London: Concrete Publications, 1929. [C&CA 1959, ICE, TCS]

Gray, W. S. and Childe, H. L. *Concrete Surface Finishes, Rendering and Terrazzo*. London: Concrete Publications Ltd, 1935. [Copy: TCS]

Clifford, Walter Woodbridge

Sutherland, H. & Clifford, W. W. *Introduction to reinforced concrete design*. New York: Wiley, 1926 [BL, ICE. Rev: *SE*, 1927]

Cochran, Jerome

Cochran, J. *A treatise on the inspection of concrete construction*. Chicago: Myron C. Clark Publishing Co., 1913 [PCA]

Cochran, J. *General specifications for concrete and reinforced concrete: including finishing and waterproofing*. New York: D. Van Nostrand Co., 1913 [Copy: Internet Archive]

Colby, Albert Ladd

Colby, A. L. *Reinforced concrete in Europe*. Easton, PA: Chemical Publishing, 1909 [BL, PCA, Peddie. Copy: Internet Archive]

Coleman, George Stephen

Goleman, G. S. *Reinforced concrete diagrams for the calculation of beams, slabs, and columns in reinforced concrete*. London: Crosby Lockwood, 1908 [ICE], and New York: Van Nostrand, 1908 [Peddie]

Coleman, T. E.

Coleman, T. E. *Estimating for reinforced concrete work*. London: Batsford, 1912 [IStructE. Rev: *C&CE*, v. 8, 1913; *Trans CI*, v. 5, 1913]

Coleman, T. E. *The civil engineer's cost book*. London: Spon, 1924 [IStructE]

Considere, Armand

Considere, A. (trans. Leon S. Moisseiff). *Experimental researches on reinforced concrete*. New York: McGraw Hill, 1903. Second edition, 1906. [C&CA 1959, BL, ICE, IStructE]

Corkhill, T.

Corkhill, T. (ed). *Brickwork, concrete and masonry: a complete work by practical specialists describing modern practice in the work of the bricklayer, concreter, mason, plasterer and slater*. (8 vols). London: Pitman, 1930 [ICE. Rev: *SE* v. 9 July 1931, p. 260; *SE* v. 10 April 1932, p. 198]

Crane, Theodore

Crane, T. *Concrete building construction*. New York: Wiley, 1927 [BL, Yrbk 1934, C&CA 1952. Copy: TCS. Rev: *SE* v6 1928]

Cross, Hardy

Cross, H. & Morgan, N. D. *Continuous frames of reinforced concrete*. New York: John Wiley & Sons, 1932 [BL, C&CA 1959, ICE]. Rev edition, 1954 [BL]

Dana, Richard Turner

Dana, R. T. & Kingsley, J. W. *Concrete computation charts*. New York: Codex Book Co., Inc., 1922.

Davenport, John Alfred

Davenport, J. A. *Graphical reinforced concrete design*. London and New York: Spon, 1911 [ICE, Peddie, Yrbk 1934. Rev: *C&CE* v. 7, p. 311)

Davison, Ralph C.

Davison, R. C. *Concrete pottery and garden furniture*. New York: Munn & Co., 1910. Second edition, 1917 [C&CA 1959. Rev: *CA*, v. 9]

Davison, Thomas Raffles

Burren, F. & Gregory, G. R. [illus. Davison, T. R.] *Moulds for cast stone and concrete*. London: Concrete Publications Ltd, 1930 [C&CA 1959. Copy: TSC]

Day, William Peyton

Leonard, J. B. & Day, W. P. *The concrete bridge: a book on why the concrete bridge is replacing other forms of bridge construction*. San Francisco, 1913 [Copy: Internet Archive. Rev: *C-CA*, v. 4, January 1914, p. 42]

DeGroot, A.

Baker, S. & DeGroot, A. *Cements and aggregates*. Scranton, PA: International Textbook Co., 1931 [Copy: TCS]

Baker, S. & DeGroot, A. *Concrete and mortar*. Scranton, PA: International Textbook Co., 1935 [Copy: TCS]

Desch, Cecil Henry

Lea, F. M. & Desch, C. H. *The chemistry of cement and concrete*. London: Edward Arnold, 1935 [BL, ICE]

Dodge, Gordon Floyd

Dodge, Gordon Floyd. *Diagrams for designing reinforced concrete structures, including diagrams for reactions and strengths of steel beams*. New York: Clark Publishing Co., 1910 and London: Spon, 1910 [Peddie]

Douglas, Walter J.

Douglas, W. J. *Practical hints for concrete constructors*. (Reprint No.12). New York: Engineering News Publishing Co., 1907 [Peddie. Copy: Internet Archive]

Dowsett, John Frederick

Dowsett, J. F. & Bartle, E. G. *Practical formwork and shuttering*. London: Library Press, 1932 [IStructE. Rev: *SE*, April 1932, v. 10, p. 198]

Dunn, William Newton

Marsh, Charles F. [and Dunn, William in the Third edition]. *Reinforced Concrete*. London: Archibald Constable, 1904. Second edition, 1905, Third ed, 1906. [C&CA 1959, BL, ICE, IStructE. Copy: TCS Third edition Rev: *C&CE*, v. 1, p. 405]

Marsh, Charles F. and Dunn, William. *Manual of reinforced concrete and concrete block construction*. London: Archibald Constable, and New York: Van Nostrand, 1909. [Rev: *CA*, v. 7, p. 461]. Fourth edition, 1922. [C&CA 1959, BL, TCS. Copies: TCS. Rev: *C&CE*, v. 6, p. 715-16]

Dunn, William. *Lectures on reinforced concrete delivered at the Institution of Civil engineers in November 1910*. London: University of London Press and Hodder & Stoughton, 1911 [ICE, IStructE, PCA, Peddie]

Dunn, William. *Diagrams for the solution of T beams in reinforced concrete*. London: University of London Press and Hodder & Stoughton, 1911 [Peddie]

Eddy, Prof. Henry Turner

Eddy, H. T. *The theory of the flexure and strength of rectangular flat plates applied to reinforced concrete floor slabs*. Minneapolis: Rogers & Co., 1913 [PCA. Rev: *C&CE*, v. 8, p. 589]

Eddy, H. T. & Turner, C. A. P. *Concrete-steel construction. Part 1 – building: a treatise upon the elementary principles of design and execution of reinforced concrete work in building*. Minneapolis: Heywood Manufacturing Co., 1914 [Rev: *C&CE*, v. 10, pp. 106–7]. Second edition, 1919.

Ekblaw, Karl John Theodore

Ekblaw, K. J. T. *Farm concrete*. New York: Macmillan, 1917 [PCA]

Etchells, Ernest Fiander

Etchells, E. F. *Mnemonic notation for engineering formulae: report of the Science Committee of the Concrete Institute*. London: E&FN Spon, 1918.

Faber, Oscar

Faber, O. and Bowie, P. G. *Reinforced concrete design* (later Vol. 1 – theory). London: Edward Arnold, 1912 [BL, ICE. Rev: *C&CE* v. 7, p. 390], and New York: Longman, Green & Co., 1912 [PCA. Rev: *CA*, v. 14, p. 314]. Second edition, 1919 [BL, C&CA 1959, ICE, Yrbk 1934. Copy: TCS] and 1924 [BL]

Faber, O. *Reinforced concrete design:* Vol. II – *practice*. London: Edward Arnold, 1920 [BL, ICE, Yrbk 1934. Rev: *C&CE* v. 16, p. 66 and p. 602] and 1924 [BL, C&CA 1959]; and Oxford University Press, 1929 [BL]

Faber, O. *Reinforced concrete simply explained*. London: Henry Frowde and Hodder & Stoughton, 1922 [Rev: *C&CE*, p. 371]. Second edition. Oxford University Press, 1926 and 1929 [Yrbk 1934]

Faber, O. *Reinforced concrete in bending and shear: theory and tests in support*. London: Concrete Publications Ltd, 1924 [C&CA 1959, ICE, TCS, Yrbk 1934. Copy: TCS]

Faber, O. and Childe, H. L. *Concrete Yearbook*. London: Concrete Publications Ltd, 1924 et seq [Yrbk 1934. Copies TCS]

Falk, Myron S.

Falk, M. S. *Cements, mortars and concrete: their physical properties*. New York: Myron C. Clark, 1904. [ICE. Rev: *EN*, 13 October 1904]

Fallon, John Tiernan

Fallon, J. T (ed). *How to make concrete garden furniture*. New York: Robert M. McBride, 1917 and London: Batsford, 1917 [Yrbk 1934]

Fleming, J. Gibson

Fleming. J. G. *Reinforced concrete*. Chatham: Royal Engineers Institute, 1910 [BL, C&CA 1959. Copy: TCS. Rev: *C&CE* 1910]

Foster, Wolcott C.

Foster, W. C. *A treatise on wooden trestle bridges and their concrete substitutes according to the present practice of American railroads*. Fourth edition. New York: Wiley, and London; Chapman & Hall, 1913 [ICE. Rev: *C&CE* v. 8, p. 883]

Fougner, Nicolay Knudtzon

Fougner, N. K. *Seagoing and other concrete ships*. (Oxford Technical Publications). London: Henry Frowde and Hodder & Stoughton, 1922 [BL, C&CA 1959, ICE, Yrbk 1934. Copy: TCS]

Fraser, Percival Maurice

Scott, A. A. H. and Fraser, P. M. *A specification for reinforced concrete work*. London: Witherby & Co., 1911 [Peddie. Rev. *C&CE* v. 6, pp. 461–462]

Gammon, John C.

Gammon, J. C. *Reinforced concrete design simplified*. London: Crosby Lockwood, 1911 [ICE, Peddie. Rev: *C&CE* v. 6, p. 881]. Second edition, 1913 [BL]. Third edition. Kingston Hill: Technical Press, 1921 [BL, C&CA 1959, Yrbk 1934. Copy: TCS]

Geen, Albert Burnard

Geen, B. *Continuous beams in reinforced concrete*. London: Chapman & Hall, 1913 [C&CA 1959, ICE, Yrbk 1934. Copy: TCS. Rev: *C&CE* v. 8, p. 512, *Trans & Notes* v. 5, pt. 1, xiv]

Gibson, William Herbert

Webb, W. L. & Gibson, W. H. *Reinforced concrete: a treatise on cement, concrete and concrete steel and their applications to modern structural work*. Chicago: American School of Correspondence, 1908 and London: Crosby Lockwood, 1927 [Yrbk 1927]

Webb, W. L. and Gibson, W. H. *Masonry and reinforced concrete*. Chicago: American School of Correspondence and London: Crosby Lockwood, 1909 [Peddie. Rev: *C&CE* v. 6, p. 154]

Webb, W. L. and Gibson, W. H. *Concrete and reinforced concrete: a condensed practical treatise on the problems of concrete construction in cement mixtures, tests, beams an d slab design, construction work, retaining walls, etc*. Chicago: American Technical Society, 1916 [BL, PCA] and 1919 and 1925.

Webb, W. L. and Gibson, W. H. *Concrete design and construction*. Chicago: American Technical Society, 1931 [BL, ICE. Rev: *SE* v. 10, p271] 1940.

Gilbert, Clyde Dee

Whipple, H. and Gilbert, C. D. *Concrete houses and how they were built*. Detroit: Concrete-Cement Age Publishing Co., 1917 [C&CA 1959. Copy: TCS]

Gilbreth, Frank Bunker

Gilbreth, F. B. *Concrete system*. New York: Engineering News, 1908 [PCA. Rev: *CA* November 1908 p. 394], Easton: Hive Publishing Co., 1908 [BL], and London: Constable, 1908 [Peddie]

Gillette, Halbert Powers

Gillette, H. P. & Hill, C. S. *Concrete construction: methods and cost*. New York: Myron Clark Publishing Co., 1908 [C&CA 1959, ICE, PCA. Rev: *EN*, 11 June 1908 and *C&EN*, 1908, v.20, p.153] and London: Spon, 1908 [BL, Peddie]

Glanville, Sir William Henry,

Scott, W. L. & Glanville, W. H. *Explanatory handbook on the code of practice for reinforced concrete as recommended by the Reinforced Concrete Structures Committee of the Building Research Board*. London: Concrete Publications Ltd, 1934 [BL, TCS]

Godfrey, Edward

Godfrey, E. *Concrete*. Pittsburg: Godfrey, 1908 [C&CA 1959. Rev: *CA*, v.7, *C&EN*, v.20, p.98 and *EN*, 14 May 1908]

Godfrey, E. *Steel and reinforced concrete in buildings*. Pittsburg: Godfrey, 1911 [Peddie; *CA*, v. 13, p. 178]

Goodrich, E. P.

Morsch, E., trans Goodrich, E. P. *Concrete-steel construction* (from *Der Eisenbetonbau*, Third ed, 1908). New York: Engineering News Publishing Co., 1909 [BL, ICE]. Second edition, 1910 [C&CA 1959, BL, TCS. Copy: TCS]

Gray, William Samuel

Gray, W. S. *Reinforced concrete reservoirs and tanks*. London: Concrete Publications Ltd, 1931 [BL, Yrbk 1936. Rev: *SE*, v. 9, November 1931, p. 382; *C&CE*, v. 26, 1931, p. 676]

Gray, W. S. *Reinforced concrete water towers, bunkers, silos and gantries*. London: Concrete Publications Ltd, 1933 [BL, C&CA 1959, ICE, Yrbk 1936. Rev: *SE*, v. 11 August 1933, p. 361]

Gray, W. S. and Childe, H. L. *Concrete Surface Finishes, Rendering and Terrazzo*. London: Concrete Publications Ltd, 1935. [Copy: TCS]

Gregory, G. R.

Burren, F. & Gregory, G. R. *Moulds for cast stone and concrete*. London: Concrete Publications Ltd, 1930 [C&CA 1959. Copy: TSC]

Hammersley-Heenan, John

Hammersley-Heenan, J. *Concrete in freezing weather and the effect of frost upon concrete*. London, 1915 [C&CA 1952]

Hanson, Edward Smith

Hanson, E. S. *Cement pipe and tile*. Chicago: Cement Era Publishing Co., 1909 [Rev: *CA*, v. 8]

Hanson, E. S. *Concrete roads and pavements*. Chicago: Cement Era Publishing Co., 1913 [BL, C&CA 1959]

Hanson, E. S. *Concrete silos: their advantages, different types, how to build them*. Chicago: Cement Era Publishing Co., 1916 [BL]

Harris, Wallace Rutherford

Harris, W. R. & Campbell, H. C. *Concrete products: their manufacture and use*. Chicago: International Trade Press, 1921 and 1924 [C&CA 1952]

Harrison, J. L.

Harrison, J. L. *Management and methods in concrete highway construction*. New York: McGraw-Hill, 1927 [ICE, Yrbk 1934]

Hatt, William Kendrick

Hatt, W. K. & Voss, W. C. *Concrete work: a book to aid the self development of workers in concrete*. New York: John Wiley & Sons, 1921 [BL, C&CA 1959] and Chapman & Hall, 1921.

Hatt, W. K. & Voss, W. C. *Answers to concrete work*. Chapman & Hall, 1922 [Yrbk 1934]

Hawkesworth, John

Hawkesworth, John. *Graphical handbook for reinforced concrete design*. New York: Van Nostrand, 1906 and London: Crosby, Lockwood & Son, 1907 [C&CA 1959, IStructE. Rev: *EN*, 16 May 1907]

Heidenreich, Eyvind Lee

Heidenreich, E. L. *Treatise on armored concrete construction*. Chicago: Cement & Engineering News, 1904 [C&CN, v.15, 1903/04]

Heidenreich, E. L. *Engineers' pocketbook of reinforced concrete*. Chicago: Myron C. Clark Publishing Co., 1909 [IStructE, PCA. Rev: *EN*, 14 January 1909], and London: Spon, 1909 [Peddie]. Second edition, 1915 [BL]

Hering, Oswald Constantin

Hering, O. C. *Concrete and stucco houses: the use of plastic materials in the building of country and suburban houses in a manner to insure the qualities of fitness durability and beauty*. New York: McBride, Nast & Co., 1912 [BL] and London: Batsford, 1912 [Yrbk 1934]. Second edition. New York: R. M. McBride, 1922 [BL]

Hill, Charles Shattuck

Buel, A. W. and Hill, C. S. *Reinforced concrete construction*. New York: Engineering News Publishing, and London: Archibald Constable & Co., 1904 [ICE]. Second ed., 1906 [C&CA 1959, BL, ICE, IStructE. Copy: TCS. Rev: *EN*, 15 November 1906, *C&CE* May 1907, p.166]

Gillette, H. P. & Hill, C. S. *Concrete construction: methods and cost*. New York: Myron Clark Publishing Co., 1908 [C&CA 1959, ICE, PCA. Rev: *EN*, 11 June 1908] and London: Spon, 1908 [BL]

Hill, C. S. *Concrete inspection*. Chicago and New York: Myron C. Clark [PCA. Rev: *CA*, v. 9 1909, p. 444]

Hill, C. S. *Winter construction methods*. McGraw Hill, 1928 [Yrbk 1934]

Hilton, Geffrey William

Hilton, G. W. *The concrete house: an explanatory treatise on how the author, during war time, largely by his own labour, erected and completed a detached, two-storied, mono-bloc, concrete house, designed for his own occupation*. London: E. & F. N. Spon, 1919 [C&CA 1959, Yrbk 1934. Copy: TCS] and New York: Spon & Chamberlain, 1919.

Hodgson, Fred T.

Hodgson, F. T. *Concrete, cements, mortars, artificial marbles, plasters and stucco: how to use and how to prepare them*. Chicago: Drake, 1906.

'Hollie'

"Hollie". *Concrete products*. London: Anglo-German Publishing Co., 1914 [Copy: TCS. Rev: *C&CE* v. 9, p. 286]

Hollister, Solomon Cady

Hool, G. A. and Johnson, N. C. [assisted by Hollister, S. C.]. *Concrete engineers' handbook: data for the design and construction of plain and reinforced concrete structures*. New York: McGraw-Hill, 1918 [C&CA 1959, IStructE, Yrbk 1934. Copy: TCS]

Hool, Prof George Albert

Hool, G. A. *Reinforced concrete construction*. Vol. I: *fundamental principles*. (Engineering education series). New York, McGraw-Hill, 1912 [Rev: *C-CA*, v. 2, p. 147; *C&CE* v. 7, p. 788; *Trans*, v. 5, pt. 1, p. xv]. Second edition, 1917 [BL, C&CA 1959, ICE. Copy: TCS]. Third edition, 1927 [BL, Yrbk 1934]

Hool, G. A. *Reinforced concrete construction*. Vol. II: *retaining walls and buildings*. New York, McGraw-Hill, 1913 [BL, C&CA 1959, ICE. Rev: *C-CA* v. 4, p. 139; *C&CE* v. 9, p. 210]. Second edition, 1927 [BL, Yrbk 1934]

Hool, G. A. *Reinforced concrete construction*. Vol. III: *bridges and culverts*. New York, McGraw-Hill, 1916 [C&CA 1959, ICE. Rev: *C&CE* v. 11, p. 267]. Second edition, 1928 [BL, IStructE, Yrbk 1934]

Hool, G. A. and Johnson, N. C. *Concrete engineers' handbook: data for the design and construction of plain and reinforced concrete structures*. New York: McGraw-Hill, 1918 [C&CA 1959, IStructE, Yrbk 1934. Copy: TCS]

Hool, G. A. and Whitney, C. S. *Concrete designers' manual*. New York: McGraw-Hill, 1921 [Rev: *C&CE* v. 16, p. 746]. Second edition,1926 [BL, C&CA 1959, Yrbk 1934]

Hool, G. A. and Kinne, W. S. (eds). *Reinforced concrete and masonry structures*. New York: McGraw-Hill, 1924 [BL, ICE, Yrbk 1934]

Hool, G. A. and Pulver, H. E. *Concrete practice: a textbook for vocational and trade schools*. New York: McGraw-Hill, 1926 [C&CA 1959, ICE, TCS, Yrbk 1934. Copy: TCS]

Houghton, Albert Allison

Houghton, A. A. *Ornamental concrete without molds*. New York: Norman W. Henley Publishing Co., 1910 [C&CA 1959. Rev: *C&CE*, v. 6, p. 634]

Houghton, A. A. *Concrete from sand molds*. New York: Norman W. Henley Publishing Co., 1910 [Rev: *C&CE* v. 6, p. 634]

Houghton, A. A. (Concrete workers' reference books). New York: Norman W. Henley Publishing Co., 1910-11 [Rev: *C&CE* v. 6, p. 634], and London: Spon

1. *Concrete wall forms*. [ICE, Peddie]
2. *Concrete floors and sidewalks*
3. *Practical silo construction* [*CA*, v. 13, p. 44]
4. *Molding concrete chimneys, slate and roof tiles* [*CA*, v. 12, p. 333]
5. *Molding and curing ornamental concrete* [*CA*, v. 12, p. 333]
6. *Concrete monuments mausoleums and burial vaults*. [BL; *CA*, v. 12, p. 333]

Houghton, A. A. *Concrete bridges, culverts and sewers*. New York: Norman W. Henley Publishing Co., 1912 [BL; *CA*, v. 14, p. 266]

Houghton, A. A. *Molding concrete bathtubs, aquariums and natatoriums*. New York, ND

Houghton, A. A. *Constructing concrete porches*. New York: Normal W. Henley Publishing Co., 1912 [*CA*, v. 14, p. 266]

Houghton, A. A. *Molding concrete flower-pots, boxes, jardiniers*. New York, 1912.

Houghton, A. A. *Molding concrete fountains and lawn ornaments*. New York, 1912.

Howe, Harrison Estell

Howe, H. E. *The new stone age*. New York: The Century Co., 1921 and London: University of London Press, 1921 [C&CA 1959]

Howe, Malverd A.

Howe, M. A. *Symmetrical masonry arches: including natural stone, plain concrete and reinforced-concrete arches*. New York: Wiley, 1906 [IStructE]. Second edition 1914 [ICE]

Hudson, Richard John Harrington

Harrington Hudson, R. J. *Reinforced concrete: a practical handbook for use in design and construction*. London: Chapman & Hall, 1922 [BL, C&CA 1959, Yrbk 1934. Copy: TCS. Rev: C&CA, 1922, p. 601]

Johnson, Lewis Jerome

Johnson, L. J. *Reinforced concrete*. New York: Moffat, 1909 [Peddie]

Johnson, Nathan Clarke

Hool, G. A. and Johnson, N. C. *Concrete engineers' handbook: data for the design and construction of plain and reinforced concrete structures*. New York: McGraw-Hill, 1918 [C&CA 1959, IStructE, Yrbk 1934. Copy: TCS]

Jones, Bernard E.

Jones, B. E. (ed) and Lakeman, A. *Cassell's Reinforced Concrete: a complete treatise on the practice and theory of modern construction in concrete-steel*. London and New York: Cassell & Co., 1913 [ICE, PCA. Rev: *C&CE* v. 8, pp. 212–213]. Second edition. London: Waverley, 1920 [BL, C&CA 1959, Yrbk 1934. Copy: TCS]

Ketchum, Milo Smith

Ketchum, M. S. *The design of highway bridges of steel, timber and concrete*. New York: McGraw Hill, 1908. Second edition [C&CA 1959, BL, ICE, Yrbk 1934. Copy: TCS]

Kingsley, J. W.

Dana, R. T. & Kingsley, J. W. *Concrete computation charts*. New York: Codex Book Co., Inc., 1922.

Kinne, William Spaulding

Hool, G. A. and Kinne, W. S. (eds). *Reinforced concrete and masonry structures*. New York: McGraw-Hill, 1924 [BL, ICE, Yrbk 1934]

Kloes, Jacobus Alida van der

Kloes, J. A. van der, and Searle, A. B. (rev. and adapt.) *A manual for masons, bricklayers, concrete workers and plasterers*. London: Churchill, 1914 [BL, ICE]

Kommers, Jesse Benjamin

Moore, H. F. & Kommers, J. B. *The Fatigue of metals: with chapters on the fatigue of wood and concrete*. New York: McGraw-Hill, 1927 [BL]

Lakeman, Albert

Jones, B. E. (ed) and Lakeman, A. *Cassell's Reinforced Concrete: a complete treatise on the practice and theory of modern construction in concrete-steel*. London: Cassell & Co., 1913 [ICE. Rev: *C&CE* v. 8, pp. 212–213]. Second edition. London: Waverley, 1920 [BL, C&CA 1959, Yrbk 1934. Copy: TCS]

Lakeman, A. *Concrete cottages: small garages and farm buildings* [later subtitled *bungalows and garages*]. London: Concrete Publications Ltd, 1918 [Rev: *C&CE* v. 13, p. 146]. Second edition, 1924 [BL, C&CA 1959. Copy: TCS]. Third edition, 1932 [C&CA 1959, Yrbk 1934]

Lakeman, A. *Elementary guide to reinforced concrete*. London: Concrete Publications Ltd, 1928 (12 editions to 1950) [BL, C&CA 1959, ICE, Yrbk 1934. Copy: TCS]

Turner, L and Lakeman, A. *Concrete construction made easy*. London: Concrete Publications Ltd, 1929 [Yrbk 1934]

Lea, Sir Frederick Measham

Lea, F. M. & Desch, C. H. *The chemistry of cement and concrete*. London: Edward Arnold, 1935 [BL, ICE]

Leffler, Burton R.

Leffler, B. R. *The elastic arch: with special reference to the reinforced concrete arch*. New York: Holt, 1907 [BL]

Leonard, John Buck

Leonard, J. B. & Day, W. P. *The concrete bridge: a book on why the concrete bridge is replacing other forms of bridge construction*. San Francisco, 1913 [Copy: Internet Archive. Rev: *C-CA*, v. 4, January 1914, p. 42]

Lesley, Robert Whitman

Lesley, R. W. *Concrete factories: an illustrated review of the principles of construction of reinforced concrete buildings*. New York: Bruce & Banning (for the Cement Age Co.), 1907. [Peddie. Rev: *C&CE* May 1906, p.166]

Lewis, Myron H.

Lewis, M. H. *Modern methods of waterproofing concrete and other structures*. Norman W. Henley Publishing Co., 1911 and 1914.

Lewis, M. H. & Chandler, A. H. *Popular handbook for cement and concrete users*. Norman W. Henley Publishing Co., 1911 [C&CA 1952, *CA*, v. 12, p. 220. Rev: *C&CE* v. 6, pp. 633–634]. Second edition. Henry Frowde and Hodder & Stoughton, 1921 [BL. Rev: *C&CE* pp. 53–56]

Lynam, C. G.

Lynam, C. G. *Growth and movement in Portland cement concrete*. London: Oxford University Press, 1934 [C&CA 1959, ICE, IStructE. Rev: *SE* December 1934, p. 512]

McCullough, Col. Ernest

McCullough, E. *Reinforced concrete: a manual of practice*. Chicago: Cement Era/ Myron C. Clark, 1908 [PCA. Rev: *CA*, v. 7, p. 79 and *C&EN* 1908, v.20, p.208], and London: Spon, 1908 [Peddie]

McCullough, E. *Practical structural design in timber, steel and concrete*. New York: United Publishers, 1918 [PCA]. Third edition, New York: Scientific Book Corporation, 1926 [ICE] and 1927 [BL]

McMillan, Franklin

McMillan, F. R. *Basic principles of concrete making*. New York: McGraw-Hill, 1929 [BL, ICE]

Manning, George Philip

Manning, G. P. *Reinforced concrete design*. London: Longmans, Green & Co., 1924 [BL, C&CA 1959, ICE, ISE]. Second edition, 1936 [C&CA 1959, ISE. Copy: TCS]. Third edition, 1966 [ISE]

Manning, G. P. *Construction in reinforced concrete: an elementary book for designers and students*. London: Pitman, 1932 [BL, C&CA 1959, ICE, ISE]. Second edition, 1947 [ISE. Copy: TCS]

Manning, G. P. *Reinforced concrete arch design: a textbook for engineers and advanced students*. London: Pitman, 1933 [BL, ICE, ISE. Copy: TCS]. Second edition, 1954 [ISE. Copy: TCS]

Marsh, Charles Fleming

Marsh, C. F. [and Dunn, W. in the Third edition]. *Reinforced Concrete*. London: Archibald Constable, 1904. Second edition, 1905, Third ed, 1906. [C&CA 1959, BL, ICE, IStructE. Copy: TCS Third edition Rev: *C&CE*, v. 1, p. 405]

Marsh, C. F. and Dunn, W. *Manual of reinforced concrete and concrete block construction*. London: Archibald Constable, and New York: Van Nostrand, 1909. [Rev: *CA*, v. 7, p. 461]. Fourth edition, 1922. [C&CA 1959, BL, TCS. Copies: TCS. Rev: *C&CE*, v.3, p.178; v. 6, p. 715–16]

Marsh, C. F. *A concise treatise on reinforced concrete: a companion to 'the reinforced concrete manual'*. London: Constable, 1909 and New York: Van Nostrand, 1909 [Peddie]. Second edition, 1911, Third ed, 1920 [C&CA 1959, BL. Copy: TCS. Rev. *C&CE*, v. 5, pp. 144–5]

Marsh, C. F. *Reinforced concrete compression member diagram*. Archibald Constable, 1911 [Peddie. Yrbk 1934. Rev: *C&CE*, v. 7, p. 71]

Marshall, Roy

Boynton, W. C. & Marshall, R. (comp). *How to use concrete*. Detroit: Concrete Publishing, 1910. [PCA. Rev: *CA*, v. 10; *C&CE* v. 5, pp. 612–13]

Martin, Nathaniel

Martin, N. (trans). *The properties and design of reinforced concrete: instructions, authorised methods of calculation, experimental results and reports by the French Government Commissions on reinforced concrete*. London: Constable & Co., Ltd, 1912 [C&CA 1959, ICE, Yrbk 1934. Copy: TCS. Rev: *C&CE*, v. 8, p. 142], and New York: Van Nostrand [Rev; *C-CA*, v. 5, p. 131]

Matthews, Ernest Romney

Adams, H. & Matthews, E. R. *Reinforced concrete construction in theory and practice: an elementary manual for students and others*. London: Longmans, Green & Co., 1911 [C&CA 1959, ICE, Peddie]. Second edition, 1920. [C&CA 1959, BL, ICE, Yrbk 1934. Copy: TCS. Rev: *C&CE* 1912]

Maurer, Edward Rose

Turneaure, F. E. and Maurer, E. R. *Principles of reinforced concrete construction*. New York: John Wiley & Sons, 1907 [Copy: TCS. Rev: *C&CE*, v. 3, p. 258]. Second edition, 1909 [BL, ICE. Rev: *C&CE*, v. 5, p. 612; *CA*, v. 9]. Third ed, 1919 [C&CA 1959, BL, ICE. Copy: TCS. Rev: *C&CE*, v. 15, p. 650]. Fourth edition, 1932 (BL, ICE, Yrbk 1934. Rev: *SE*, v. 11, p. 102]

Mehta, N. L.

Mehta, N. L. *Notes on reinforced concrete structures*. Madras: Higginbothams, 1916 [BL]

Melan, Josef

Melan, J. and Steinman, D. B. (trans). *Plain and reinforced concrete arches*. New York: John Wiley & Sons, 1915 [BL, ICE] and 1917 [BL]. London: Chapman & Hall, 1917 [Yrbk 1934. Copy: TCS. Rev: *C&CE*, v. 10, pp. 424–425].

Mensch, L. J.

Mensch, L. J. *Architects' and engineers' handbook of reinforced concrete constructions*. Chicago: Cement & Engineering News, 1904 [PCA]

Mensch, L. J. *The reinforced concrete pocket book*. Mensch: San Francisco, 1909 [PCA. Rev: *CA*, v. 9]

Middleton, George Alexander Thomas

Middleton, G. A. T. *The effects of reinforced concrete building*. London: Francis Griffiths, 1909 [ICE] and New York: Spon, 1909 [Peddie]

Moore, Herbert Fisher

Moore, H. F. & Kommers, J. B. *The Fatigue of metals: with chapters on the fatigue of wood and concrete*. New York: McGraw-Hill, 1927 [BL]

Morgan, Newlin Dolbey

Cross, H. & Morgan, N. D. *Continuous frames of reinforced concrete*. New York: John Wiley & Sons, 1932 [BL, C&CA 1959, ICE]. Rev edition, 1954 [BL]

Morsch, Emil

Morsch, E., trans Goodrich, E. P. *Concrete-steel construction* (from *Der Eisenbetonbau*, Third ed, 1908). New York: Engineering News Publishing Co., 1909 [BL, ICE, PCA] and London: Constable, 1909. Second edition, 1910 [C&CA 1959, BL, TCS. Copy: TCS]

Nichols, Charles Eliot

Thomas, M. E. and Nichols, C. E. *Reinforced concrete design tables: a handook for engineers and architects in designing reinforced concrete structures*. New York: McGraw-Hill, 1917 [PCA Yrbk 1934]

Onderdonk, Francis S.

Onderdonk, F. S. *The ferro-concrete style: reinforced concrete in modern architecture*. New York: Architectural Book Publishing Co., 1928 [BL, C&CA 1959, ICE. Copy: TCS]

O'Rourke, Charles Edward

Urquhart, L. C. & O'Rourke, C. E. *Design of concrete structures*. New York: McGraw-Hill, 1923 [BL]. Second edition, 1926. Third edition, 1935. Fourth edition, 1940. Sixth edition, 1958.

Ostrup, John Christian

Ostrup, J. C. *Standard specifications for structural steel, timber, concrete and reinforced concrete.* New York: McGraw-Hill, 1911. [Rev: *C&CE*, 1911, v. 6, p. 153]

Palliser, Charles

Palliser, C. *Practical concrete block making.* New York: Industrial Publications, 1908 [C&CA 1959. Rev: *CA*, v. 8]

Perrot, Emile G.

Ballinger, W. F. & Perrot, E. G. *Inspector's handbook of reinforced concrete.* New York: Engineering News Publishing Co., 1909 [PCA], McGraw-Hill, 1909 [ICE. Rev: *CA*, v. 10], and London: Constable, 1909 [Peddie]

Piggott, J. T.

Piggott, J. T (ed). *Reinforced concrete calculations in a nutshell.* London: Spon, 1911 [Yrbk 1934]

Pond, Clinton de Witt

Pond, Clinton de Witt. *Concrete construction for architects: a concise treatise on the design of reinforced concrete slabs, beams, girders, columns and footings, and a description of the actual design of a concrete building involving the use of flat slab construction.* New York: C. Scribner, 1923 [C&CA 1952]

Porter, Harry F.

Porter, H. F. *Concrete its composition and use: a clear, detailed, complete statement of the fundamental principles of the basic process of the concrete industry, including essential up-to-date, proven tables and data for the users of concrete.* Cleveland: Concrete Engineering, 1909. [IStructE. Rev: *CA*, v. 10; *C&CE*, 1910, v. 5, p. 163]

Post, Chester Leroy

Post, C. L. *Building superintendence for reinforced concrete structures.* American Technical Press, 1917. Second edition. American Technical Press and London: Crosby Lockwood, 1927 [Yrbk 1934]

Pulver, Harry E.

Hool, G. A. and Pulver, H. E. *Concrete practice: a textbook for vocational and trade schools.* New York: McGraw-Hill, 1926 [C&CA 1959, ICE, TCS, Yrbk 1934. Copy: TCS]

Radford, William Addison

Radford, W. A. *Cement houses and how to build them: perspective view and floor plans of concrete block and cement plaster houses.* Chicago and New York: Radford Architectural Co., 1909 [Copy: Internet Archive]

Radford, W. A. *Cement and how to use it: a working manual of up-to-date practice in the manufacture and testing of cement: the proportioning, mixing and depositing of concrete …* Chicago: Radford Architectural Co., 1910 [BL]

Radford, W. A. *Radford's cyclopaedia of cement construction: a general reference work on up-to-date practice in the manufacture and testing of cements, the selection of concreting materials, tolls and machinery; the proportioning, mixing and depositing of concrete, plain, ornamental and reinforced.* 5 vols. Chicago: Radford Architectural Co., 1911 [Peddie]

Ransome, Ernest Leslie

Ransome, E. and Saurbrey, A. *Reinforced concrete buildings: a treatise on the history, patents, design and erection of the principal parts entering into a modern reinforced concrete building.* New York: McGraw-Hill Book Co., 1912 [Copy: TCS]

Reid, Homer Austin

Reid, H. A. *Concrete and reinforced concrete construction.* New York: Myron C. Clark Publishing Co., 1907 [C&CA 1959, PCA. Copy: TCS]; 1908 [BL, ICE]

Reuterdahl, Arvid

Reuterdahl, A. *Theory and design of reinforced concrete arches: a treatise for engineers and technical students*. Chicago: Myron C. Clark, 1908 [BL, ICE. Rev: *EN*, 17 December 1908], and London: Spon, 1908 [Peddie]

Reynolds, Charles Edward

Reynolds, C. E. *Reinforced concrete designers' handbook*. London: Concrete Publications Ltd, 1932 [BL, C&CA 1959. Copy: TCS. Rev: *SE* v. 10, p. 446, October 1932]

Rice, Harmon H.

Rice, H. H. and Torrance, W. M. *The manufacture of concrete blocks and their use in building construction*. New York: Engineering News Publishing Co., 1906 and London: A. Constable & Co., 1906 [C&CA 1959. Rev: *EN*, 16 August 1906]

Rice, H. H. *Concrete block manufacture: processes and machines*. New York: John Wiley & Sons, 1906 [Rev: *EN*, 16 August 1906; *CA*, v. 3, p. 3] and London: Chapman & Hall, 1906 [Rev: *C&CE* v.2 May 1907, p.166]

Rings, Frederick

Rings, F. *Reinforced concrete; theory and practice*. London: B. T. Batsford Ltd, 1910 [C&CA 1959, ICE, IStructE. Copy: TCS. Rev: *C&CE* v. 5, p. 455], and New York: Van Nostrand [Peddie]. Second edition, 1918 [BL, ICE, IStructE]

Rings, F. *Ready reckoners for reinforced concrete design*, London: the author, 1910 [ICE, Peddie. Rev: *C&CE*, v. 6, p. 462]

Rings, F. *Reinforced concrete bridges*. London: Constable, 1913 [BL, C&CA 1959, ICE, IStructE, Yrbk 1934. Copy: TCS. Rev: *C&CE*, v. 8, p. 441, *Trans CI*, v. 5, p. xiv], and New York: Van Nostrand [Rev: *C-CA*, v. 5, p. 37]

Sabin, Louis Carlton

Sabin, L. C. *Cement and concrete*. New York: McGraw Publishing Co., and London: Constable, 1905 [BL, ICE, IStructE. Rev: *CA*, v. 2, p. 210; *C&CE*, v.2 May 1907, p.165]; and Boston: Stanhope Press, 1907 [*EN*, 18 July 1907]

Saurbrey, Alexis

Ransome, E. and Saurbrey, A. *Reinforced concrete buildings: a treatise on the history, patents, design and erection of the principal parts entering into a modern reinforced concrete building*. New York: McGraw-Hill Book Co., 1912 [Copy: TCS]

Scott, Augustine Alban Hamilton

Scott, A. A. H. and Fraser, P. M. *A specification for reinforced concrete work*. London: Witherby & Co., 1911 [Peddie. Rev: *C&CE* v. 6, pp. 461–462]

Scott, A. A. H. *Reinforced concrete in practice*. London: Scott, Greenwood & Son, 15 [ICE]. Second edition, 1925 [BL, C&CA 1959, ICE, Yrbk 1934. Copy: TCS]

Scott, Ernest A.

Scott, E. A. *Arrol's reinforced concrete reference book of examples, general formulae, model regulations, and designing table as used by the reinforced concrete department of Sir William Arrol and Company, Ltd. Glasgow*. London: E. & F. N. Spon, 1930 [BL, C&CA 1959, ICE. Copy: TCS. Rev: *C&CE* v26, p. 244; *SE* December 1930, p. 435]

Scott, William Leslie

Scott, W. L. *Reinforced concrete bridges: the practical design of modern reinforced concrete bridges including notes on temperature and shrinkage effects*. London: Crosby Lockwood, 1925 [BL, C&CA 1959]. Second edition, 1928 [BL, ICE]. Third edition, London: Technical Press, 1931 [BL, ICE]

Scott, W. L. & Glanville, W. H. *Explanatory handbook on the code of practice for reinforced concrete as recommended by the Reinforced Concrete Structures Committee of the Building Research Board*. London: Concrete Publications Ltd, 1934 [BL, TCS]

Searle, Alfred Broadhead

Searle, A. B. *Cement, concrete and bricks*. London: Constable & Co., 1913 [Rev: *C&CE* v. 8, p. 885]. Second edition, 1926 [BL, ICE]

Kloes, J. A. van der, and Searle, A. B. (rev. and adapt.) *A manual for masons, bricklayers, concrete workers and plasterers*. London: Churchill, 1914 [BL, ICE]

Seaton, Roy Andrew

Seaton, R. A. *Concrete construction for rural communities*. (Agricultural Engineering Series). New York: McGraw-Hill, 1916 [BL, PCA, Yrbk 1934. Rev: *C&CE* v. 12]

Singleton-Green, John

Singleton-Green, J. *Reinforced concrete roads*. Contractor's Record, 1930 [BL, ICE]

Singleton-Green, J. *Concrete engineering:* Vol. I, *practical concrete*. Charles Griffin & Co., 1933 [BL, C&CA 1959, ICE. Rev: *SE* v. 10, p299]

Singleton-Green, J. *Concrete engineering:* Vol. II, *properties of concrete*. Charles Griffin & Co., 1935 [BL, C&CA 1959, ICE]

Slater, Willis A.

Talbot, A. N. and Slater, W. A. *Tests of reinforced concrete flat slab structures*. (Bulletin No.84 of the University of Illinois Engineering Experiment Station). London: Chapman & Hall, 1917 [Rev: *C&CE* v. 12, p. 203]

Smith, Maj. R. A. B.

Smith R. A. B. *Design and construction of concrete roads*. London: Concrete Publications Ltd., 1934 [BL, C&CA 1959, ICE, IStructE]

Smulski, Edward

Taylor, F. W., Thompson, S. E. & Smulski, E. (*A treatise on*) *concrete plain and reinforced*. Fourth edition. New York: John Wiley, 1925 [BL, C&CA 1959, ICE, IStructE, TCS, Yrbk 1934. Copy: TCS. Rev: *SE*, 1926, p. 67]. Fifth edition, 1932. (Originally published in 1905)

Spicer, C. W. J.

Scott, W. L. *Reinforced concrete bridges: the practical design of modern reinforced concrete bridges including notes on temperature and shrinkage effects*. London: Crosby Lockwood, 1925 [BL, C&CA 1959]. Second edition, 1928 [BL, ICE]. Third edition, London: Technical Press, 1931 [BL, ICE]

Steinman, David Barnard

Melan, J. and Steinman, D. B. (trans). *Plain and reinforced concrete arches*. New York: John Wiley & Sons, 1915 [BL, ICE] and 1917 [BL]. London: Chapman & Hall, 1917 [Yrbk 1934. Copy: TCS. Rev: *C&CE*, v. 10, pp. 424–425].

Stroyer-Nielsen, (Jens Peter) Rudolf

Stroyer, R. N. *Concrete structures in marine work*. London: Knapp, Drewett & Son Ltd, 1934 [BL, ICE, IStructE]

Sutherland, Hale

Sutherland, H. & Clifford, W. W. *Introduction to reinforced concrete design*. New York: Wiley, 1926 [BL, ICE. Rev: *SE*, 1927]

Talbot, Arthur Newell

Talbot, A. N. and Slater, W. A. *Tests of reinforced concrete flat slab structures*. (Bulletin No.84 of the University of Illinois Engineering Experiment Station). London: Chapman & Hall, 1917 [Rev: *C&CE* v. 12, p. 203]

Taylor, Frederick Winslow

Taylor, F. W. and Thompson, S. E. *(A treatise on) concrete plain and reinforced*. New York: John Wiley, 1905 [BL, C&CA 1959, ICE. Rev: *CA*, v. 2, p. 610]. Second edition, 1910 [BL. Rev: *CA*, v. 10; *C&CE*, 1910, p. 855]. Third ed, 1919 [BL, ICE]. Fourth edition, 1925 [BL, C&CA 1959, ICE, IStructE, TCS, Yrbk 1934. Copy: TCS. Rev: *SE*, 1926, p. 67]

Taylor, F. W. *Concrete costs*. New York: John Wiley & Sons, 1912. [BL, ICE, Yrbk 1934. Rev. *C&CE*, 1912, p. 553]

Thomas, M. Edgar

Thomas, M. E. and Nichols, C. E. *Reinforced concrete design tables: a handook for engineers and architects in designing reinforced concrete structures*. New York: McGraw-Hill, 1917 [PCA, Yrbk 1934]

Thompson, Sanford Eleazer

Taylor, F. W. and Thompson, S. E. *(A treatise on) concrete plain and reinforced*. New York: John Wiley, 1905 [BL, C&CA 1959, ICE. Rev: *CA*, v. 2, p. 610]. Second edition, 1910 [BL. Rev: *CA*, v. 10; *C&CE*, 1910, p. 855]. Third ed, 1919 [BL, ICE]. Fourth edition, 1925 [BL, C&CA 1959, ICE, IStructE, TCS, Yrbk 1934. Copy: TCS. Rev: *SE*, 1926, p. 67]

Toch, Maximillian

Toch, M. *The protection and decoration of concrete*. New York: Van Nostrand, 1931 [ICE]

Torrance, W. M.

Rice, H. H. and Torrance, W. M. *The manufacture of concrete blocks and their use in building construction*. New York: Engineering News Publishing Co., 1906 and London: A. Constable & Co., 1906 [C&CA 1959. Rev: *EN*, 16 August 1906]

Trautwine, John Cresson

Trautwine, J. C. Jr. and J. C. III. *Concrete*. New York: John Wiley & Sons, 1909 [Copy: TCS], and London: Chapman & Hall, 1909 [Peddie]

Travers Morgan, Reginald

Travers Morgan, R. *Tables for reinforced concrete floors and roofs*. London: Chapman & Hall, 1925 [BL, C&CA 1959, Yrbk 1934]

Turneaure, Frederick Eugene

Turneaure, F. E. and Maurer, E. R. *Principles of reinforced concrete construction*. New York: John Wiley & Sons, 1907 [Copy: TCS. Rev: *C&CE*, v. 3, p. 258]. Second edition, 1909 [BL, ICE. Rev: *C&CE*, v. 5, p. 612; *CA*, v. 9]. Third ed, 1919 [C&CA 1959, BL, ICE. Copy: TCS. Rev: *C&CE*, v. 15, p. 650]. Fourth edition, 1932 (BL, ICE, Yrbk 1934. Rev: *SE*, v. 11, p. 102]

Turner, Claude Allen Porter

Turner, C. A. P. *Concrete-steel construction. Part 1 – building: a practical treatise for the constructor and those commercially engaged in the industry*. Minneapolis: Farnham Printing & Stationery Co., 1909 [PCA, Peddie]

Eddy, H. T. & Turner, C. A. P. *Concrete-steel construction. Part 1 – building: a treatise upon the elementary principles of design and execution of reinforced concrete work in building*. Minneapolis: Heywood Manufacturing Co., 1914. Second edition, 1919.

Turner, Leslie

Turner, L. *Concrete construction made easy*. London: Concrete Publications Ltd, 1929 [ICE, Yrbk 1934]

Twelvetrees, Walter Noble

Twelvetrees, W. N. *Concrete-steel: a treatise on the theory and practice of reinforced concrete construction*. London: Whittaker & Co., 1905 [BL, C&CA 1959, IStructE. Copy: TCS. Rev: *EN*, 14 Sep 1905; *CA*, v. 2, p. 686]

Twelvetrees, W. N. *Concrete-steel buildings*. London: Whittaker & Co., 1907 [ICE. Rev: *EN*, 15 August 1907; *C&CE*, v.2 Sep 1907, p.320], and New York: Macmillan, 1907 [Peddie]

Twelvetrees, W. N. *Simplified methods of calculating reinforced concrete beams*. London: Sir Isaac Pitman, 1909 [ICE], and New York: Macmillan, 1907 [Peddie]. Second edition, 1921 [C&CA 1959, Yrbk 1934. Copy: TCS]

Twelvetrees, W. N. *The practical design of reinforced concrete beams and columns*. London: Whittaker & Co., 1911 [C&CA 1959, ICE. Copy: TCS. Rev: *C&CE*, v. 6, p. 799], and New York: Macmillan, 1911 [Peddie, *CA*, v. 14, p. 164]

Twelvetrees, W. N. *A treatise on reinforced concrete*. London: Sir Isaac Pitman, 1920 [C&CA 1959, Yrbk 1934]

Twelvetrees, W. N. *Concrete and reinforced concrete*. London: Sir Isaac Pitman & Sons, 1922 [C&CA 1959, Yrbk 1934]

Twelvetrees, W. N. *Concrete making machinery*. London: Scott, Greenwood & Son, 1925 [C&CA 1959, ICE]

Tyrrell, H. G.

Tyrrell, H. G. *Concrete bridges and culverts*. Chicago: Myron C. Clarke, 1909 [ICE]

Urquhart, Leonard Church

Urquhart, L. C. & O'Rourke, C. E. *Design of concrete structures*. New York: McGraw-Hill, 1923 [BL]. Second edition, 1926. Third edition, 1935. Fourth edition, 1940. Sixth edition, 1958.

Voss, Walter Charles

Hatt, W. K. & Voss, W. C. *Concrete work: a book to aid the self development of workers in concrete*. New York: John Wiley & Sons, 1921 [BL, C&CA 1959] and Chapman & Hall, 1921.

Hatt, W. K. & Voss, W. C. *Answers to concrete work*. Chapman & Hall, 1922 [Yrbk 1934]

Warren, Frank Dinsmore

Warren, F. D. *A handbook on reinforced concrete: for architects engineers and contractors*. New York: D. Van Nostrand Co., 1906 [Rev: *CA*, v. 2, p. 888]. Second edition, 1907 [Moisseiff]

Warren, William Henry

Warren, W. H. *Engineering construction. Part II: in masonry and concrete*. London: Longmans Green, 1921 [Yrbk 1934]

Watson, Wilbur J.

Watson, W. J. *General specifications for concrete work: as applied to building construction*. New York: Engineering News Publishing Co., 1908 [Rev: *Engineering News*, 20 February 1908] and New York: McGraw-Hill, 1915 [PCA]

Watson, W. J. *General specifications for concrete bridges*. New York: Engineering News Publishing Co., 1908 [Rev: *Engineering News*, 12 March 1908; *CA*, v. 6, p. 429]. Second edition, 1910 [Rev: *CA*, v. 13].]. Third edition, McGraw-Hill, 1916.

Webb, Walter Loring

Webb, W. L. & Gibson, W. H. *Reinforced concrete: a treatise on cement, concrete and concrete steel and their applications to modern structural work*. Chicago: American School of Correspondence, 1908 and London: Crosby Lockwood, 1927 [Yrbk 1927]

Webb, W. L. and Gibson, W. H. *Masonry and reinforced concrete*. Chicago: American School of Correspondence and London: Crosby Lockwood, 1909 [Peddie. Rev: *C&CE* v. 6, p. 154]

Webb, W. L. and Gibson, W. H. *Concrete and reinforced concrete: a condensed practical treatise on the problems of concrete construction in cement mixtures, tests, beams an d slab design, construction work, retaining walls, etc*. Chicago: American Technical Society, 1916 [BL, PCA] and 1919 and 1925.

Webb, W. L. and Gibson, W. H. *Concrete design and construction*. Chicago: American Technical Society, 1931 [BL, ICE. Rev: *SE* v. 10, p271] 1940.

Whipple, Harvey

Whipple, H. *Concrete stone manufacture*. Concrete-Cement Age Publishing Co., 1915 [Rev: *C&CE* v. 10, p. 259]. Second edition, 1918.

Whipple, H. and Gilbert, C. D. *Concrete houses and how they were built*. Detroit: Concrete-Cement Age Publishing Co., 1917 [C&CA 1959. Copy: TCS]

Williamson, James

Williamson, J. *Calculating diagrams for design of reinforced concrete sections*. London: Constable & Co., 1919 [BL, C&CA 1959, ICE, TCS, Yrbk 1934. Copy: TCS]

Winn, Col. John

Winn, J. *Notes on steel concrete construction*. Chatham: Royal Engineers Institute, 1903. [IStructE]

Whitney, Charles S.

Hool, G. A. and Whitney, C. S. *Concrete designers' manual: tables and diagram for the design of reinforced concrete structures*. New York: McGraw-Hill, 1921 [Rev: *C&CE* v. 16, p. 746]. Second edition,1926 [BL, C&CA 1959, Yrbk 1934]

Wynn, Albert Edward

Wynn, A. E. & Andrews, E. S. *Modern methods of concrete making*. London: Concrete Publications Ltd, 1926 [ICE, IStructE]. Second edition, 1928. Third edition, 1935 [C&CA 1959]. 1944 [TCS]

Wynn, A. E. *Design and construction of formwork for concrete structures*. London: Concrete Publications Ltd, 1926 [BL, ICE. Rev: *SE*, 1927]. Second edition, 1930 [ICE]. Third edition,1933 [C&CA 1959], revised 1939 [IStructE]. Fourth edition, 1951 [IStructE], revised 1956 [IStructE]

Wynn, A. E. *Making precast concrete for profit: how to keep costs and determine profits*. London: Concrete Publications Ltd, 1930 [C&CA 1959. Copy: TCS]

Wynn, A. E. *Estimating and cost keeping for concrete structures*. London: Concrete Publications Ltd., 1930 [BL, C&CA 1959, ICE, IStructE. Rev: *SE* v. 9, February 1931]. Second edition, 1946, revised, 1949 [IStructE]

Yerbury, Frank R.

Bennett, T. P. *Architectural design in concrete*. London: Ernest Benn, 1927 [C&CA 1959, ICE. Copy: TCS

Appendix C – Chronology

The following table compares the date of first (or earliest known) editions issued by British and American publishers. There is an abundance of titles in the years 1908–1912, while the clear dominance of the USA in these early years is reversed by the 1920s.

United Kingdom of Great Britain & Ireland/N. Ireland	**United States of America**
	1902
	Cain, W. *Theory of steel-concrete arches and of vaulted arches*. New York: Van Nostrand, 1902.
1903	**1903**
Winn, J. *Notes on steel concrete construction*. Chatham: Royal Engineers Institute, 1903.	Considere, A. (trans. Leon S. Moisseiff). *Experimental researches on reinforced concrete*. New York: McGraw Hill, 1903.
1904	**1904**
Marsh, Charles F. *Reinforced Concrete*. London: Archibald Constable, 1904.	Buel, A. W. and Hill, C. S. *Reinforced concrete construction*. New York: Engineering News and London: Archibald Constable & Co., 1904.
	Falk, M. S. *Cements, mortars and concrete: their physical properties*. New York: Myron C. Clark, 1904.
	Heidenreich, E.L. *Treatise on armored concrete construction*. Chicago: Cement & Engineering News, 1904.
	Mensch, L. J. *Architects' and engineers' handbook or reinforced concrete*

United Kingdom of Great Britain & Ireland/N. Ireland	United States of America
	constructions. Chicago: Cement & Engineering News, 1904.
1905	**1905**
Twelvetrees, W. N. *Concrete-steel: a treatise on the theory and practice of reinforced concrete construction*. London: Whittaker & Co., 1905.	Buel, A. W. and Hill, C. S. *Reinforced concrete construction*. London: Archibald Constable & Co., 1905. Sabin, L. C. *Cement and concrete*. New York: McGraw Publishing Co., 1905. Taylor, F. W. and Thompson, S. E. (*A treatise on*) *concrete plain and reinforced*. New York: John Wiley, 1905.
1906	**1906**
	Brayton, L. F. *Brayton standards: a pocket companion for the uniform design of reinforced concrete*. Minneapolis, 1906. Hawkesworth, John. *Graphical handbook for reinforced concrete design*. New York: Van Nostrand, 1906 and London: Crosby, Lockwood & Son, 1907. Hodgson, F. T. *Concrete, cements, mortars, artificial marbles, plasters and stucco: how to use and how to prepare them*. Chicago: Drake, 1906. Howe, M. A. *Symmetrical masonry arches: including natural stone, plain concrete and reinforced-concrete arches*. New York: Wiley, 1906. Rice, H. H. and Torrance, W. M. *The manufacture of concrete blocks and their use in building construction*. New York: Engineering News, 1906. Rice, H. H. *Concrete block manufacture: processes and machines*. New York: John Wiley & Sons, 1906. Warren, F. D. *A handbook on reinforced concrete: for architects engineers and contractors*. New York: D. Van Nostrand Co., 1906.

United Kingdom of Great Britain & Ireland/N. Ireland	United States of America
1907	**1907**
Twelvetrees, W. N. *Concrete-steel buildings*. London: Whittaker & Co., 1907.	Douglas, W. J. *Practical hints for concrete constructors*. New York: Engineering News Publishing Co., 1907. Leffler, B. R. *The elastic arch: with special reference to the reinforced concrete arch*. New York: Holt, 1907. Lesley, R. W. *Concrete factories*. New York: Bruce & Banning (for the Cement Age Co.), 1907. Reid, H. A. *Concrete and reinforced concrete construction*. New York: Myron C. Clark Publishing Co., 1907. Turneaure, F. E. and Maurer, E. R. *Principles of reinforced concrete construction*. New York: John Wiley & Sons, 1907.
1908	**1908**
Coleman, G. S. *Reinforced concrete diagrams for the calculation of beams, slabs, and columns in reinforced concrete*. London: Crosby Lockwood, 1908.	Andrews, H. B. *Practical reinforced concrete standards for the designing of reinforced concrete buildings*. Boston: Simpson Bros., 1908. Balet, J. W. *Analysis of elastic arches: three-hinged, two-hinged and hingeless, of steel, masonry and reinforced concrete*. New York: Engineering News Publishing, 1908. Gilbreth, F. B. *Concrete system*. Easton: Hive Publishing Co., 1908. Gillette, H. P. & Hill, C. S. *Concrete construction: methods and cost*. New York: Myron Clark Publishing Co., 1908. Godfrey, E. *Concrete*. Pittsburg: Edward Godfrey, 1908. Ketchum, M. S. *The design of highway bridges of steel, timber and concrete*. New York: McGraw Hill, 1908. McCullough, E. *Reinforced concrete: a manual of practice*. Chicago: Myron C. Clark, 1908.

United Kingdom of Great Britain & Ireland/N. Ireland	United States of America
	Palliser, C. *Practical concrete block making*. New York: Industrial Publications, 1908. Reuterdahl, A. *Theory and design of reinforced concrete arches: a treatise for engineers and technical students*. Chicago: Myron C. Clark, 1908. Watson, W. J. *General specifications for concrete work: as applied to building construction*. New York: Engineering News Publishing Co., 1908. Watson, W. J. *General specifications for concrete bridges*. New York: Engineering News Publishing Co., 1908. Webb, W. L. & Gibson, W. H. *Reinforced concrete: a treatise on cement, concrete and concrete steel and their applications to modern structural work*. Chicago: American School of Correspondence, 1908.
1909	**1909**
Marsh, C. F. and Dunn, W. *Manual of reinforced concrete and concrete block construction*. London: Archibald Constable, 1909. Marsh, C. F. *A concise treatise on reinforced concrete: a companion to 'the reinforced concrete manual'*. London: Constable, 1909. Middleton, G. A. T. *The effects of reinforced concrete building*. London: Francis Griffiths, 1909. Twelvetrees, W. N. *Simplified methods of calculating reinforced concrete beams*. London: Sir Isaac Pitman, 1909.	Ballinger, W. F. & Perrot, E. G. *Inspector's handbook of reinforced concrete*. New York and London: McGraw-Hill, 1909. Brett, A. (ed). *Reinforced concrete field handbook*. Cleveland: Technical Publishing Co., 1909. Colby, A. L. *Reinforced concrete in Europe*. Easton, PA: Chemical Publishing, 1909. Hanson, E. S. *Cement pipe and tile*. Chicago: Cement Era Publishing Co., 1909. Heidenreich, E. L. *Engineers' pocketbook of reinforced concrete*. Chicago: Myron C. Clark Publishing Co., 1909. Hill, C. S. *Concrete inspection*. Chicago and New York: Myron C. Clark, 1909. Johnson, L. J. *Reinforced concrete*. New York: Moffat, 1909.

United Kingdom of Great Britain & Ireland/N. Ireland	United States of America
	Mensch, L. J. *The reinforced concrete pocket book*. Mensch: San Francisco, 1909.
	Morsch, E., trans Goodrich, E. P. *Concrete-steel construction* (from *Der Eisenbetonbau*, third ed, 1908). New York: Engineering News Publishing Co., 1909.
	Porter, H. F. *Concrete its composition and use: a clear, detailed, complete statement of the fundamental principles of the basic process of the concrete industry, including essential up-to-date, proven tables and data for the users of concrete*. Cleveland: Concrete Engineering, 1909.
	Radford, W. A. *Cement houses and how to build them: perspective view and floor plans of concrete block and cement plaster houses*. Chicago: Radford Architectural Co., 1909.
	Trautwine, J. C. Jr. and J. C. III. *Concrete*. New York: John Wiley & Sons, 1909.
	Turner, C. A. P. *Concrete-steel construction. Part I – building: a practical treatise for the constructor and those commercially engaged in the industry*. Minneapolis: Farnham Printing & Stationery Co., 1909.
	Tyrrell, H. G. *Concrete bridges and culverts*. Chicago: Myron C. Clarke, 1909.
	Webb, W. L. and Gibson, W. H. *Masonry and reinforced concrete*. Chicago: American School of Correspondence, 1909.
1910	**1910**
Becher, Heinrich. *Reinforced concrete in sub- and superstructure*. London: Constable, 1910.	Andrews, H. B. *Design of reinforced concrete slabs, beams and columns, confirming to the recommendations of the Joint Committee on Concrete and*
Fleming. J. G. *Reinforced concrete*.	

United Kingdom of Great Britain & Ireland/N. Ireland	United States of America
Chatham: Royal Engineers Institute, 1910. Rings, F. *Reinforced concrete; theory and practice*. London: B. T. Batsford Ltd, 1910. Rings, F. *Ready reckoners for reinforced concrete design*, 1910.	*Reinforced Concrete*. Boston: Andrews, 1910. Boynton, W. C. & Marshall, R. (comp). *How to use concrete*. Detroit: Concrete Publishing, 1910. Davison, R. C. *Concrete pottery and garden furniture*. New York: Munn & Co., 1910. Dodge, Gordon Floyd. *Diagrams for designing reinforced concrete structures, including diagrams for reactions and strengths of steel beams*. New York: Clark Publishing Co., 1910. Houghton, A. A. *Ornamental concrete without molds*. New York: Norman W. Henley Publishing Co., 1910. Houghton, A. A. *Concrete from sand molds*. New York: Norman W. Henley Publishing Co., 1910. Radford, W. A. *Cement and how to use it:* Chicago: Radford Architectural Co., 1910.
1911	**1911**
Adams, H. & Matthews, E. R. *Reinforced concrete construction in theory and practice: an elementary manual for students and others*. London: Longmans, Green & Co., 1911. Second edition, 1920. Cantell, M. T. *Reinforced concrete construction: elementary course*. London: Spon, 1911. Davenport, J. A. *Graphical reinforced concrete design*. London: Spon, 1911. Dunn, William. *Lectures on reinforced concrete delivered at the Institution of Civil engineers in November 1910*. London: University of London Press and Hodder & Stoughton, 1911. Dunn, William. *Diagrams for the solution of T beams in reinforced concrete*.	Brooks, J. P. *Reinforced concrete: mechanics and elementary design*. New York: McGraw Hill, 1911. Godfrey, E. *Steel and reinforced concrete in buildings*. Pittsburg: Godfrey, 1911. Houghton, A. A. (Concrete workers' reference books). New York: Norman W. Henley Publishing Co., 1911 1. *Concrete wall forms.* 2. *Concrete floors and sidewalks* 3. *Practical silo construction* 4. *Molding concrete chimneys, slate and roof tiles* 5. *Molding and curing ornamental concrete* 6. *Concrete monuments mausoleums and burial vaults.*

United Kingdom of Great Britain & Ireland/N. Ireland	United States of America
London: University of London Press, 1911. Gammon, J. C. *Reinforced concrete design simplified*. London: Crosby Lockwood, 1911. Marsh, C. F. *Reinforced concrete compression member diagram*. Archibald Constable, 1911. Piggott, J. T (ed). *Reinforced concrete calculations in a nutshell*. London: Spon, 1911. Scott, A. A. H. and Fraser, P. M. *A specification for reinforced concrete work*. London: Witherby & Co., 1911. Twelvetrees, W. N. *The practical design of reinforced concrete beams and columns*. London: Whittaker & Co., 1911.	Lewis, M. H. *Modern methods of waterproofing concrete and other structures*. Norman W. Henley Publishing Co., 1911. Lewis, M. H. *Popular handbook for cement and concrete users*. Norman W. Henley Publishing Co., 1911. Ostrup, J. C. *Standard specifications for structural steel, timber, concrete and reinforced concrete*. New York: McGraw-Hill, 1911. Radford, W. A. *Radford's cyclopaedia of cement construction*. 5 vols. Chicago: Radford Architectural Co., 1911.
1912	**1912**
Andrews, E. S. *Elementary principles of reinforced concrete construction: a text-book for the use of students, engineers, architects and builders*. London: Scott, Greenwood & Son, 1912. Cantell, M. T. *Reinforced concrete construction: advanced course*. London: Spon, 1912. Coleman, T. E. *Estimating for reinforced concrete work*. London: Batsford, 1912. Faber, O. and Bowie, P. G. *Reinforced concrete design* (later Vol. I – theory). London: Edward Arnold, 1912. Martin, N. (trans). *The properties and design of reinforced concrete: instructions, authorised methods of calculation, experimental results and reports by the French Government Commissions on reinforced concrete*. London: Constable & Co., Ltd, 1912.	Hering, O. C. *Concrete and stucco houses: the use of plastic materials in the building of country and suburban houses in a manner to insure the qualities of fitness durability and beauty*. New York: McBride Nast, 1912. Hool, G. A. *Reinforced concrete construction*. Vol. I: *fundamental principles*. (Engineering education series). New York, McGraw-Hill, 1912. Houghton, A. A. *Concrete bridges, culverts and sewers*. New York: Norman W. Henley Publishing Co., 1912. Houghton, A. A. *Molding concrete flower-pots, boxes, jardiniers*. New York, 1912. Houghton, A. A. *Molding concrete fountains and lawn ornaments*. New York, 1912. Ransome, E. and Saurbrey, A. *Reinforced concrete buildings: a treatise on the history, patents, design and erection of the principal parts entering into a*

United Kingdom of Great Britain & Ireland/N. Ireland	United States of America
	modern reinforced concrete building. New York: McGraw-Hill Book Co., 1912. Taylor, F. W. *Concrete costs*. New York: John Wiley & Sons, 1912.
1913	**1913**
Ball, J. D. W. *Reinforced concrete railway structures*. (Glasgow Text Books of Civil Engineering). London: Constable & Co., 1913. Geen, B. *Continuous beams in reinforced concrete*. London: Chapman & Hall, 1913. Jones, B. E. (ed) and Lakeman, A. *Cassell's Reinforced Concrete: a complete treatise on the practice and theory of modern construction in concrete-steel*. London: Cassell & Co., 1913. Rings, F. *Reinforced concrete bridges*. London: Constable, 1913. Searle, A. B. *Cement, concrete and bricks*. London: Constable & Co., 1913.	Cochran, J. *A treatise on the inspection of concrete construction*. Chicago: Myron C. Clark Publishing Co., 1913. Cochran, J. *General specifications for concrete and reinforced concrete: including finishing and waterproofing*. New York: D. Van Nostrand Co., 1913. Eddy, H. T. *The theory of the flexure and strength of rectangular flat plates applied to reinforced concrete floor slabs*. Minneapolis: Rogers & Co., 1913. Foster, W. C. *A treatise on wooden trestle bridges and their concrete substitutes according to the present practice of American railroads*. Fourth edition. New York: Wiley, 1913. Hool, G. A. *Reinforced concrete construction*. Vol. II: *retaining walls and buildings*. New York, McGraw-Hill, 1913. Hanson, E. S. *Concrete roads and pavements*. Chicago: Cement Era Publishing Co., 1913. Leonard, J. B. & Day, W. P. *The concrete bridge: a book on why the concrete bridge is replacing other forms of bridge construction*. San Francisco, 1913.
1914	**1914**
"Hollie". *Concrete products*. London: Anglo-German Publishing Co., 1914. Kloes, J. A. van der, and Searle, A. B. (rev. and adapt.) *A manual for masons, bricklayers, concrete workers and plasterers*. London: Churchill, 1914.	Eddy, H. T. & Turner, C. A. P. *Concrete-steel construction. Part 1 – building: a treatise upon the elementary principles of design and execution of reinforced concrete work in building*. Minneapolis: Heywood Manufacturing Co., 1914.

United Kingdom of Great Britain & Ireland/N. Ireland	United States of America
1915	**1915**
Andrews, E. S. *Regulations of the London County Council relating to reinforced concrete (London Building Act, 1909, Amendment) and steel framed buildings (LCC General Powers Act, 1909): a handy guide, containing the full text, with explanatory notes, diagrams and worked examples*. London: B. T. Batsford Ltd, 1915. Hammersley-Heenan, J. *Concrete in freezing weather and the effect of frost upon concrete*. London, 1915. Scott, A. A. H. *Reinforced concrete in practice*. London: Scott, Greenwood, 1915.	Melan, J. and Steinman, D. B. (trans). *Plain and reinforced concrete arches*. New York: John Wiley & Sons, 1915. Whipple, H. *Concrete stone manufacture*. Concrete-Cement Age Publishing Co., 1915.
1916	**1916**
Mehta, N. L. *Notes on reinforced concrete structures*. Madras: Higginbothams, 1916.	Hanson, E. S. *Concrete silos: their advantages, different types, how to build them*. Chicago: Cement Era Publishing Co., 1916. Hool, G. A. *Reinforced concrete construction*. Vol. III: *bridges and culverts*. New York: McGraw-Hill, 1916. Seaton, R. A. *Concrete construction for rural communities*. (Agricultural Engineering Series). New York: McGraw-Hill, 1916. Webb, W. L. and Gibson, W. H. *Concrete and reinforced concrete: a condensed practical treatise on the problems of concrete construction in cement mixtures, tests, beams an d slab design, construction work, retaining walls, etc.* Chicago: American Technical Society, 1916.
1917	**1917**
	Campbell, H. C. and Beyer, W. F. *Practical concrete work for the school and home*. Peoria, Illinois: Manual Arts

United Kingdom of Great Britain & Ireland/N. Ireland	United States of America
	Press, 1917 and London: Batsford, 1917. Ekblaw, K. J. T. *Farm concrete*. New York: Macmillan, 1917. Fallon, J. T (ed). *How to make concrete garden furniture*. New York: Robert M. McBride, 1917. Post, C. L. *Building superintendence for reinforced concrete structures*. American Technical Press, 1917. Talbot, A. N. and Slater, W. A. *Tests of reinforced concrete flat slab structures*. (Bulletin No. 84 of the University of Illinois Engineering Experiment Station). London: Chapman & Hall, 1917. Thomas, M. E. and Nichols, C. E. *Reinforced concrete design tables: a handbook for engineers and architects in designing reinforced concrete structures*. New York: McGraw-Hill, 1917. Whipple, H. and Gilbert, C. D. *Concrete houses and how they were built*. Detroit: Concrete-Cement Age Publishing Co., 1917.
1918	**1918**
Cantell, M. T. *Reinforced concrete construction: part I*. London: Spon, 1918? Second ed., 1920. Etchells, E. F. *Mnemonic notation for engineering formulae: report of the Science Committee of the Concrete Institute*. London: E & FN Spon, 1918. Lakeman, A. *Concrete cottages: small garages and farm buildings* [later subtitled *bungalows and garages*]. London: Concrete Publications Ltd, 1918.	Abrams, D. A. *Design of concrete mixtures*. (Bulletin No. 1). Portland Cement Association, 1918. Hool, G. A. and Johnson, N. C. [assisted by Hollister, S. C.]. *Concrete engineers' handbook: data for the design and construction of plain and reinforced concrete structures*. New York: McGraw-Hill, 1918.
1919	**1919**
Ballard, F. *Concrete for house, farm and*	

United Kingdom of Great Britain & Ireland/N. Ireland	United States of America
estate. London: Crosby Lockwood, 1919. Second edition, 1921. Third edition, 1925.	
Hilton, G. W. *The concrete house: an explanatory treatise on how the author, during war time, largely by his own labour, erected and completed a detached, two-storied, mono-bloc, concrete house, designed for his own occupation*. London: E. & F. N. Spon, 1919.	
Williamson, J. *Calculating diagrams for design of reinforced concrete sections*. London: Constable & Co., 1919.	
1920	**1920**
Faber, O. *Reinforced concrete design:* Vol. II – *practice*. London: Edward Arnold, 1920.	
Twelvetrees, W. N. *A treatise on reinforced concrete*. London: Sir Isaac Pitman, 1920.	
1921	**1921**
Andrews, E. S. *Detail design in reinforced concrete: with special reference to the requirements of the reinforced concrete regulations of the L. C. C.* London: Pitman, 1921.	Baxter, L. H. *Elementary concrete construction*. Milwaukee: Bruce Publishing Co., 1921.
Cantell, M. T. *Reinforced concrete construction: part II*. Second ed. London: Spon, 1921.	Harris, W. R. & Campbell, H. C. *Concrete products: their manufacture and use*. Chicago: International Trade Press, 1921.
Warren, W. H. *Engineering construction. Part II: in masonry and concrete*. London: Longmans Green, 1921.	Hatt, W. K. & Voss, W. C. *Concrete work: a book to aid the self development of workers in concrete*. New York: John Wiley & Sons, 1921.
	Hool, G. A. and Whitney, C. S. *Concrete designers' manual: tables and diagram for the design of reinforced concrete structures*. New York: McGraw-Hill, 1921.
	Howe, H. E. *The new stone age*. New York: The Century Co., 1921.

United Kingdom of Great Britain & Ireland/N. Ireland	United States of America
1922	**1922**
Faber, O. *Reinforced concrete simply explained*. London: Henry Frowde and Hodder & Stoughton, 1922.	Dana, R. T. *Concrete computation charts*. New York: Codex Book Co., Inc., 1922.
Fougner, N. K. *Seagoing and other concrete ships*. (Oxford Technical Publications). London: Henry Frowde and Hodder & Stoughton, 1922.	Hatt, W. K. & Voss, W. C. *Answers to concrete work*. Chapman & Hall, 1922.
Harrington Hudson, R. J. *Reinforced concrete: a practical handbook for use in design and construction*. London: Chapman & Hall, 1922.	
Twelvetrees, W. N. *Concrete and reinforced concrete*. London: Sir Isaac Pitman & Sons, 1922.	
1923	**1923**
Badder, H. C. *Impervious concrete*. London: Educational Publishing, 1923.	Pond, Clinton de Witt. *Concrete construction for architects*. New York: C. Scribner, 1923.
	Urquhart, C. E. & O'Rourke, C. E. *Design of concrete structures*. New York: McGraw-Hill, 1923.
1924	**1924**
Coleman, T. E. *The civil engineer's cost book*. London: Spon, 1924.	Hool, G. A. and Kinne, W. S. (eds). *Reinforced concrete and masonry structures*. New York: McGraw-Hill, 1924.
Faber, O. *Reinforced concrete in bending and shear: theory and tests in support*. London: Concrete Publications Ltd, 1924.	
Faber, O. and Childe, H. L. *Concrete Yearbook*. London: Concrete Publications Ltd, 1924.	
Manning, G. P. *Reinforced concrete design*. London: Longmans, Green & Co., 1924.	
1925	**1925**
Scott, W. L. *Reinforced concrete bridges: the practical design of modern reinforced concrete bridges including*	

United Kingdom of Great Britain & Ireland/N. Ireland	United States of America
notes on temperature and shrinkage effects. London: Crosby Lockwood, 1925. Travers Morgan, R. *Tables for reinforced concrete floors and roofs*. London: Chapman & Hall, 1925. Twelvetrees, W. N. *Concrete making machinery*. London: Scott, Greenwood & Son, 1925.	
1926	**1926**
Wynn, A. E. & Andrews, E. S. *Modern methods of concrete making*. London: Concrete Publications Ltd, 1926. Wynn, A. E. *Design and construction of formwork for concrete structures*. London: Concrete Publications Ltd, 1926.	Hool, G. A. and Pulver, H. E. *Concrete practice: a textbook for vocational and trade schools*. New York: McGraw-Hill, 1926. McCullough, E. *Practical structural design in timber, steel and concrete*. Third edition. New York: Scientific Book Corporation, 1926. Sutherland, H. & Clifford, W. W. *Introduction to reinforced concrete design*. New York: Wiley, 1926.
1927	**1927**
Bennett, T. P. *Architectural design in concrete*. London: Ernest Benn, 1927. Childe, H. L. *Manufacture and uses of concrete products and cast stone*. (Concrete Series, No. 11). London: Concrete Publications Ltd, 1927.	Crane, T. *Concrete building construction*. New York: Wiley, 1927. Harrison, J. L. *Management and methods in concrete highway construction*. New York: McGraw-Hill, 1927. Moore, H. F. & Kommers, J. B. *The Fatigue of metals: with chapters on the fatigue of wood and concrete*. New York: McGraw-Hill, 1927.
1928	**1928**
Andrews, E. S. and Wynn, A. E. *Modern methods of concrete making*. Second edition. London: Concrete Publications Ltd, 1928. Cantell, M. T. *Practical designing in reinforced concrete: part I*. London: Spon, 1928. Lakeman, A. *Elementary guide to*	Bauer, E. E. *Plain concrete*. New York: McGraw-Hill, 1928. Hill, C. S. *Winter construction methods*. McGraw Hill, 1928. Onderdonk, F. S. *The ferro-concrete style: reinforced concrete in modern architecture*. New York: Architectural Book Publishing Co., 1928.

United Kingdom of Great Britain & Ireland/N. Ireland	United States of America
reinforced concrete. London: Concrete Publications Ltd, 1928.	
1929	**1929**
Childe, H. L. *Precast concrete factory operation*. London: Concrete Publications, 1929.	McMillan, F. R. *Basic principles of concrete making*. New York: McGraw-Hill, 1929.
Turner, L and Lakeman, A. *Concrete construction made easy*. London: Concrete Publications Ltd, 1929.	
1930	**1930**
Burren, F. & Gregory, G. R. *Moulds for cast stone and concrete*. London: Concrete Publications Ltd, 1930.	
Corkhill, T. (ed). *Brickwork, concrete and masonry: a complete work by practical specialists describing modern practice in the work of the bricklayer, concreter, mason, plasterer and slater*. (8 vols). London: Pitman, 1930.	
Scott, E. A. *Arrol's reinforced concrete reference book of examples, general formulae, model regulations, and designing table as used by the reinforced concrete department of Sir William Arrol and Company, Ltd. Glasgow*. London: E. & F. N. Spon, 1930.	
Singleton-Green, J. *Reinforced concrete roads*. Contractor's Record, 1930.	
Wynn, A. E. *Making precast concrete for profit: how to keep costs and determine profits*. London: Concrete Publications Ltd, 1930.	
Wynn, A. E. *Estimating and cost keeping for concrete structures*. London: Concrete Publications Ltd., 1930.	
1931	**1931**
Baker, S. & DeGroot, A. *Cements and aggregates*. Scranton, PA: International Textbook Co., 1931.	Toch, M. *The protection and decoration of concrete*. New York: Van Nostrand, 1931.
	Webb, W. L. and Gibson, W. H.

United Kingdom of Great Britain & Ireland/N. Ireland	United States of America
Gray, W. S. *Reinforced concrete reservoirs and tanks*. London: Concrete Publications Ltd, 1931.	*Concrete design and construction*. Chicago: American Technical Society, 1931.
1932	**1932**
Dowsett, J. F. & Bartle, E. G. *Practical formwork and shuttering*. London: Library Press, 1932.	Cross, H. & Morgan, N. D. *Continuous frames of reinforced concrete*. New York: John Wiley & Sons, 1932.
Manning, G. P. *Construction in reinforced concrete: an elementary book for designers and students*. London: Pitman, 1932.	
Reynolds, C. E. *Reinforced concrete designers' handbook*. London: Concrete Publications Ltd, 1932.	
1933	**1933**
Adams, H. C. *Elements of reinforced concrete design*. London: Concrete Publications, 1933.	
Cantell, M. T. *Practical designing in reinforced concrete: part II*. London: Spon, 1933.	
Chettoe, C. S. and Adams, H. C. *Reinforced concrete bridge design*. London: Chapman & Hall, 1933.	
Gray, W. S. *Reinforced concrete water towers, bunkers, silos and gantries*. London: Concrete Publications Ltd, 1933.	
Manning, G. P. *Reinforced concrete arch design: a textbook for engineers and advanced students*. London: Pitman, 1933.	
Singleton-Green, J. *Concrete engineering:* Vol. I, *practical concrete*. Charles Griffin & Co., 1933.	
1934	**1934**
Lynam, C. G. *Growth and movement in Portland cement concrete*. London: Oxford University Press, 1934.	

United Kingdom of Great Britain & Ireland/N. Ireland	United States of America
Scott, W. L. & Glanville, W. H. *Explanatory handbook on the code of practice for reinforced concrete as recommended by the Reinforced Concrete Structures Committee of the Building Research Board.* London: Concrete Publications Ltd, 1934.	
Smith R. A. B. *Design and construction of concrete roads.* London: Concrete Publications Ltd., 1934.	
Stroyer-Nielsen, (J.P) Rudolf. *Concrete structures in marine work.* London: Knapp, Drewett & Son Ltd, 1934.	
1935	**1935**
Cantell, M. T. *Practical designing in reinforced concrete: part III.* London: Spon, 1935.	Baker, S. & DeGroot, A. *Concrete and mortar.* Scranton, PA: International Textbook Co., 1935.
Gray, W. S. and Childe, H. L. *Concrete Surface Finishes, Rendering and Terrazzo.* London: Concrete Publications Ltd, 1935.	
Lea, F. M. & Desch, C. H. *The chemistry of cement and concrete.* London: Edward Arnold, 1935.	
Singleton-Green, J. *Concrete engineering:* Vol. II, *properties of concrete.* Charles Griffin & Co., 1935.	

Appendix D – Publishers

This section identifies those publishers who brought to market the work of the various authors described in the forgoing survey and considers the principal firms in more detail.

Commercial Publishers

The following table lists British and American publishers respectively, arranged by date of first publication of the titles covered in the survey. In the column to the right, the number of titles first issued by the publisher is given. This number excludes reciprocally published titles from the parallel jurisdiction: that is to say, work by an American author published in the USA, and issued separately in Great Britain by a British publisher, would be listed only in the American column – unless the original 'home' publisher has not been established.

British Publishers	No.	American Publishers	No.
Archibald Constable, 1904	**13**	Van Nostrand, 1902	**5**
Whitaker & Co., 1905	3	McGraw-Hill, 1903	**20**
Crosby, Lockwood & Son, 1907	**5**	Myron C. Clark, 1904	**8**
Francis Griffiths, 1909	1	Cement & Engineering News, 1904	1
Sir Isaac Pitman, 1909	**7**	John Wiley, 1905	**11**
University of London, 1910	2	Drake, 1906	1
Hodder & Stoughton, 1910	3	Engineering News Publishing, 1906	**5**
B. T. Batsford, 1910	3	Holt, 1907	1
Longman, Green & Co., 1911	3	Bruce & Banning, 1908	1
E & FN Spon, 1911	**13**	Hive Publishing Co., 1908	1
Witherby & Co., 1911	1	Industrial Publications, 1908	1
Scott, Greenwood & Son, 1912	3	American School of Correspondence, '08	2
Edward Arnold, 1912	3	Simpson Bros Corporation, 1908	1

British Publishers	No.
Chapman & Hall	**4**
Cassell & Co., 1913	1
Anglo-German Publishing Co., 1914	1
Churchill, 1914	1
Concrete Publications Ltd, 1918	20
Henry Frowde, 1922	**2**
Educational Publishing, 1923	1
Ernest Benn, 1927	1
Contractors Record, 1930	1
Library Press, 1932	1
Charles Griffin & Co., 1933	2
Oxford University Press, 1934	1

American Publishers	No.
Chemical Publishing, 1909	1
Concrete Engineering, 1909	1
Farnham Printing & Stationery Co., 1909	1
Concrete Publishing, 1910	1
Macmillan, 1917	1
Munn & Co., 1910	1
Norman W. Henley Publishing Co., 1910	**13**
McBride Nast, 1912	1
Rogers & Co., 1913	1
Cement Era Publishing, 1913	2
Heywood Manufacturing Co., 1914	1
Concrete-Cement Age Publishing Co., 1915	2
American Technical Society, 1916	2
American Technical Press, 1917	1
Manual Arts Press, 1917	1
Bruce Publishing Co., 1921	1
Codex, 1922	1
Scientific Book Corporation, 1926	1
Architectural Book Publishing Co., 1928	1
International Textbook Co., 1931	2

The Principal British Publishers

CONCRETE PUBLICATIONS LTD (20 TITLES, 1918–1935)

The proprietors of the journal *Concrete & Constructional Engineering* – E. O. Sachs, backed by the Associated Portland Cement Manufacturers – formalised their business as Concrete Publications Ltd prior to Sachs' death in 1919 and started to market books as a spin-off from periodical publishing. The first title was *Concrete cottages* by Albert Lakeman, prepared in 1918 for the Concrete Utilities Bureau – a market development organisation controlled by APCM and housed in the same premises as CPL. A second edition was issued under the company's own imprint in 1924. *Concrete roads* followed in 1920 – a propagandist volume for the Concrete Utilities Bureau drawn from journalistic accounts of contemporary construction projects. It was reissued in 1923 and ultimately replaced by the work of Major R. A. B. Smith in 1934.

These tentative forays into book publishing were soon to be developed by H. L.Childe – the managing editor appointed in 1922, as the foundation of a new strategic direction for Concrete Publications Ltd. He established the 'Concrete Series' an imprint that was to become hugely popular in the years to come. It was expressed typographically simply as the words 'Concrete Series' set within a circle. This motif changed later to the words Concrete Series written in a circle, between two rings, in a manner reminiscent of the early APCM logos. It appeared as a spine label on CPL blue cloth bindings until the 1960s.

The 1923 edition of *Concrete roads* contains the first known reference to 'the Concrete Series', and was followed a year later by the revised *Concrete cottages*. Also in 1924 *A Hundred Years of Portland Cement* by A. C. Davis was prepared to commemorate the centenary of the patent awarded to Joseph Aspdin for his invention of Portland cement. Other books followed throughout the 1920s, including titles by Childe himself. The early volumes were given a series number in 1924 with *C&CE* and the newly launched *Concrete Yearbook* first:

- *Concrete & Constructional Engineering* (journal, launched 1906)
- *Concrete Yearbook* (annual, launched 1924)
- *Concrete roads* (Anon., 1923)
- *Concrete cottages* (Lakeman, 1924)
- *A Hundred Years of Portland Cement* (Davis, 1924)
- *Reinforced Concrete Beams in Bending and Shear* (Faber, 1925)
- *Elementary Guide to Reinforced Concrete* (Lakeman, 1925)
- Concrete for the Builder (magazine, launched 1926)
- *Design and Construction of Formwork for Concrete Structures* (Wynn, 1926)

Childe, Turner, Burren & Gregory, Gray, Reynolds, H. C. Adams, Scott & Glanville and Smith complete the list of authors up to 1935, most of them regular contributors to *C&CE* whose texts were trailed or serialised in the journal. The publisher's list was extensive and the sales volumes impressive by the time the firm was taken over by the Cement & Concrete Association in 1969. For a history of the firm see *Concrete Publications Ltd and its legacy to the concrete industry* (Trout, 2003).

Archibald Constable (14 titles, 1904–1919)

Publisher of the first British author on reinforced concrete (Charles Marsh), Constable appears to have capitalised on its early involvement in this emerging sector by securing the British rights to the American writers Buel & Hill, Louis Sabin and Rice & Torrance, even the German, Heinrich Becher. Successive editions of Marsh followed in 1909 and 1911, and in 1912–13 Constable added Martin, Bull, Rings and Searle to its list. After the war Constable had lost its lead and the only new 'concrete' author was Williamson in 1919. The firm – which had started out in Edinburgh in 1801 as a literary publisher, noted for its editions of Sir Walter Scott – issued a wide range of fiction and non-fiction by the early years of the 20th Century. It continues today as Constable & Robinson.

E. & F. N. SPON (13 TITLES, 1911-1935)

Established in 1799 by the *émigré* Frenchman Baron de Spon, E. & F. N. Spon is one of Britain's oldest technical publishers and has taken an interest in concrete over many years. Henry Reid's very early *Practical treatise on concrete and how to make it* (dating from 1869) was brought out by E. & F. N. Spon, as was John Newman's *Notes on concrete and works in concrete* – the second edition of which (in 1894) presages the subject of this survey. The first edition of Thomas Potter's *Concrete: its use in building* (1877) was published by Spon, even if the 1908 edition was issued by Batsford. Of similar vintage were several related books on cement: Reid's *A practical treatise on the manufacture of Portland cement* (1868) – the first book in the English language in which the process of manufacture is fully described; John Grant's *Experiments on the strength of cement* (1875); A. H. Heath' s *A manual on lime and cement, their treatment and use in construction* (1893); and D. B. Butler's *Portland cement: its manufacture, testing and use* (1899).

In the opening years of the new century, Spon was the sole British agent for the McGraw Publishing Co. of New York, which might explain why no new titles on concrete came out in the Spon name until 1911, shortly after the American firm's merger with Hill. In 1911 Spon published Piggott, Davenport and Cantell – the latter becoming a prolific feature of the Spon list, with new titles appearing in 1912, 1918 1921, 1928, 1933 and 1935. At the end of the First World War, Spon issued Etchells' work on notation, and the slim volume by Hilton bears all the marks of wartime economy production. Coleman's book on concrete costing in 1924 parallels Spon's own annual price books which are still so much a feature of the firm's output today. Spon's last concrete title before 1935 was *Arrol's reinforced concrete reference book* by Ernest Scott – very much in line with the firm's approach – and the imprint has remained consistently one of the most closely associated with concrete right up to the present. In more recent years E. & F. N. Spon became part of the Routledge Group, which itself was acquired by Taylor & Francis in 1998, and now appears as 'Spon Press'.

SIR ISAAC PITMAN (7 TITLES, 1909–1933)

Sir Isaac Pitman (1813–1897) is best known as the inventor of the system of shorthand named after him, and for the colleges providing training in office skills. He was also a prominent Swedenborgian and vice-president of the Vegetarian Society. More relevantly for our purposes, he founded a publishing company that turned its attention to (among other things) concrete. Twelvetrees was the firm's first concrete author in 1909. After the First World War there was a flurry of interest in concrete with two new titles by Twelvetrees in 1920 and 1922, and A. S. Andrews in 1921. Similarly, ten years later there was another burst with three books by Corkhill and Manning in 1930, 1932 and 1933.

CROSBY LOCKWOOD & SON (5 TITLES, 1907–1925)

Lockwood & Co. was established in the early 19th Century, becoming Crosby Lockwood & Co.in 1850s. The firm specialised in technical subjects, as well as school-books and popular literature, until 1972. An early interest in concrete was expressed by the publication of Dobson's *Foundations and concrete works*, dating from the mid 19th Century, which by 1893 had reached its sixth edition. A revised edition – updated to cover reinforced concrete – was published in 1903 and thereafter the firm published a steady, if not extensive succession of new authors on concrete: Hawkesworth 1907; Coleman, 1908; Gammon, 1911; Ballard, 19'19; and W. L. Scott, 1925.

CHAPMAN & HALL (4 TITLES, 1913–1933)

Founded in the first half of the 19th Century by Edward Chapman and William Hall Chapman was perhaps best known for publishing Charles Dickens in the 1840s and 1860s, and William Thackeray. The firm was then sold to the novelist Anthony Trollope and his son Henry. Between 1902 and 1930 (under the management of Arthur Waugh), Chapman & Hall moved away from its concentration on literature and published books on technology. The first of its concrete authors was Burnard Geen in 1913; a year which also saw Chapman & Hall publishing the British edition of W. C. Foster's *A treatise on wooden trestle bridges and their concrete substitutes according to the present practice of American railroads* (a long-established title published in New York by John Wiley). Chapman & Hall had entered into an agreement with Wiley to act as its sole agent in Great Britain and Europe, and Foster's was only one of many such titles. Similarly Chapman & Hall was responsible for bringing Talbot and Slater's important University of Illinois Bulletin No.84 to the British market in 1917, during the lean war years. A further American title followed in 1922 when Hatt & Voss were published in 1922, alongside the home-grown Harrington-Hudson. British classics by Travers Morgan and Chettoe & Adams were issued in 1925 and 1933 respectively.

In the 1930s Chapman & Hall merged with Methuen and joined E. & F. N. Spon in 1955 to form Associated Book Publishers. Having passed through Thomson's ownership between 1987 and 1998, Chapman & Hall is currently used as an imprint for science and technology books by Taylor & Francis.

EDWARD ARNOLD (3 TITLES, 1912–1935)

With only three concrete authors, Edward Arnold's publishing activity was spread over three decades: Faber (1912 and 1920) in the 1910s and 1920s; Lea & Desch (1935) in the 1930s.

B. T. BATSFORD (3 TITLES, 1910–1915)

Thomas Potter's *Concrete: its use in building and the construction of concrete walls, floors, etc.* was originally published by Spon in 1877, but with that firm's preoccupation with American authors in the opening years of the new century, Batsford took on the revised edition of 1908. Rings (in 1910) was one of the pioneering British authors and was followed in the Batsford list by Coleman (1912) and Andrews (1915). Batsford was notable however, in publishing a greater number of American authors for the British market than home-grown writers. Hering (1912), Campbell (1917), Fallon (1917) and Baxter (1921) were notably orientated towards art and architecture, with their American editions issued by publishers less associated with the mainstream concrete literature: McBride Nast; Manual Arts; Robert McBride and Bruce. Batsford published no British author on concrete after the First World War. This period in the firm's history was covered in detail by the centenary book – Hector Bolitho's *A Batsford Century* in 1943.

HODDER & STOUGHTON (3 TITLES, 1910–1922)

Hodder & Stoughton's involvement in concrete publishing appears to have been only in partnership with the university presses. William Dunn was published with the University of London Press in 1910, while in 1922 both Faber and Fougner were published with Henry Frowde of the Oxford University Press. All three authors were important, progressive writers.

LONGMAN, GREEN & CO. (3 TITLES, 1911–1924)

This venerable firm (established in 1724) published across a wide range of subjects in its time. Its presence in the literature of concrete in the early 20th Century was limited to Adams & Mathews in 1911, Warren in 1921, and Manning in 1924. More recently though, it was the publisher of Adam Neville's standard work, *The properties of concrete*. No longer an independent publishing house, Longman is now an imprint of Pearson Education, specialising in English, History, Philosophy, Politics and Religion.

SCOTT, GREENWOOD & SON (3 TITLES, 1912–1925)

Thomas Greenwood (1851–1908) – the driving force in this publishing firm – was best known as a passionate advocate of the public library and to a lesser extent, a novelist. Having worked briefly in the hatting trade, and then a library, Greenwood (with Hoseason Smith) established the printing and publishing firm Smith, Greenwood & Co. in 1875 and immediately set up the *Hatters' Gazette* and the *Pottery Gazette*. Soon after the firm changed to Scott, Greenwood & Son. Other trade journals followed such as *Oil & Colour Trades Journal*, along with many books on technical subjects. The presence of authors E. S. Andrews (1912), A. A. H. Scott (1915) and W. N. Twelvetrees (1925) on the list was entirely in keeping with the industrial nature of much of this firm's output.

WHITTAKER & CO. (3 TITLES, 1905–1911)

The only involvement in reinforced concrete by this house was in publishing three titles by Noble Twelvetrees between 1905 and 1911 – one of the very earliest of British authors in this survey. Twelvetrees later transferred his loyalties to Sir Isaac Pitman and Scott, Greenwood & Son.

The Principal American Publishers

McGraw-Hill (20 titles, 1908–1928)

The McGraw-Hill Book Company was formed in 1909 as a merger of the book departments of the McGraw Publishing Co. and the Hill Publishing Co. Both James H. McGraw and John A. Hill had become involved in publishing in the 1880s – each editing a railway trade journal. For fifteen years or so they followed parallel careers in technical and trade publishing until McGraw established his publishing house in 1899 and Hill in 1902. Both firms continued independently, whilst acting as partners in the combined book business – Hill acting as president and McGraw as vice-president. On Hill's sudden death in 1916, a complete merger resulted in the establishment of McGraw-Hill Publishing Company, with offices in the Hill Building, Tenth Avenue, New York. Similarly Hill's *Engineering News* and McGraw's *Engineering Record* became McGraw-Hill's *Engineering News-Record*.

McGraw-Hill expanded in the 1920s, acquiring New Falls Paper Co. and A. W. Shaw, and setting up offices in California and London. James McGraw retired as president in 1928, but remained as chairman of the board until 1935. The company's strengths were in science and engineering, and an increasing range of textbooks aimed at college students. Of the concrete authors on the firm's list, Hool was the most prolific – with his seven titles (and numerous revised editions) based on university extension courses. Similarly both Ketchum and Urquhart were university professors and each wrote more widely than on concrete alone. McGraw-Hill's output in this subject averaged one a year over the period surveyed, but has gone on to include many more in the years since.

Norman W. Henley Publishing Co. (13 titles, 1910–1912)

Although Henley published 11 titles, the figure is misleading as an expression of the extent of the firm's interest in concrete. Output was limited to only two authors (Houghton and Lewis), the bulk of the titles was from one series, and the period of activity just three years. The subject matter concentrated on ornamental concrete and the approach was one of practical guidance for the popular market. It seems the firm aimed at a niche not served by the more mainstream technical publishers.

John Wiley & Sons (11 titles, 1905–1932)

Established in 1807 when Charles Wiley opened a print shop in Manhattan, the firm became famous in the 19th Century for publishing some of the great names of American literature. During the later years of the century the firm – known as John Wiley & Sons from 1876 – gradually shifted its focus to scientific, technical and engineering subjects, including civil engineering, architecture and construction. In 1904 the family firm incorporated, with John's sons (Charles's grandsons) taking the roles of president, vice-president and secretary. The new corporation entered into an agreement with Chapman & Hall to act as its sole agent in Great Britain and Europe. The next few years saw an increase in business management titles and books aimed at the higher education sector. In 1929 sales exceeded one million for the first time.

MYRON C.CLARK (9 TITLES, 1904–1913)

Little is known about this firm, which was an important publisher in our field during the early years of the period, though it appears to have been linked with the magazine *Cement Era*. With offices in New York and Chicago (the emerging capital of concrete), Myron C. Clark included on its list several noteworthy authors including Reid, Gillette, McCullough and Reuterdahl. By 1918 texts for which the copyright was held by Myron C. Clark were being published by McGraw-Hill.

D.VAN NOSTRAND (5 TITLES, 1902–1931)

Now known as Van Nostrand Reinholdt, this New York firm has the distinction of publishing the first book of our survey – Cain's *Theory of steel-concrete arches and of vaulted arches* in 1902. Subsequent authors included Hawkesworth (1906), Warren (1906), Cochran (1913) and Toch (1931).

ENGINEERING NEWS (5 TITLES, 1906–1909)

Engineering News was a forerunner of the well-known American engineering journal, *Engineering News-Record* (known today as *ENR*). It started out as *The Engineer & Surveyor* in 1874, then renamed successively *The Engineer, Architect & Surveyor*, then *Engineering News & American Railway Journal* (under the editorship of John A. Hill), and finally, *Engineering News*. Hill established a book-publishing firm in 1902 [see McGraw-Hill above] but continued *Engineering News* as a standalone business at 220 Broadway. The eponymous journal was of course, the principal purpose of the enterprise, but the Engineering News Publishing Co. also had its own book department publishing reprints or compilations of articles from the journal, and acting as a mail order bookseller for technical titles. Of the latter service it was proud to claim that they: '[carry] in stock books of all leading publishers and can supply promptly and at publishers' price any technical book in print.'

Of the authors' work covered by the present survey, Buel & Hill's book first appeared in 1904 (published in the UK by Constable), but with copyright vested in Engineering News. Other authors of books published by Engineering News included Rice & Torrance (1906), Douglas (1907), Balet (1908), Watson (1908) and Goodrich, whose translation of Morsch appeared in 1909. Also in 1909, concrete publishing to date was summarised by Moisseiff in his *Review of the literature of reinforced concrete*.

In 1909, when the Hill Publishing Company combined its book department with that of the McGraw Publishing Co. to form McGraw-Hill, the firms' respective journals – *Engineering News* and *Engineering Record* – remained separate businesses until the full merger following Hill's death in 1916.

Technical and Trade Journals

Given the close relationship between book authors and the periodical press – with books promoted by or serialised in the journals, or articles reprinted separately or worked up into longer volumes – it is worth noting the specialist journals covering concrete. Furthermore, several authors in this survey earned their livelihood as editors of periodicals – most conspicuously H. L. Childe and, at different times, Halbert Gillette and Harvey Whipple.

The trade and technical journals below are listed in date order. The selection includes those specific to cement and concrete, but makes no attempt to be comprehensive in listing general engineering, construction or architecture titles. Those that are included have a particular interest for our subject.

United Kingdom of Great Britain & Ireland (and dominions)

Publisher (place)	Title	Dates
Concrete Publications Ltd	*Concrete & Constructional Engineering*	1906–1966
St Bride's Press for LG Mouchel	*Ferro-concrete* (monthly)	1909–1939
Monetary Times Printing Co.	*Canadian Cement & Concrete Review*	1910–1911
Trussed Concrete Steel Co.	*Kahncrete Engineering* (bi-monthly on examples of the Kahn System)	1914–1935
Concrete Publications Ltd	*Concrete for the Builder* *Concrete Building & Concrete Products*	1926–1928 1929–
H. E. Ormerod (Bombay) for Cement Marketing Co. of India	*The Indian Concrete Journal*	1927–present
British Reinforced Concrete Co.	*The Concrete Way*	July 1928–1937
(Birmingham)	*Cement, Lime & Gravel*	1928–
Industrial Constructions Ltd	*The Concrete Age*	1929–1931
	Cast Stone Architecture & Concrete Design	1931–
Trussed Concrete Steel Co.	*Truscon Review*	April 1933– 1935

United States of America

Publisher (place)	Title	Dates
(Chicago)	*Cement & Engineering News*	1896–1924 to *Rock Products*
(New York)	*Cement*	1900–1913
(Chicago)	*Rock Products & Building Matls*	1902–present
(Chicago)	*Cement Era*	1903–1917
Concrete Pub. Co. (Detroit) later Concrete Publishing Corp. (Chicago)	*Concrete* *Concrete-Cement Age* *Concrete*	March 1904–1912 July 1912–December 1915 January 1916–February 1961
R. W. Lesley (NY)	*Cement Age*	June 1904–12 + *Concrete*
(Atlanta)	*Concrete Age*	1905–1923
(Chicago)	*Cement World*	1907–1917
Tech.Publishing (Cleveland)	*Concrete Engineering*	1907–1910 to *Cement Age*
Technical Literature Co. (NY)	*The Engineering Digest*	1908–1909
Hill Publishing Co. (NY)	*Engineering News*	1875–1917
McGraw Publishing Co.	*Engineering Record*	1877–1917
McGraw-Hill Publishing Co.	*Engineering News-Record*	1917–present
(Chicago)	*Concrete Builder*	1918– [suspended 1923–31]
(Chicago)	*Concrete Products*	1918–1936
(New York)	*Concrete Craft*	1919

Chicago is conspicuous as the centre of specialist publishing for concrete in America, in the same way as New York is the natural centre for commercial book publishing.

Appendix E – Professional honours

Many of the authors' careers we have examined culminated in election to the presidency of one or more professional institutions with which they were associated, or the award of an institutional medal in recognition of a lifetime's achievement. While making no claim to be comprehensive, we list the recipients of the more prominent honours.

Presidents of the Concrete Institute (later the Institution of Structural Engineers)

2nd	Sir Henry Tanner	1910–1912
4th	Prof Henry Adams	1914–1916
7th	E. Fiander Etchells	1920–1923
15th	Ewart S. Andrews	1934–1935
16th	Oscar Faber	1935–1936
30th	Leslie Turner	1949–1950

Presidents of the Society of Engineers

W. Noble Twelvetrees	1919
Burnard Geen	1920

Presidents of the Societé des Ingenieurs Civils de France, British Section

4th	W. Noble Twelvetrees	1922–1923
8th	E. Fiander Etchells	1926–1927

Presidents of the American Concrete Institute

4th	William K. Hatt	1917–1919
9th	Duff A. Abrams	1930–1931
10th	S. C. Hollister	1932–1933
13th	Franklin R. McMillan	1936
32nd	Charles S. Whitney	1955

Honorary Members of the American Concrete Institute

Honorary Membership was instituted in 1932, with nine significant figures in the development of concrete technology recognised – including the five below, the Austrian engineer / author Fritz Emperger, and three early ACI presidents:

1932 William K. Hatt
1932 Robert W. Lesley
1932 Arthur N. Talbot
1932 Sanford E. Thompson
1932 Frederick E. Turneaure

Other important figures in the later literature were duly honoured with Arthur Lord, William Lerch and T. C. Powers among the first few in the post-war years. Of the authors in our survey the following were honoured:

1943 Franklin R. McMillan
1959 Hardy Cross
1963 S. C. Hollister
1966 Sir William Glanville
1967 Frederick M. Lea

Gold Medal of the Institution of Structural Engineers

Rarely bestowed in its early days, three of 'our' authors were deserving recipients:

1922 Henry Adams
1958 Hardy Cross
1962 Sir William Glanville

The Wason Medal (ACI)

Instituted in 1917 by past President Leonard Wason, the medal was for 'the most meritorious paper'. Authors from our survey whose writing has justified the award include:

1919 Duff A. Abrams
1920 Willis A. Slater
1929 Franklin R. McMillan
1933 Charles S. Whitney (and again in 1953)
1936 Hardy Cross

The Henry C. Turner Medal (ACI)

An open award made no more than once a year for notable achievement in, or service to, the concrete industry, the Henry C. Turner Medal commemorates the name of a former ACI president. Recipients include by-now familiar names:

1928 Arthur N. Talbot
1929 William K. Hatt
1930 Frederick E. Turneaure
1932 Duff A. Abrams
1956 Franklin R. McMillan

The Benjamin Garver Lamme Award

Instituted in 1928 by the American Society for Engineering Education, the Lamme Award was granted to several of our authors whose careers lay in the universities:

1932 Arthur N. Talbot
1934 Edward S. Maurer
1937 Frederick E. Turneaure
1944 Hardy Cross
1942 S. C. Hollister

The Telford Premium

Named for a bequest by the Institution of Civil Engineers' first president, 1820–1834, the Telford Medal and four prizes are presented each year for outstanding papers submitted to the Institution. Recipients include:

G. S. Coleman 1923/4
Oscar Faber 1928/9
C. S. Chettoe 1942
Frederick M. Lea 1943

Nominal Index

D

E

F

G

H

Thomas, M. Edgar
1917/USA 134

Thompson, Sanford Eleazer (1867–1949)
1905/USA 17

Toch, Maximillian
1931/USA 199

Torrance, W. M.
1906/USA 25

Trautwine, John Cresson (1850–1924)
1909/USA 64

Travers Morgan, Reginald (1891–1940)
1925/UK 173

Turneaure, Frederick Eugene (1866–1951)
1907/USA 32

Turner, Claude Allen Porter (1869–1955)
1909/USA 65

Turner, Leslie (1891–1971)
1929/UK 188

Twelvetrees, Walter Noble (1853–1941)
1905/UK 19

Tyrrell, H. G. (b.1867)
1909/USA 67

U

Urquhart, Leonard Church (b.1886)
1923/USA 160

V

Voss, Walter Charles (b.1887)
1921/USA 151

W

Warren, Frank Dinsmore (1879–1930)
1906/USA 25

Warren, William Henry (1852–1926)
1921/UK 151

Watson, Wilbur J. (1871–1939)
1908/USA 48

Webb, Walter Loring (1863–1941)
1908/USA 49

Whipple, Harvey (b.1884)
1915/USA 127

Whitney, Charles S.
1921/USA 153

Williamson, James
1919/UK 147

Winn, Lt. Col. John (1860–1928)
1903/UK 3

Wynn, Albert Edward (1888–1956)
1926/USA 174

Y

Yerbury, Frank R. (1885–1970)
1927/UK 180